6th International Conference Heat and Mass Transfer and Hydrodynamics in Swirling Flows 2017

AA006080

Journal of Physics: Conference Series Volume 980

Novosibirsk, Russia
21 – 23 November 2017

ISBN: 978-1-5108-6044-5
ISSN: 1742-6588

Printed from e-media with permission by:

Curran Associates, Inc.
57 Morehouse Lane
Red Hook, NY 12571

proceedings.com

Some format issues inherent in the e-media version may also appear in this print version.

Copyright© (2017) by the Institute of Physics
All rights reserved. The material featured in this book is subject to
IOP copyright protection, unless otherwise indicated.

Printed by Curran Associates, Inc. (2018)

For permission requests, please contact the Institute of Physics
at the address below.

Institute of Physics
Dirac House, Temple Back
Bristol BS1 6BE UK

Phone: 44 1 17 929 7481
Fax: 44 1 17 920 0979

techtracking@iop.org

Additional copies of this publication are available from:

Curran Associates, Inc.
57 Morehouse Lane
Red Hook, NY 12571 USA
Phone: 845-758-0400
Fax: 845-758-2633
Email: curran@proceedings.com
Web: www.proceedings.com

Table of contents

Volume 980

6th International Conference Heat and Mass Transfer and Hydrodynamics in Swirling Flows

21–23 November 2017, Novosibirsk, Russian Federation

Accepted papers received: 20 February 2018
Published online: 19 March 2018

Preface

Papers

Heat and mass transfer in swirling flows

IOP Conf. Series: Journal of Physics: Conf. Series **980** (2018) 011001 doi:10.1088/1742-6596/980/1/011001

Preface for the Sixth International Conference «Heat and Mass Transfer and Hydrodynamics in Swirling Flows», Novosibirsk, November 21-23, 2017.

The conference took place in Novosibirsk at the Institute of Thermophysics SB RAS from November 21 to November 23, 2017.

The topics were mainly focused on hydrodynamics, heat and mass transfer in swirling flows, but the spectrum of problems discussed was much wider and included thermohydrodynamics on micro- and nanoscale, the application of vortex and heat exchange intensifiers to enhance the heat transfer, swirling flames, various devices based on these types of flows for the power engineering applications and some advances in experimental methods to tackle these problems.

The number of participants reached 77 reseachers including scientists from academia and industry with a good share of young members coming from 14 cities in Russia, Belarus, USA, Germany and Belgium. A large number of insightful keynote lectures have been presented by Profs. K. Oberleithner (Germany), M. Vanierschot (Belgium), Ph. Ligrani (USA), N.V. Kornev (Germany), A.A. Khalatov (Ukraine), S.V. Alekseenko, E.E. Son, A,Yu. Varaksin, V.I. Terekhov, E.M. Smirnov, S.K. Nemirovskii, I.A. Popov, S.A. Isaev, V.M. Dulin (Russia). As a results there were 14 keynote and 6 topical lectures followed by 56 oral presentations. The proceedings in this volume contain 43 papers grouped by sections where they were presented.

The scientific committee appreciates the enormous work of the editorial board and reviewers during the preparation of this volume. We would like to express our sincere gratitude to all authors for their contributions.

Content from this work may be used under the terms of the Creative Commons Attribution 3.0 licence. Any further distribution of this work must maintain attribution to the author(s) and the title of the work, journal citation and DOI.
Published under licence by IOP Publishing Ltd

Peer review statement

All papers published in this volume of *Journal of Physics: Conference Series* have been peer reviewed through processes administered by the proceedings Editors. Reviews were conducted by expert referees to the professional and scientific standards expected of a proceedings journal published by IOP Publishing.

Content from this work may be used under the terms of the Creative Commons Attribution 3.0 licence. Any further distribution of this work must maintain attribution to the author(s) and the title of the work, journal citation and DOI.
Published under licence by IOP Publishing Ltd

IOP Conf. Series: Journal of Physics: Conf. Series **980** (2018) 012001 doi:10.1088/1742-6596/980/1/012001

Swirl effect on flow structure and mixing in a turbulent jet

Z D Kravtsov[1,2], D K Sharaborin[1,2], V M Dulin[1,2]

[1]Kutateladze Institute of Thermophysics, 1 Ak. Lavrentyev Avenue, Novosibirsk, 630090, Russia
[2]Novosibirsk State University, 2 Pirogov Street, Novosibirsk, 630090, Russia

E-mail: zakhar.d.kravtsov@gmail.com

Abstract. The paper reports on experimental study of turbulent transport in the initial region of swirling turbulent jets. The particle image velocimetry and planar laser-induced fluorescence techniques are used to investigate the flow structure and passive scalar concentration, respectively, in free air jet with acetone vapor. Three flow cases are considered, viz., non-swirling jets and swirling jets with and without vortex breakdown and central recirculation zone. Without vortex breakdown, the swirl is shown to promote jet mixing with surrounding air and to decrease the jet core length. The vortex core breakdown further enhances mixing as the jet core disintegrates at the nozzle exit.

1. Introduction

Swirling jets are widely used in combustors to stabilize the flame by means of a swirl-induced recirculation zone [1]. The swirl intensifies processes of turbulent heat and mass transfer and increases efficiency of mixers and burners. This is achieved via formation of longitudinal vortices in jet flows. For jets with swirl rate above the critical value, a vortex breakdown occurs, leading to unsteady flow dynamics for high swirl rates. The vortex breakdown associated with formation of stagnation point is followed by a recirculation region downstream [2]. The recirculation zone often triggers unsteady jet rotation, corresponding to vortex core precession [3]. It has been shown in several papers that the swirl has a positive effect on the mixing processes [4]. However, there is still a lack of experimental studies of flow structure and mixing in turbulent swirling jets with detailed information on the instantaneous velocity and concentration fields.

The planar laser-induced fluorescence (PLIF) technique is widely used to measure concentration distribution [5, 6]. This approach is based on registration of local fluorescence of passive admixture molecules in the measurement plane, excited by laser radiation. The PLIF method is an almost non-intrusive technique. Another technique, the particle image velocimetry (PIV) is used to measure the instantaneous velocity fields. The method is based on double illumination of tracer particles in the measurement plane and evaluation of their local displacement from image correlation analysis. The aim of the present work was to study the turbulent transport and mixing in the initial region of a turbulent jet for various swirl rates, using PIV and PLIF. The paper reports on swirl effect on the mean concentration and velocity fields.

Content from this work may be used under the terms of the Creative Commons Attribution 3.0 licence. Any further distribution of this work must maintain attribution to the author(s) and the title of the work, journal citation and DOI.
Published under licence by IOP Publishing Ltd

2. Experimental setup

The swirling jet flow was organized by a contraction nozzle with changeable vane swirlers mounted inside. Details on the nozzle may be found in [7]. Swirlers with different inclination angles of the vanes had swirl rate S (the definition is taken from [1]) equal to 0.41 and 1.0.

$$S = \frac{2}{3}\left(\frac{1-(d_1/d_2)^3}{1-(d_1/d_2)^2}\right)\tan(\psi) \quad (1)$$

Here, d_1 = 7 mm is the diameter of the centre body supporting the vanes, d_2 = 27 mm is the external diameter of the swirler, and ψ is the vanes inclination angle relative to the axis. The swirler was mounted inside the nozzle as demonstrated in Figure 1 (bottom left corner). The nozzle exit diameter d was 15 mm. The Reynolds number, defined as Re = U_0d/v (U_0 is the bulk flow velocity of the jet; v is the air kinematic viscosity), was 5 000.

The air flow was supplied from air-pressure line, and its flow rate was precisely controlled by mass flow meters (Bronkhorst). An acetone seeder was used to introduce acetone vapour into the main air flow. The seeder was a container with liquid acetone, placed in a heated water bath to provide thermal stabilization. The air flow was bubbled through the liquid acetone. The temperature of air with acetone vapour at the nozzle exit was monitored by a thermocouple. The surrounding air was seeded by a fog machine.

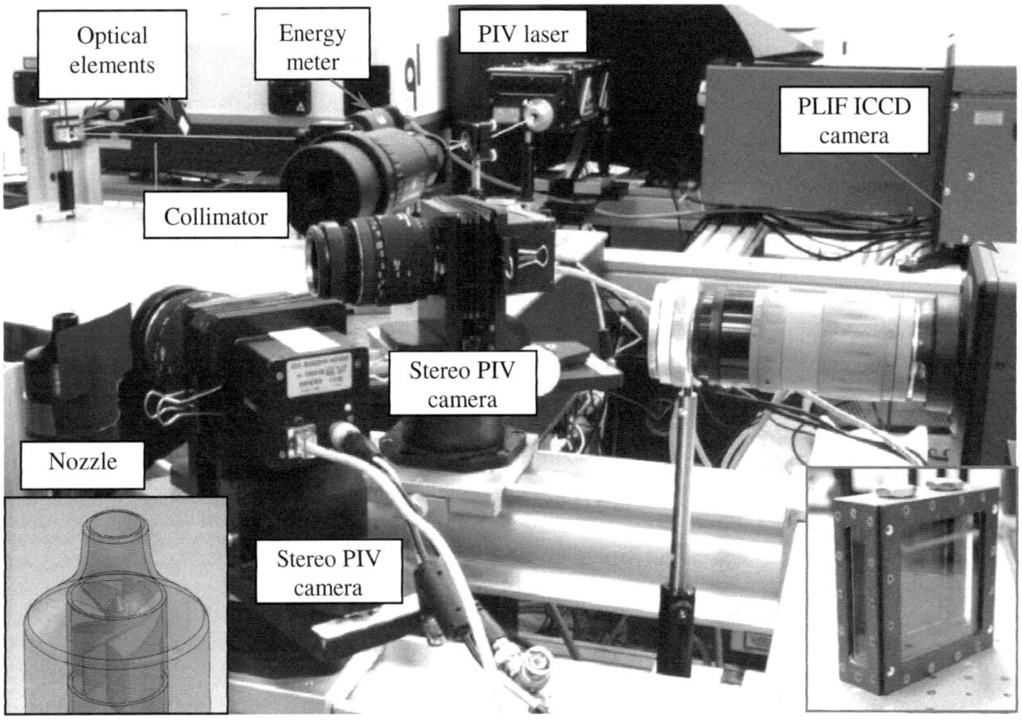

Figure 1. Photograph of the PIV/PLIF setup and calibration cuvette

Figure 1 shows the photographs of PIV/PLIF equipment and beam alignment optics. Double-head Nd:YAG laser (Quantel, EverGreen) was used to illuminate the particles tracer for PIV measurements. The laser beam was converted into a collimated laser sheet with the width of 50 mm and thickness below 0.8 mm by using a system of cylindrical and spherical lenses. On average it was 53 mJ after the laser sheet optics (measured by a power meter Coherent LabMax). Particle images were captured by a pair of CCD cameras (Bobcat ImperX). A narrow band-pass optical filter (532 nm ± 10 nm) was used. The laser and camera were synchronized via TTL signal. The PIV images (2048×2048 pixels is size) were processed using in-house software "ActualFlow". The images were pre-processed to remove background (minimal intensity for each pixel). During four iterations of an adaptive cross-correlation algorithm the interrogation area size was reduced from 64×64 to 32×32 pixels. The spatial overlap rate between the interrogation areas was 50%. The velocity fields were validated by signal-to-noise criterion.

The fourth harmonic (266 nm) of a pulsed Nd:YAG laser (Quantel, Briliant B) was used for acetone PLIF excitation. A small portion of laser radiation was directed to a power meter, using a semi-transparent mirror. The energy of each laser pulse was measured for exact comparison of the instantaneous fluorescence images. The energy of the laser radiation in the wavelength range of 266 nm was approximately 60 mJ. The laser beam was aligned with that of the PIV system, using a dichroic mirror, and transformed into a collimated sheet. Image intensifier (LaVision IRO) based on a UV-sensitive photocathode S20 (multialkali) and multi-channel amplifier were used to register the fluorescence signal of the acetone vapor. The quantum efficiency of the photocathode was about 25% at the considered wavelength range (280-320 nm). The image intensifier was equipped with a quartz lens (LaVision UV-lens, f # 2.8, 100 mm) and an optical band-pass filter (300-320 nm). The optical filter was used to remove the unwanted intensity at the laser wavelength, including reflections from the nozzle. The signal was recorded by sCMOS camera (16-bit images with resolution of 2560 × 2160 pixels). The images were processed by DaVis software from LaVision.

PLIF data were processed using a number of routines. The correction associated with non-uniform spatial sensitivity of the detector was performed using an image of a white paper list. The paper was placed away from the focus plane of the optics and was uniformly illuminated by a diffusive light source. The sensitivity between the center and the corners differed up to 20%. The energy distribution across the laser beam was not uniform either. To take into account this effect, an image of the laser sheet intensity inside a calibration cuvette, uniformly filled with acetone vapor, was captured. The PLIF data were averaged over the cells in order to correspond to the velocity fields.

3. Results

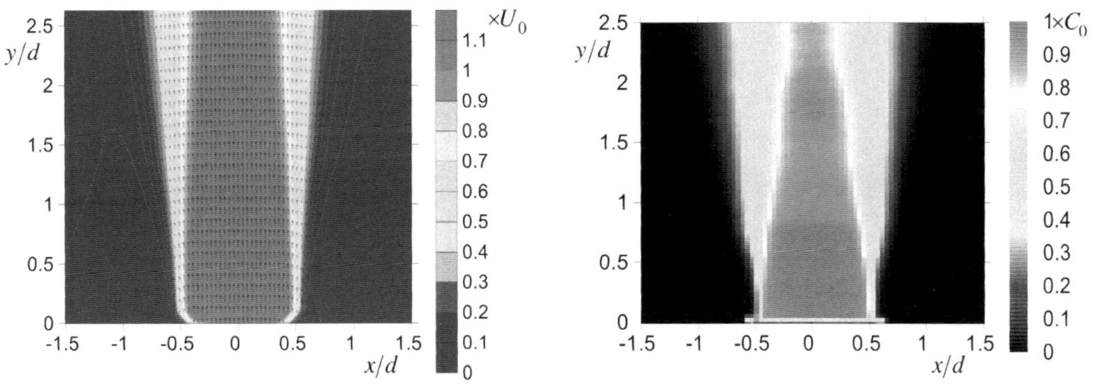

Figure 2. Normalized mean velocity (left) and concentration (right) fields for the jet at S=0

The time-averaged two-component velocity field is shown by the vectors in Figure 2 (left). The averaging is based on 3 000 snapshots of the velocity fields. The axial velocity component is shown by color. The spatial distribution of the concentration, normalized by the value at the nozzle exit, is shown in Figure 2 (right). For the case without swirl, the concentration distribution is seen to have the shape of a cone. The concentration in the jet core falls downstream. For the non-swirling jet there is a conical jet core, surrounded by the mixing layer.

Figure 3 (left) shows the time-averaged velocity field for the swirling jet with swirl rate $S = 0.41$. The normal-to-plane component is shown by color. For this swirl rate there is no recirculation zone at the jet center. The normalized concentration distribution is shown in Figure 2 (right). The swirl intensifies mixing and results in a greater opening angle, and the concentration core is frustum of a cone.

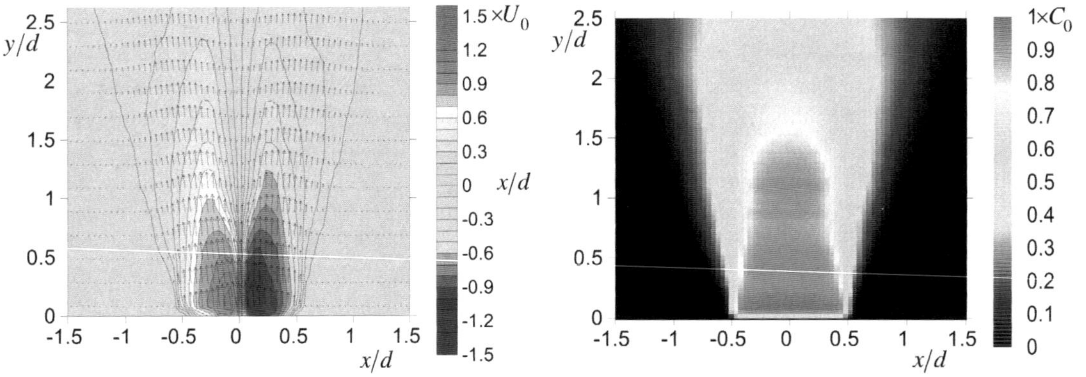

Figure 3. Normalized mean velocity (left) and concentration (right) fields for the jet at S=0.41

Figure 4 (left) shows the averaged velocity field for the jet with the swirl number $S = 1$. There is a bubble-type recirculation zone at the jet axis. It results in a much greater opening angle of the jet. The mean concentration of the scalar supplied by the jet is shown in Figure 4 (right). The concentration core disintegrates quickly, as the jet comes out of the nozzle. Thus, the concentration is typically below 70% downstream $y/d = 0.5$.

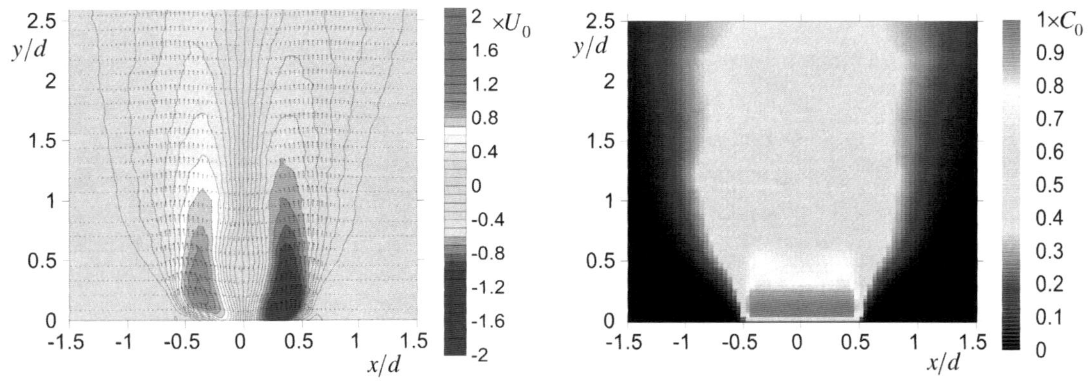

Figure 4. Normalized mean velocity (left) and concentration (right) fields for the jet at $S = 1$

Profiles of the axial component of mean velocity and the mean concentration are shown in Figure 5 for the cross-section at $y/d = 1.5$. In general, the swirl provides a wider jet and a region of the jet velocity deceleration around the jet axis. For the swirling jet with vortex breakdown the axial velocity reaches negative values, the jet is considerably wider, and the concentration profile is much flatter. Figure 5 also shows the axial velocity data from the study [8], where the jet was produced by tangential inlets inside the nozzle. One can observe that shapes of the axial velocity profiles are similar despite that the jets were produced by nozzles with fundamentally different configurations.

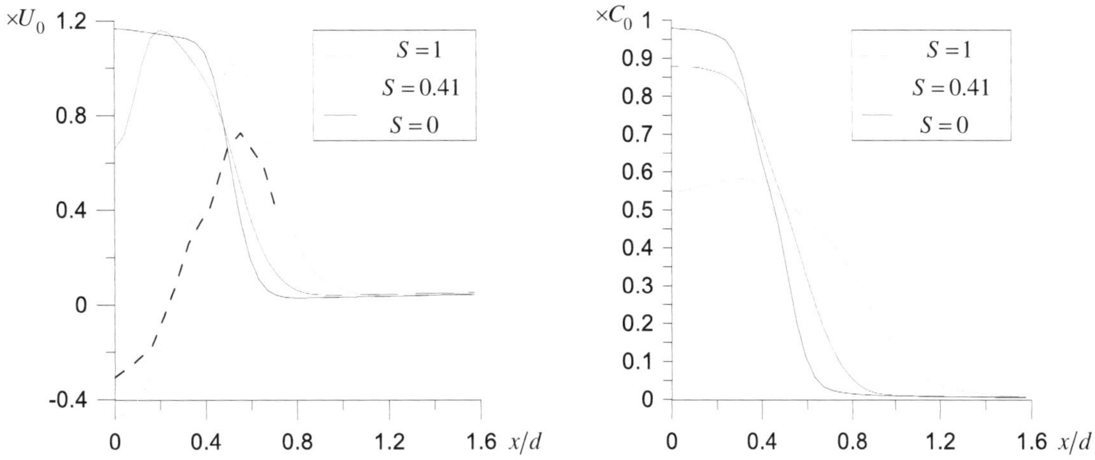

Figure 5. Profiles of the axial component of mean velocity (left) and concentration (right) at $y/d = 1.5d$. The dashed line corresponds to the data of [8] for $S = 1.0$, Re = 24.400

4. Conclusions

The flow structure and mixing in the initial region of a swirling turbulent jet were experimentally studied using the planar optical methods for velocity and concentration measurements. The Reynolds number of the jet was 5000. Three cases of the swirl rate were considered, viz., non-swirling jet, low-swirl jet without central recirculation zone and high-swirl jet with bubble-type vortex breakdown and central recirculation zone. The swirl was observed to result in a wider jet flow and to intensify mixing. For the case of vortex breakdown, the mean concentration decreases immediately downstream the nozzle exit and is found to be below 70% for the distance of $0.5d$ from the nozzle.

Acknowledgments

Financial support was obtained from the Russian Foundation for Basic Research (grant № 16-08-01236).

References

[1] Gupta A K, Lilley D G and Syred N 1984 *Swirl Flows* (Abacus Press, Kent, U.K)
[2] Wang S and Rusak Z 1997 The dynamics of a swirling flow in a pipe and transition to axisymmetric vortex breakdown. *J Fluid Mech* **340** 177–223
[3] Su L K 1998 Measurements of the three-dimensional scalar dissipation rate in gas-phase planar turbulent jets. *Annular Report of Center for Turbulence Research* 35-46
[4] Wang G H, Clemens N T , Barlow R S and Varghese P L 2007 A system model for assessing scalar dissipation measurement accuracy in turbulent flows. *Meas. Sci. Technol.* **18** 1287-1303
[5] Mourtazin D and Cohen J 2007 The effect of buoyancy on vortex breakdown in a swirling jet. *J Fluid Mech.* **571** 177-189
[6] Webster D R, Roberts P J W and Ra'ad L 2001 Simultaneous DPTV/PLIF measurements of a turbulent jet. *Exp. in Fluids* **30**(1) 375–382
[7] Alekseenko S V, Dulin V M, Kozorezov Y S and Markovich D M 2008 Effect of axisymmetric forcing on the structure of a swirling turbulent jet. *Int. J. Heat Fluid Flow* **29** 1699-1715
[8] Martinelli F, Cozzi F and Coghe A 2012 Phase-locked analysis of velocity fluctuations in a turbulent free swirling jet after vortex breakdown. *Exp. fluids* **53**(2) 437-449

IOP Conf. Series: Journal of Physics: Conf. Series **980** (2018) 012002 doi:10.1088/1742-6596/980/1/012002

Investigating the structure of a vortex flow in the closed polygonal containers

I Yu Podolskaya[1,2], G V Bakakin[1], I V Naumov[1]

[1]Kutateladze Institute of Thermophysics SB RAS, Novosibirsk 630090, 1 Lavrentyev Ave., Russia
[2]Novosibirsk State University, Novosibirsk 630090, 1 Pirogova Str., Russia

Email: naumov@itp.nsc.ru

Abstract. The structure of confined vortex flow generated by a rotating lid in a closed container with polygonal cross-section geometry (eight, six and five angles) has been investigated numerically for different height/radius aspect ratios h from 3.0 to 4.5 and for Reynold numbers ranging from 1500 to 3000. The critical Reynolds numbers at which the flow becomes unsteady were determined numerically by STAR-CCM+ computational fluid dynamics software for pentagonal and hexagonal cross-section configurations. The obtained results were compared with the flow structure in the closed cylindrical container. The boundary of a nonstationarity in polygonal containers is found to shift to the region of smaller aspect ratio and smaller Reynolds numbers with a decrease in the number of angles in the cross-section of the container relative to the boundary in a cylindrical container. It is additionally established that the structure of the flow in the near-axis region remains similar to the vortex structure in the cylinder, therefore the shape of the container does not influence the near-axis region.

1. Introduction

Development of the vortex devices in chemical, biological and power technologies requires investigating the characteristics of confined vortex flows, defining the regimes of emergence and destruction of self-organized vortex structures and the steady/unsteady regimes of the vortex flow to improve heat and mass transfer control. One of the key parameters influencing transfer processes is vortex breakdown phenomenon that appears in many engineering applications.

It is well known [1, 2] that the onset of vortex breakdown (VB) in a classical closed cylinder depends on aspect ratio h (ratio of cylinder high H to its radius R), Reynolds number $Re = \Omega R^2/\nu$ (where Ω is the angular velocity of lid rotation, and ν is the liquid kinematic viscosity) and boundary conditions, when the rotating lid drives the fluid around the container axis.

It is necessary not only to determine the regime parameters at which vortex breakdown occurs but also to control vortex breakdown formation. In work [3] it was experimentally found that the scenario of VB development is the same for cylinders with both low and high aspect ratio, and it remains independent of stationary-nonstationary transition boundary for the main vortex flow at h, ranging from 1 to 6. VB control without an input of significant additional energy but with a change in

Content from this work may be used under the terms of the Creative Commons Attribution 3.0 licence. Any further distribution of this work must maintain attribution to the author(s) and the title of the work, journal citation and DOI.
Published under licence by IOP Publishing Ltd

container geometry is a very attractive way of manipulating the flow environment. A non-intrusive control method of VB, based on co-rotation and counter-rotation of a small disk embedded in the non-rotating end wall was successfully applied [4, 5]. They showed that the aspect ratio has a dominant effect on the VB appearance, and that varying the size of the rotating disk one can control the VB behavior. They found that the co-rotating disk increases the swirl decay downstream. As a result, the VB onset occurs at a smaller Re than that with no control. By contrast, for the counter-rotating disc the bubble appears at larger Re. The chamber geometry may also have its influence. However, the majority of previous studies focused on the case of a closed cylindrical container. Other geometries have not been well explored experimentally, and only a few numerical studies considered the problem where the container consists of an inclined or wave sidewall [6].

In various technical applications the configuration of the container may also differ from the axisymmetric cylindrical one, for example, a polygonal cross-section geometry. Recent papers [7, 8] have shown that in the near-axis region of VB location the flow structure remains axisymmetric (Fig.1) in polygonal containers. The VB region for the octagonal cross-section was detected to almost correspond to the one for cylindrical container but these areas for square and cylindrical containers do not overlap in entire range of aspect ratios [8].

It should be added that even in the axisymmetric cylindrical configuration, multiplex structures of helical vortices: doublets, triplets and even quadruplets [9, 10], are formed with increasing aspect ratio, which causes the transition to the unsteady flow regime. Recent numerical [11] and experimental [12] investigations of the boundary of the unsteady flow regime formation have shown that the critical Reynolds number ($Re \sim 3000$) is independent from the aspect ratio, exceeding 5.

The purpose of this work is to expand the investigation of influence of polygonal geometry on the vortex flow especially on the unsteadiness boundary at regime parameters (h, Re) where the vortex multiplets was formed. The pentagonal and octagonal cross-section configurations have been chosen in this study (Fig.1) since the vortex breakdown region for the octagonal cross-section almost corresponds to the one for cylindrical container and these areas for square and cylindrical containers do not overlap in the entire range of aspect ratios [7, 8].

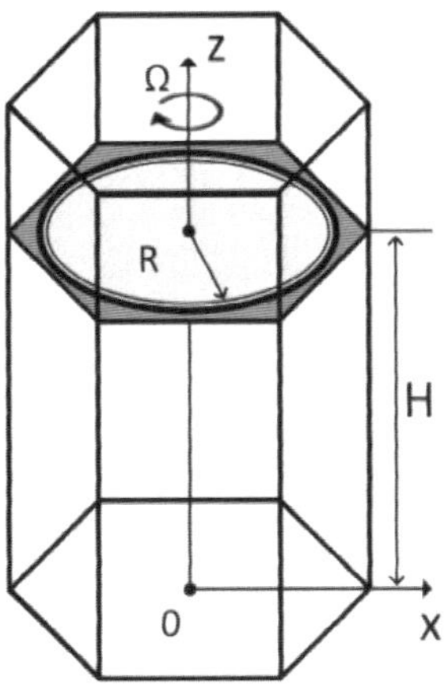

(a)

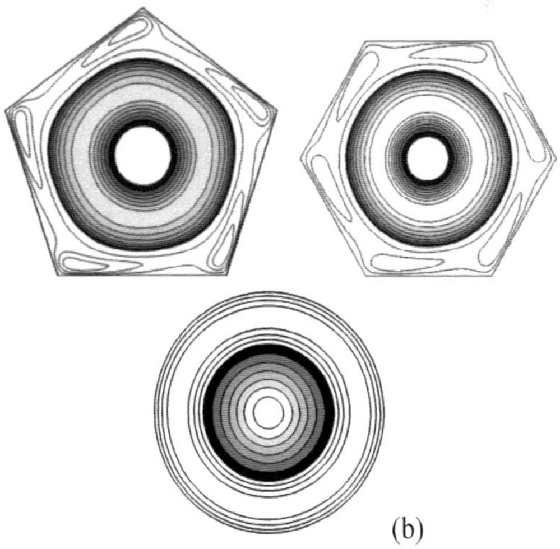

(b)

Figure 1. Schematic of the problem (a). Distribution of axial velocity in (x, y) plane, z = 0.3H, 0.5H and 0.25H for pentagonal, hexagonal and cylindrical configuration, respectively (b).

2. Problem formulation and technique of numerical simulations

The topology of confined vortex flow generated by a rotating disk in closed containers with polygonal cross-sections (pentagonal and hexagonal) was analyzed numerically by STAR CCM+ computational fluid dynamics software at h ranging from 3.0 to 4.5. The schematic of the problem geometry is presented in figure 1. The reference point of the coordinate system is set in the center of each polygon bottom. The x-axis passes through the non-rotating bottom plane. The z-axis coincides with the vertical container axis. Contrary to the cylindrical geometry, where the rotating lid is also an upper end wall of the container, in the polygonal geometry, the rotating disk is embedded in the upper end wall since its rotation is impossible due to the non-circular geometry.

The critical Reynolds numbers at which the flow becomes unsteady were determined experimentally and numerically by STARCCM+ computational fluid dynamics software. The assumption that the flow was incompressible and steady was made to compute the three-dimensional flow field in the containers with rotating end walls. The well-established STAR CCM+ commercial package is based on the finite volume method. In numerical calculation, the rotating disk radius was 50 mm. The pressure-based segregated flow model was employed to solve the governing equations. The standard no-slip boundary condition was imposed on all surfaces.

The linkage between the momentum and continuity equations was achieved using a predictor–corrector approach. The second-order upwind scheme was applied to the convective term. The SIMPLE solver algorithm was used to couple the continuity and momentum equations. To neglect total diffusion at flow boundaries near the container walls, the flow boundary diffusion property of segregated flows was used. Double precision (15 d.p.) was used for all the calculations to neglect round-off errors. Detailed description of numerical simulation and its verification is presented in papers [7, 8]. These parameters of numerical simulation allowed determining both the Reynolds numbers at which vortex breakdown bubble occurs, the position of this vortex breakdown bubble on the axis and the size of the recirculation zone, which were confirmed experimentally [8]. Therefore, it is possible to use these numerical parameters to determine the critical Reynolds numbers of the unsteadiness boundary and regimes of multiplets flow in polygonal cross-section configuration of containers.

3. Results and discussion

The structure of confined vortex flow generated by a rotating lid in a closed container with polygonal cross-section geometry has been investigated numerically for different height/radius aspect ratios h from 3.0 to 4.5 and for Reynolds numbers ranging from 1500 to 3000 to determine boundaries of multiplets formation and the onset of unsteady flow regime. In contrast to cylindrical configuration, the polygonal container geometry a priori assumes the dependence on the azimuthal coordinate.

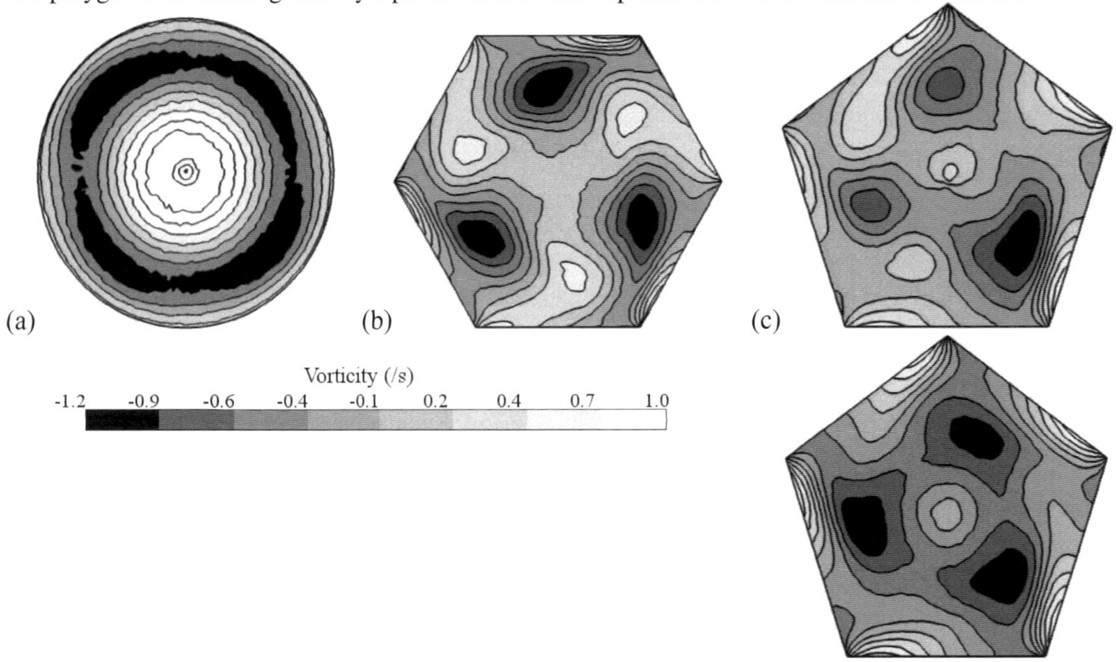

Figure 2. Numerical calculation of radial vorticity at horizontal cross-section at $z = H/4$ in cylindrical (a), hexagonal (b) and pentagonal (c, d) container configurations with aspect ratio h equal to 3.5 at Reynolds number Re equal to 2100 (a, b, c) and 1900 (d).

Figure 2 shows the distribution of radial vorticity at horizontal plane at $z = H/4$ (the area where vortex multiplets have maximal amplitude of oscillations [9, 11]) with aspect ratio $h = 3.5$. Figure 3 shows the distribution of the axial vorticity in cylindrical (a), hexagonal (b) and pentagonal (c) container configurations with aspect ratio h equal to 3.5 at Reynolds number equal to 2100.

At $Re = 2100$ the flow remains steady in the cylinder and there are no multiplets (a); in the hexagonal container it is clearly seen that the triplet occurs (b); in the pentagonal container there are several wave modes (c) that become obvious at larger Re in the cylinder [2, 12]. With a decrease of Re to 1900 (d) only one mode contributes to the distribution of radial vorticity, and the flow pattern looks like the one in the cylindrical configuration [10]. It is seen that for a fixed Re in the cylinder there are no multiplets and in hexagonal and pentagonal configurations triplets occur that is also evident from the diagram on figure 4. The triplet in the pentagonal configuration has a lower intensity compared to the one in the hexagonal configuration due to the appearance of additional oscillation modes. As for the cylinder, the triplet vortex structure is primarily formed in the hexagonal and pentagonal configurations.

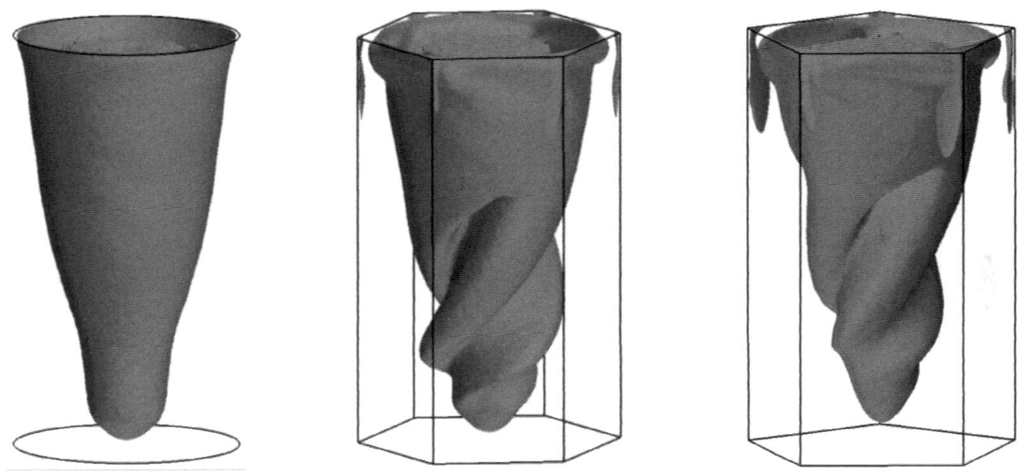

(a) (b) (c)

Figure 3. Isosurface of axial vorticity equals 3.0 in cylindrical (a), hexagonal (b) and pentagonal (c) container configurations with aspect ratio $h = 3.5$ at Reynolds number $Re = 2100$.

The obtained results are compared with the flow structure in the closed cylindrical container. Figure 4 illustrates VB boundaries in the cylinder (dashed line) and the pentagonal cross-section (solid line) in the (h, Re) plane. Open circles correspond to critical Reynolds numbers where unsteadiness appears in the cylinder; triangles are used in hexagonal and solid circles are used in pentagonal cross-section of containers. Figure 4 shows the trend of unsteadiness boundary shifting to the area of smaller aspect ratio and lower Reynolds numbers with decreasing angle numbers in the container cross-section. The same tendency is observed for the recirculation zone of vortex breakdown [7, 8], where VB onset shifts to smaller aspect ratio but to higher Reynolds numbers. The critical Reynolds number is seen to rapidly decrease (especially in the cylinder [2, 9-10] when the boundary of unsteadiness overlaps the VB onset boundary. It probably means that the vortex breakdown can make the flow stable that may be used in technical applications. In addition, figure 4 illustrates that the unsteadiness

IOP Conf. Series: Journal of Physics: Conf. Series **980** (2018) 012002 doi:10.1088/1742-6596/980/1/012002

boundary starts tending to the horizontal line at aspect ratio over 4.0 and is independent from Reynolds number, that was observed in the cylindrical container [11, 12], where the critical Reynolds number was found to be nearly *h*-independent for $h > 5.0$.

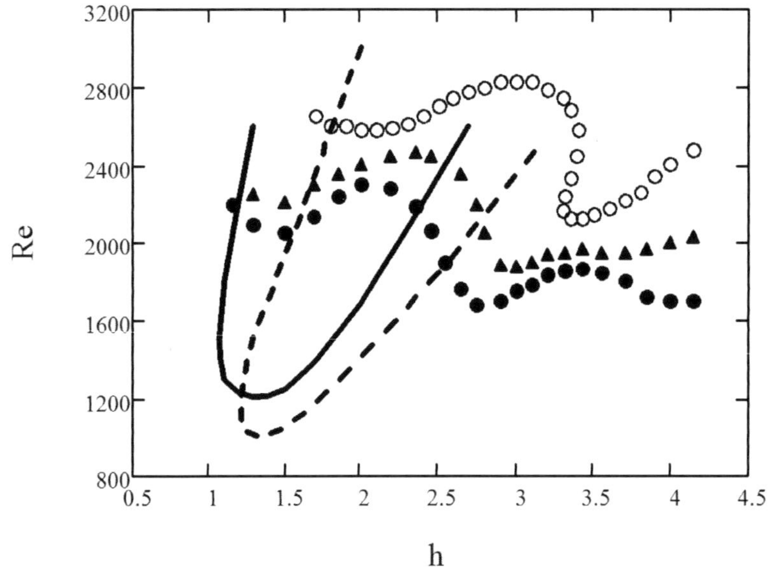

Figure 4. VB boundaries in the cylinder (dashed line) and pentagonal cross-section (solid line). Open circles correspond to critical Reynolds numbers in the cylinder, triangles – in hexagonal, and solid circles – in pentagonal cross-section configuration of containers.

Figure 5 illustrates the distribution of the axial vorticity for same intensity in cylindrical (a), hexagonal (b) and pentagonal (c) container configurations with aspect ratio *h* equal to 3.5 and Reynolds number equal to 2300, 2100, and 1900, respectively, being the critical Re for those configurations.

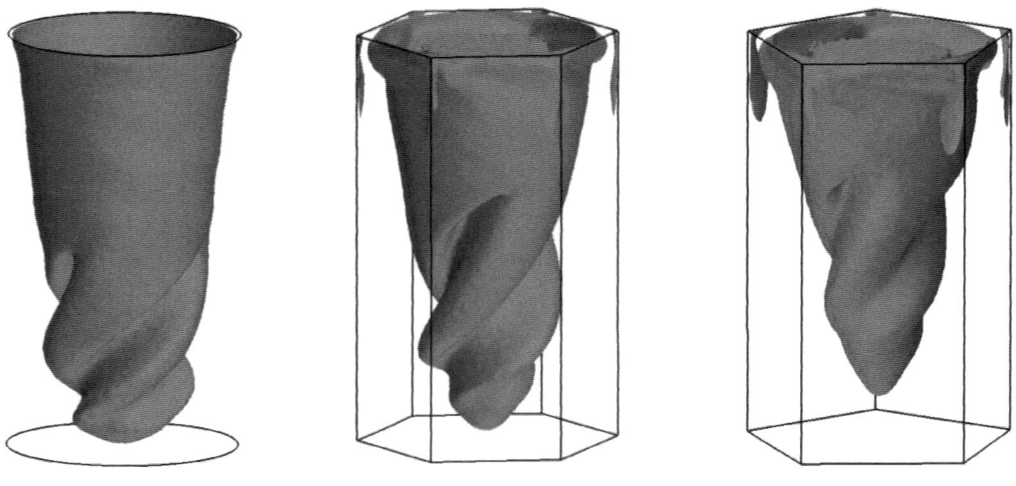

(a) (b) (c)

Figure 5. Isosurface of axial vorticity equals 3.0 in cylindrical container at Re = 2300 (a), in hexagonal container at Re = 2100 (b) and in pentagonal one at Re = 1900 (c) for aspect ratio h = 3.5.

It is seen that the structure of multiplets is of the same type. In addition, despite the smaller rotating energy (smaller Reynolds number) the multiplet is formed much lower downstream along the container axis in polygonal configuration (closer to the rotating lid) compared to the cylinder. Thus more intensive mixing occurs in the middle of the container rather than near the bottom. The same tendency was observed for the recirculation zone of vortex breakdown in the polygonal container geometry [7, 8]. In this case, the effect of the number of angles in the cross-section of polygonal container on the flow pattern in the axial region of the container is not observed. We can assume that the behavior of three-dimensional instability in pentagonal and hexagonal configurations is similar to that in the cylinder at larger Reynolds numbers.

4. Conclusions

The initial axisymmetry-breaking instability of the flow in closed containers with polygonal geometry for different aspect ratios was investigated for steady flow conditions. The influence of the container configuration on the closed vortex flow structure was studied, and the unsteadiness boundary was determined for aspect ratios ranging from 3.0 to 4.5. The flow structure in containers with polygonal cross-section and without axial symmetry was compared with a vortex flow in axisymmetric configurations. The results allow revealing regularities of formation both VB and vortex multiplets depending on Re, h and cross-section geometry of the closed container.

The new insight into some practical aspects of container flows enables a more detailed investigation of the flow structure for non-axisymmetric geometry of closed container. It has been found that reducing the number of cross-section angles from eight to five shifts the instability onset to lower Reynolds numbers and smaller aspect ratios. The results show that at steady conditions there are no asymmetric distortions of the flow topology due to the non-axisymmetric effect of the polygonal geometry of the used containers (neither in horizontal nor in vertical cross-section).

The self-organizing coherent vortex structures have been found and analysed. The vortex flow structure in the near-axis area remains similar to that in the cylinder; therefore, the polygon form of the container does not exert considerable impact on the multiplet structure. We can assume that the onset of the three-dimensional instability based on appearance of vortex multiplets in pentagonal, hexagonal and octagonal configurations is similar to that in the cylinder at increasing Reynolds number.

5. Acknowledgments

The study was funded by the Russian Foundation for Basic Research within the research project № 18-08-00508.

References

[1] Escudier M P 1984 *Exp. Fluids* **2** 89-196
[2] Sørensen J N, Naumov I V and Mikkelsen R 2006 *Exp. Fluids* **41** 425-440
[3] Naumov I V, Mikkelsen R F and Okulov V L 2014 *Thermophys. Aeromech.* **21** 767-770
[4] Mununga L, Lo Jacono D, Sørensen J N, Leweke T, Thompson M C and Hourigan K 2014 *J. Fluid Mech.* **738** 5-33
[5] Yu P, Lee T P, Zeng Y and Low HT 2007 *Phys. Fluids* **19** 057104
[6] Yu P and Meguid S A 2009 *Phys. Fluids* **21** 017104
[7] Naumov I V, Dvoynishnikov S V, Kabardin I K and Tsoy MA 2015 *Phys. Fluids* **27** 124103
[8] Naumov I V and Podolskaya I Yu 2017 *J. Fluid Mech.* **820** 263-283
[9] Naumov I V, Okulov V L, and Sørensen J N 2014 *Thermophys. Aeromech.* **17** 551-558
[10] Sørensen J N, Naumov I V and Okulov V L 2011 *J. Fluid Mech.* **683** 430-441
[11] Herrada M A, Shtern V N and Torregrosa M M 2015 *J. Fluid Mech.* **766** 560-610
[12] Sørensen J N, Gelfgat A Y, Naumov I V and Mikkelsen R 2009 *Phys. Fluids* **21** 054102

IOP Conf. Series: Journal of Physics: Conf. Series **980** (2018) 012003 doi:10.1088/1742-6596/980/1/012003

Kelvin waves on helical vortex tube in swirling flow

M A Tsoy[1,2], **S G Skripkin**[1], **P A Kuibin**[1], **S I Shtork**[1] **and S V Alekseenko**[1]

[1] Kutateladze Institute of Thermophysics, Siberian Branch of Russian Academy of Sciences, 1 Ac. Lavrentyev Ave., Novosibirsk, 630090, Russia

[2] Email: miketsoy@gmail.com

Abstract. Kelvin waves formed on precessing helical vortex tube after vortex reconnection were experimentally visualized. The experiments were performed in a swirling flow of viscous fluid in optically transparent conical diffuser. Vortex visualization was facilitated by cavitation in the vortex core and carried out using high-speed video cameras. Kelvin waves of zero, first and second modes $m = 0, 1, 2$ were formed in packets as a result of reconnection due to vortex tube self-intersection. Kelvin waves were observed to form both on the helical vortex tube and the detached vortex ring and to propagate along the axis of the vortex tubes.

1. Introduction

The stable wave modes supported by a straight vortex tube of uniform vorticity and circular cross-section which propagate along the axis of the vortex tube were mentioned more than 130 years ago by Lord Kelvin. Azimuthal wavenumber m describes the azimuthal structure of the wave. The perturbation of a vortex tube by a single wave mode takes the form of m helices.

Kelvin waves are of immense interest in quantum vortex dynamics [1-5]. A fundamental question in quantum turbulence is the nature of dissipation in the zero-temperature limit where the effects of friction vanish. A cascade of Kelvin waves generated on vortex lines as a result of occasional close approach and by individual vortex lines reconnection events transfers energy to higher wave numbers where it is radiated as sound. These effects highlight the importance of Kelvin waves and reconnections in the transfer of energy within a turbulent vortex tangle. Arendt, Fritts and Andreassen (1997) showed that Kelvin modes form a basis for the perturbations localized in the vortex core. This means that any perturbation in the core of a Rankine vortex can be expressed by a combination of Kelvin wave modes [6]. There are just a few experimental works dedicated to Kelvin waves dynamics in viscous fluid. The latest of such works which includes the observation of large-amplitude 'kink' waves on the vortex cores produced by an oscillating grid in a rotating fluid was published by Maxworthy and Hopfinger more than 30 years ago [7]. It can be stated (to our knowledge), that Kelvin waves on helical vortex tube were observed for the first time.

2. Experimental setup

Research was carried out at the Institute of Thermophysics in Novosibirsk on a closed hydrodynamic loop. The test rig, experimental equipment and mean flow velocity measurements are described in detail in previous publications [8, 9]. This test rig generates a swirling flow (S = 0.47) in a conical diffuser with throat diameter of 100 mm and allows visualizing a precessing helical vortex and the process of vortex reconnection at Reynolds number Re ~ 10^5. The diffuser is made of acrylic glass, which allows obtaining the optical access to the flow. The connected vacuum pump creates rarefaction in the loop, thus allowing visualizing vortex structures by the vapor phase formed as a result of cavitation in the vortex core.

The structure of the base flow in the conical diffuser is shown in figure 1. The main flow is directed along the walls from top to bottom and swirled in clockwise direction (view from above). A reverse flow region is formed in the axial zone. Hollow vortex tube folds in the left-handed (winding of the

Content from this work may be used under the terms of the Creative Commons Attribution 3.0 licence. Any further distribution of this work must maintain attribution to the author(s) and the title of the work, journal citation and DOI.
Published under licence by IOP Publishing Ltd

helix is opposite to the flow swirling direction) helix rotating in the same direction with main swirling flow. For current experimental conditions the frequency of the helical vortex precession is 58 Hz, which corresponds to a period of full revolution of the helix equal to 17 ms. Every 2-3 periods of precession due to unstable interaction of nearby tube coils via the velocity field that they induce on each other helical tube deforms, and helix coils approach each other, leading to self-intersection of the tube. Vortex tubes with opposite vorticity vectors are oriented parallel to each other (see figure 2 t = 0). Over 0.7 ms, the tubes are bent and flattened in reconnection zone, and a reconnection process takes place within 0.8 ms. Bended bridges are formed and begin to move away from each other due to induced motion.

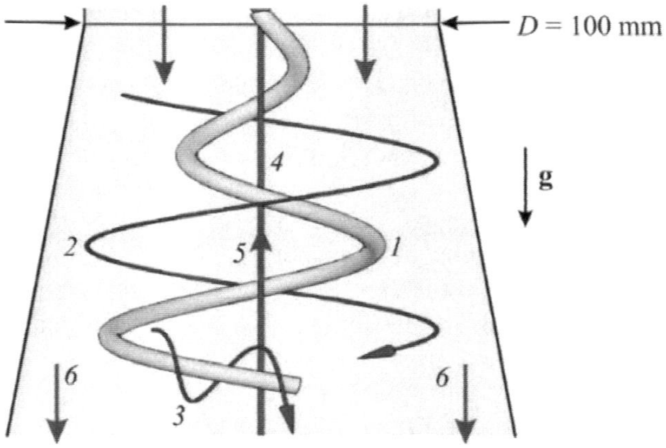

Figure 1. Structure of the base flow in the conical diffuser: 1 – helical vortex tube, 2 – fluid rotation, 3 – trajectories of fluid near the vortex, 4 – diffusor axis, 5 – axial reverse flow, 6 – flow direction.

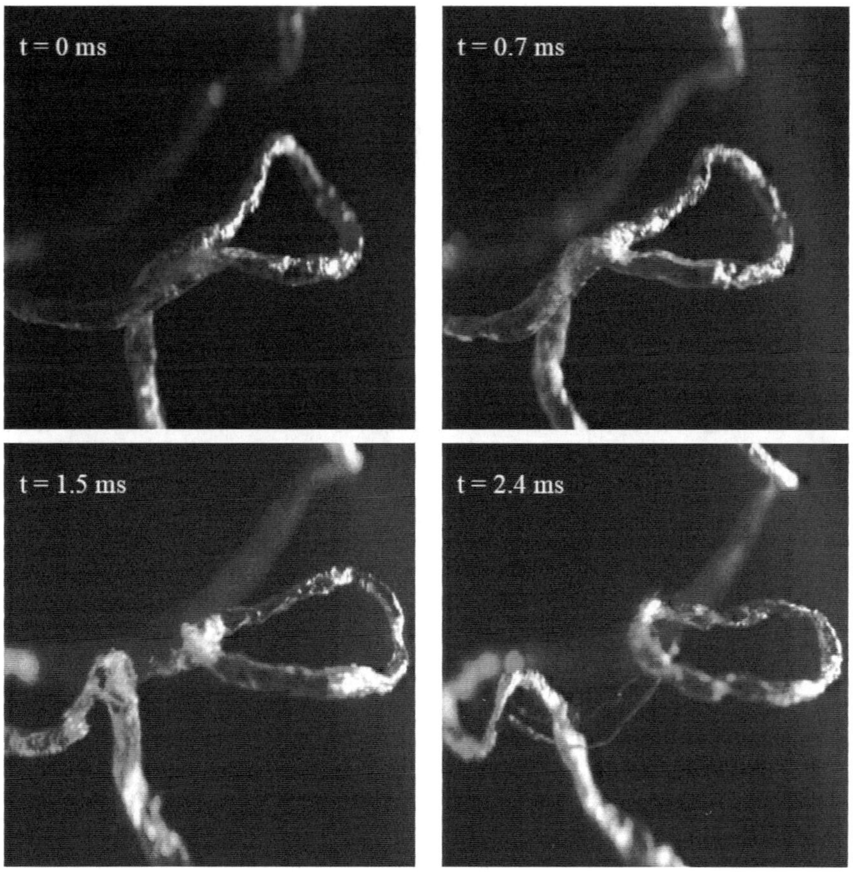

Figure 2. Visualization of reconnection process with detached vortex ring formation, 7000 fps and an exposure time 50 μs [4].

3. Kelvin wave observation

After the reconnection, strong perturbations of vortex tubes remain: sharp bending and deformation of bridges, leading to formation of Kelvin waves packets, containing various modes travelling on the tube. Despite the fact that the vortex tube in our experiments is helical ,our observations coincide with Arendt's et al. finding that any initial perturbation of such a tube evolves as a sum of Kelvin waves, so that any localized disturbance propagates away from its original location as a wave packet [2]. As a result of reconnection depending on the ratio of various magnitudes of bending, compression and flattening perturbations of the vortex tube, one of the modes turns out to be more pronounced than others, but with time another mode may become dominant due to the difference in dispersion rates. At a sufficiently short time between successive reconnection events (comparable to the lifetime of the Kelvin wave), several Kelvin wave packets are generated and simultaneously propagate on the vortex tube, see figure 3(b).

The Kelvin wave of zero mode $m = 0$ is visualized on the helical tube as a steep axisymmetric decrease of the diameter of vortex tube cavity, propagating along the tube upward, i.e. in the direction opposite to the tube vorticity vector according to vortex lines' twist inducing an axial flow, see figure 3. In its turn the tube moves downstream with velocity of the same order and in result $m = 0$ wave can be observed for several periods of helical tube precession without dispersion. Usually such waves are separated from the helical tube by another reconnection event or are transferred with helix motion out of test section downstream. This mode is induced by axisymmetric deformation of the vortex tube, associated with a sharp local change in the diameter and vorticity of vortex tube core in the reconnected section of the tube. It should be noted that such waves are not clearly observed on the separated vortex ring.

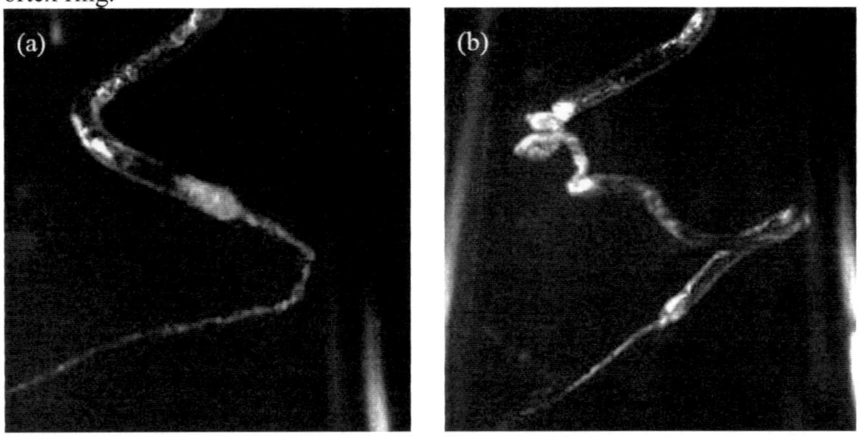

Figure 3. Visualization of Kelvin waves: (a) - single zero mode Kelvin wave on helical vortex tube, (b) – zero and first mode simultaneously, exposure time - 200 μs.

Prior to reconnection, the connecting parts of the helical tube approach each other at an almost perpendicular orientation, leading to a sharp kink of both reconnected vortex tubes after reconnection (see figure 2) which in turn induces $m = 1$ Kelvin waves in the form of rotating helical deformation of the vortex tube centerline without changes in tube diameter. As one can see on figure 4(a) after reconnection the first mode waves are observed to propagate both on the helical tube and on the separated ring. On the helical tube left-handed (helix winding direction is opposite to vortex tube rotation direction) wave travels up along the tube in the direction opposite to the tube vorticity vector, as shown in figure 4(b). On the separated vortex ring $m = -1$ right-handed wave (helix winding direction coincides with vortex tube rotation direction) travels along the tube in the direction opposite

IOP Conf. Series: Journal of Physics: Conf. Series **980** (2018) 012003 doi:10.1088/1742-6596/980/1/012003

to vorticity vector. Both an axial propagation and an azimuthal rotation of waves are induced by self-advection of a curved vortex tube. An important consequence of the self-induced motion of the waves $m = 1$ and $m = 2$ is that the direction of propagation is determined by the vorticity direction in the tube and the direction of the helical rotation of the wave packet. Such Kelvin waves may be also formed outside the reconnection process when two tube sections approach close enough to interact but not close enough to reconnect. The most frequently observed modes combination in our experiment is $m = 1$ and $m = 0$ shown in figure 4(c).

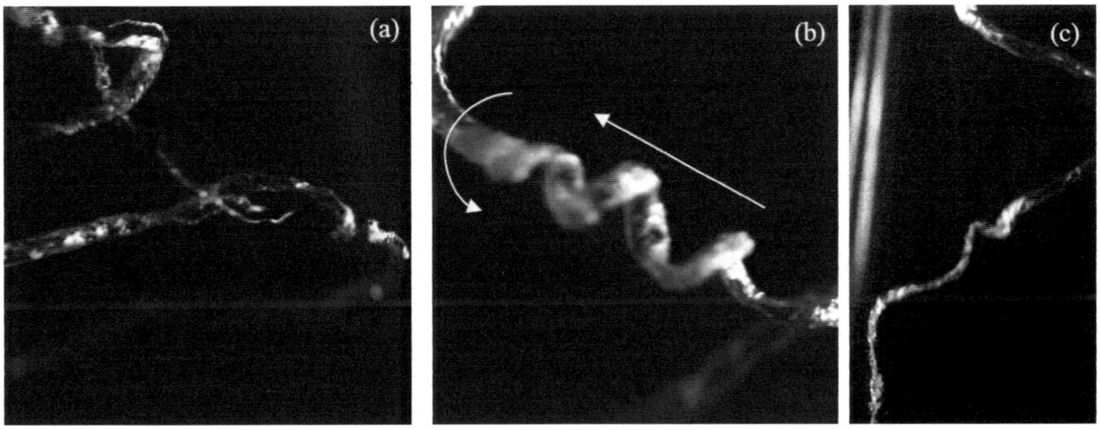

Figure 4. Visualization of Kelvin waves: (a) - m=1 waves generated simultaneously on vortex ring and helical vortex, (b) - single m=1 wave, (c) - wave packet containing m=0 and m=1 modes. Exposure times : (a), (b) – 50 μs, (c) - 200 μs.

One can clearly observe in figure 5 that $m = 1$ Kelvin waves' packets disperse while propagating - their amplitude decreases, its period increases and the vortex tube restores its unperturbed state.

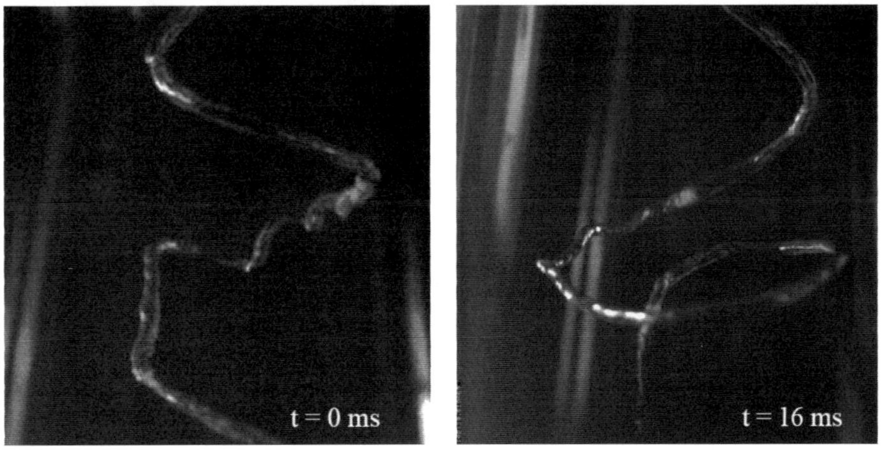

Figure 5. Visualization of $m = 1$ wave packet dispersion, exposure time - 200 μs.

Non-axisymmetric deformation of the tube - its flattening without displacement from the tube's centerline leads to the appearance of $m = 2$ Kelvin waves. With sufficient initial deformation, such a wave splits the vortex tube into two intertwined left-handed vortex tubes of smaller diameter (see figure 6). Since vorticity is divided in half it is hard to visualize an intertwined vortex pair with cavitation because of a lower pressure drop of the vortex core. It may be noted that vorticity is not evenly divided in half, and one vortex of the pair may be more pronounced than another. The curvature of these intertwined tubes (taking into account the influences on each other) self-induces both the tangential and axial components of motion - rotation of the double-helix pattern around the tube and axial movement up along the vortex tube as well as in the $m = 1$ mode. This mode is the least common in our experiments and was not observed on the separated vortex ring.

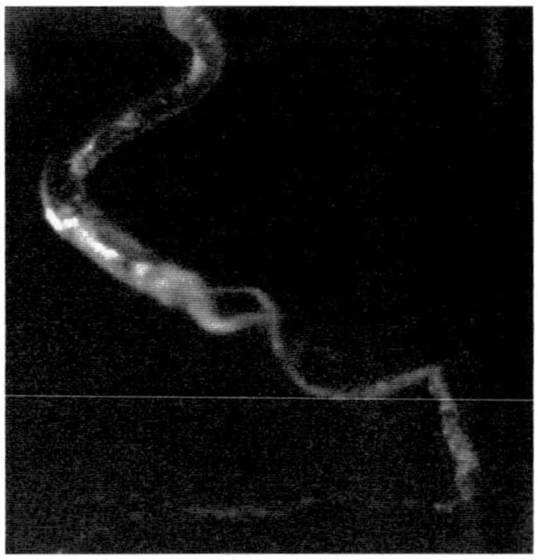

Figure 6. m = 2 twist wave on the helical vortex tube.

4. Conclusion

This work for the first presents time an experimental visualization of varying modes of Kelvin waves formed on the helical vortex tube as a result of the vortex reconnection. The $m = 0$ and $m = 2$ modes propagate along the helical vortex tube in direction opposite to the tube vorticity vector. The $m = 1$ mode forms the left-handed helix, propagating along the helical vortex tube in the direction opposite to the tube vorticity vector, and the right-handed helix, propagating on the vortex ring in the direction coinciding with the tube vorticity vector. It should be noted that modes $m = 0$ and $m = 2$ lead to a much lower pressure drop in the vortex core and in the vortex core cavity size, and the single $m = 1$ mode does not affect the size of the vortex core cavity noticeably. Further work should be devoted to a quantitative research of Kelvin waves' properties.

Acknowledgments
This research was supported by the Russian Foundation for Basic Research, Grant No. 16-08-01191.

References

[1] Fonda E, Meichle D P, Ouellette N T, Hormoz S and Lathrop D P 2014 Direct observation of Kelvin waves excited by quantized vortex reconnection *Proc. Natl. Acad. Sci.* **111** 4707–10
[2] Paoletti M S, Fisher M E and Lathrop D P 2010 Reconnection dynamics for quantized vortices *Phys. D Nonlinear Phenom.* **239** 1367–77
[3] Kivotides D, Vassilicos J C, Samuels D C and Barenghi C F 2001 Kelvin Waves Cascade in Superfluid Turbulence *Phys. Rev. Lett.* **86** 3080–3
[4] Van Gorder R A 2014 Decay of helical Kelvin waves on a quantum vortex filament *Phys. Fluids* **26** 75101
[5] Tsubota M, Araki T and Nemirovskii S K 2000 Dynamics of vortex tangle without mutual

friction in superfluid ^{4}He *Phys. Rev. B* **62** 11751–62
[6] Arendt S, Fritts D C and Andreassen O 1997 The initial value problem for Kelvin vortex waves *J. Fluid Mech.* **344** 181–212
[7] Maxworthy B T and Hopfinger E J 1985 Wave motions on vortex *J. Fluid Mech.* **151** 141–65
[8] Alekseenko S V, Kuibin P A, Shtork S I, Skripkin S G and Tsoy M A 2016 Vortex reconnection in a swirling flow *JETP Lett.* **103** 455–9
[9] Skripkin S G, Tsoy M A, Kuibin P A and Shtork S I 2017 Study of pressure shock caused by a vortex ring separated from a vortex rope in a draft tube model *J. Fluids Eng.* **139** 81103

Hydraulic resistance of tubes with internal helical finning designed by deforming cutting

I A Popov[1], A V Shchelchkov[1], A N Skrypnik[1], D V Ryzhkov[1], N N Zubkov[2], Yu V Zhukova[3], S A Sverchkov[3]

[1]Kazan National Research Technical University named after A.N. Tupolev – KAI, Kazan, Russia
[2]Bauman Moscow State Technical University, Moscow, Russia, zoubkovn@bmstu.ru
[3] A.V. Luikov Heat and Mass Transfer Institute of NAS of Belarus, Minsk, Belarus,

Email: skrart555@gmail.com

Abstract: An object of investigation is a tube having helical fins on the internal surface and different geometric sizes. Investigation methods: experiments to obtain quantitative results for hydraulic resistance of tubes with internal helical finning and to verify computational algorithm; numerical simulation to visualize the flow structure in the tube. The hydraulic resistance of tubes with internal helical finning was studied over a wide range of operating and geometric parameters: $Re_D=2\cdot10^3...2.5\cdot10^5$, under the variation of the angle of swirling $\alpha=14^\circ$-87°, the relative height of a protrusion $h/d=(25\text{-}87{,}5)\cdot10^{-3}$, and the relative axial pitch $p/d=0.16$-12.73. It has been revealed that the hydraulic resistance of tubes with helical finning increases by a factor of 1.1 to 11.7. The numerical simulation results show that with an increase in the angle of helical swirling, in the near-wall layers the share of the circumferential velocity component increases and the share of the longitudinal component decreases. And since the finning height exceeds the boundary layer thickness, the hydraulic resistance grows.

1. Introduction

In forced heat carrier flow in the tube from the tube inlet, dynamic boundary layers start forming at the walls; their thickness gradually grows with increasing the distance from the tube inlet. At a time, thermal boundary layers are formed, which hinder heat transfer between the tube and heat carrier. The thickness of temperature and dynamic layers is related as $\delta_{temp}/\delta_{dyn} \sim Pr^{-0.5}$. At some distance from the tube inlet, boundary layers merge and flow becomes stabilized.

Heat transfer enhancement techniques can be divided into three types. The first type is active techniques requiring external power supply (induced vibration, acoustic action, boundary layer scraping), the second type is passive methods not requiring external power supply, and the third type is combined methods assuming the use of two or more active /passive methods.

Many of heat transfer enhancement techniques include surface modification such as different kinds of surface roughness on the tube inside or internally finned tubes. The thermal and hydraulic characteristics of tubes having helical fins on the internal surface – internal helical finning (for example, specially made ridging or helically corrugated internal surface) have been widely investigated over the last 30 years [1-11]. The installation of such designs allows a flow to be swirled in effort to make disturbances in the near-wall layers of heat carrier Such designs provide flow swirling producing disturbances in the near-wall layers of heat carrier [1-11]. Most of the published experimental works are devoted to the possibility of industrial use of such tubes in shell-tube heat exchangers. This will allow either decreasing the mass and size (metal consumption) of heat exchangers, or increasing heart load per surface area at fixed overall sizes of the latter. The tubes with helical fins on the internal surface increase heat transfer coefficient due to disturbances in the near-wall layers of heat carrier. However, at a time heat carrier flowing inside the tube becomes turbulent [4, 10-11].

Content from this work may be used under the terms of the Creative Commons Attribution 3.0 licence. Any further distribution of this work must maintain attribution to the author(s) and the title of the work, journal citation and DOI.
Published under licence by IOP Publishing Ltd

As shown in [2-9], thermal and hydraulic characteristics of tubes with internal helical finning are insufficiently studied for the standard flow conditions of viscous liquids in industrial heat exchangers. The studies of hydraulic resistance of tubes with single-threaded internal helical finning [2-9] have been made over the range of $Re_D=3\cdot10^3...1.2\cdot10^5$ for the transient and turbulent flow regions of heat carrier. Dimensional geometric parameters of internal helical finning – the ratio of a helical swirling pitch p to a tube diameter D and of a helical finning height h to a tube diameter D – varied within p/D=0.14-1.2 and $h/D=(6\text{-}88)\cdot10^3$. The attempt to extend the range of the operating characteristics, namely, to make investigations at high Prandtl numbers Pr_f=10-90 and low Reynolds numbers Re_D ranging from $2\cdot10^3$ is outlined elsewhere in [10]. The maximum values of thermal efficiency reach Nu/Nu_0=2-2.1 at a comparable growth of hydraulic resistance ξ/ξ_0=1.8-2.4, as shown in [8].

A short review brings to a conclusion that, despite a significant amount of experimental works on tubes with single-threaded internal helical finning, additional studies should be made of flow structures over wide ranges of Reynolds and Prandtl numbers, as well as of geometric tube parameters. In particular, the flow structure visualization has not been realized in such tubes because of technical difficulties. To get information about the flow structure in tubes with internal helical finning, numerical simulation methods can be adopted.

It should be emphasized that, to obtain adequate results by numerical simulation methods, first, the computational algorithm must be verified by problems having physical analogs; second, it is necessary to correctly use the designed computational grids taking into account the geometry features of the object of investigation and the boundary layer flow; third, the correct approaches to close the Navier–Stokes equations must be correctly used, i.e., for the equations to be closed, the turbulence model must be correctly chosen.

Thus, the objective of this study is to conduct joint experimental and numerical research of hydraulic resistance and flow structure in heat carrier flow in tubes with internal helical finning. Experimental investigation will allow obtaining characteristic curves for hydraulic resistance of tubes with internal helical finning for laminar, transient and turbulent flow regimes, as well as determining optimal geometric parameters of tubes with internal helical finning, depending on operating parameters at minimum hydraulic loss ξ/ξ_0. Numerical simulation results will enable visualization of the flow structure in tubes having different helical fins on the internal surface and will qualitatively explain the effects obtained.

2. Object of investigation

An object of investigation is a tube having helical fins on the internal surface and different geometric sizes. The basic geometric parameters of the object of investigation are shown in Fig. 1. The present work studies six tubes having single-threaded helical fins made on the internal surface by the deforming cutting technique, combining at a time the processes of deforming and cutting the surface layers of the tube [12-13]. The copper tubes of inner diameter D=16 mm and length L=800 mm were investigated; their relative length was L/D=50, which is indicative of the fact that studies of heat transfer in such tubes must allow for the influence of the thermal entry length.

The basic geometric parameters of tubes are cited in Table 1. Water at 20° (Pr=6.97) served as heat carrier.

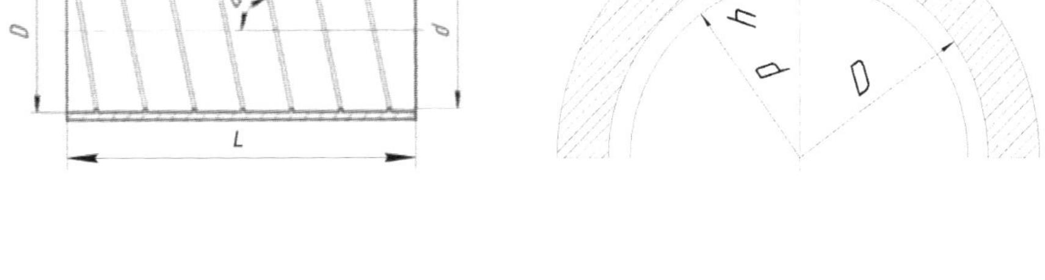

a) *b*)

Figure 1. Longitudinal (*a*) and transverse (b) sections of tubes with internal helical finning: *D* – diameter of a starting smooth tube, *d* – deformed tube diameter taken from the height of an intensifier (helical protrusion), *p* – helical swirling pitch of an intensifier, *h* – height of an intensifier (helical protrusion), α – helical swirling angle, $\phi=h^2/(p\cdot d)$– dimensionless parameter responsible for the influence of created swirling on flow characteristics.

Table 1. Basic geometric parameters of tubes under study

α°	*p*, mm	*h*, mm	*d*/*D*	$h/D\cdot 10^3$	*p*/*D*
14	198	1.4	0.825	87.5	12.73
32	80	0.6	0.925	37.5	5
46	48	0.7	0.913	43.75	3
61	28	0.7	0.913	43.75	1.75
76	12	0.4	0.95	25	0.75
87	2,5	0.7	0.913	43.75	0.16

3. Research techniques

Test bench for investigation of hydraulic resistance of tubes with internal helical finning.

The hydraulic scheme of the test bench (Fig. 2) is designed in the form of an open loop with a system of forced supply of heat carrier to the working (measuring) section. The test bench comprises the facilities of water storage *2* and water supply *3*, *4* to working section *12*, a piping, a measuring system of bulk (mass) flow rate by reference flowmeters *6*,*7* and a control system.

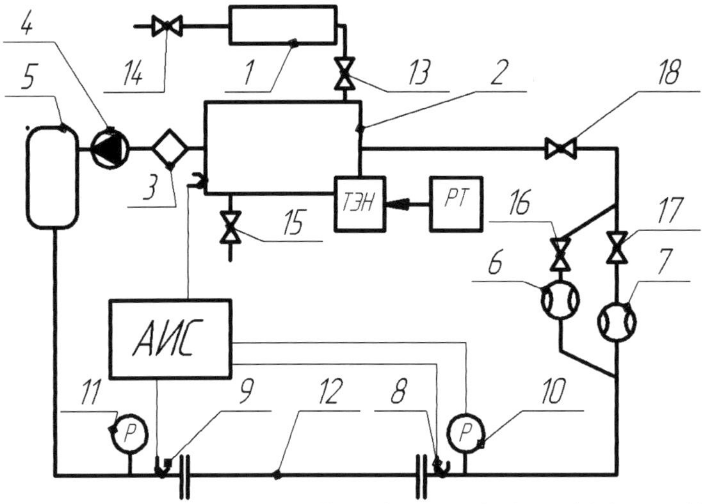

Figure 2. Test bench for investigation of hydraulic resistance of tubes with internal helical finning: 1– distiller, 2 – tank-heater, 3 – filter, 4 – high pressure pump, 5 – receiver, 6,7 – flowmeters, 8, 9 – thermocouples, 10,11 – pressure sensors, 12 – working section, AIS – automated information system, TEH – tubular electric heater, TC – temperature controller, 13, 14, 15, 16, 17 – valves.

Working section *12* for experimental study of hydraulic resistance of tubes with internal helical finning is a channel with its inlet and outlet located on the axis. Heat carrier temperature at the working section inlet and outlet was controlled by chromel-alumel thermocouples *8, 9*. To measure static pressure, pressure taps with 0.8 mm diameter were envisaged in connecting pipes at inlet *10* and outlet *11*, respectively. Pressure drop on the working section was measured by OWEN pressure sensors for isothermal conditions of heat carrier flow.

Hydraulic resistance coefficient was determined by formula (1):

$$\xi = \frac{2\Delta p}{\rho w^2}\frac{D}{L} \quad (1)$$

where Δp is the pressure drop on the working section, ρ is the heat carrier density, and w is the bulk velocity of heat carrier. A relative error in determining hydraulic resistance did not exceed 6.5%.

Numerical simulation

Numerical simulation was performed with the use of gasdynamic solver ANSYS Fluent 14.5. Steady-state Reynolds-averaged Navier–Stokes equations (Reynolds equations) and the continuity equation were solved. The Reynolds equations were closed by the κ-ω Menter shear stress transfer model [14]. A medium moving in a tube was assumed to be incompressible, and its thermophysical properties (density and viscosity) were assumed constant and independent of temperature and pressure.

A computational grid consisted of tetrahexagonal and hybrid elements closely packed near the tube finning. Total capacity of the computational grid was from 4 mln cells for a tube without finning to 8.5 mln cells for a tube with finning. A minimal size of a computational cell was 0.2×10^{-3} m.

Mass flow rate and operating pressure were assigned at the computational domain inlet; a tube inner surface was assumed to be smooth, on which no-slip conditions were realized. Outflow boundary conditions were set at the computational domain outlet. Gravity was directed perpendicular to the incoming flow. In the course of numerical simulation, boundary conditions were predetermined to be consistent with experiment conditions.

4. Analysis of the results obtained

Analysis of the experimental data on hydraulic resistance of tubes with internal helical finning

Experiments on hydraulic resistance of tubes with single-threaded internal helical finning were performed within the turbulent regime of forced water flow over the Reynolds number range $Re_D=2\cdot10^3...2.5\cdot10^5$ (Fig. 3). The obtained experimental results are compared with the data on the tube without internal helical fining and with those calculated by the formula $\xi_0=0.3164/Re^{0.25}$ (Blasius' law).

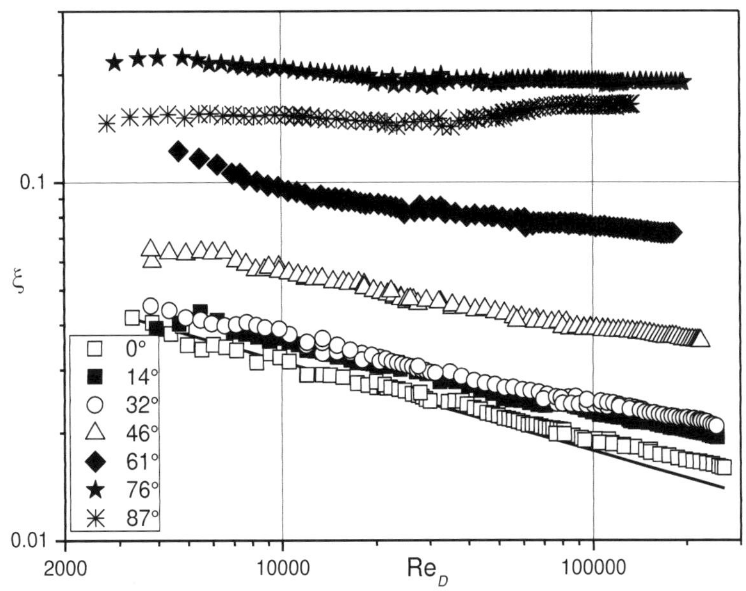

Figure 3. Experimental data on hydraulic resistance: line – the calculation by the relation $\xi=0.3164/Re^{0.25}$ for a tube without internal finning

To evaluate the adequacy of the data obtained, the results for the experimental turbulent regime of the authors of this article, presented in Fig. 4, were compared with the data [10]. Comparison was made for a tube at an angle of swirling $\alpha=46°$ and for tube No. 08 [10], as well as for a tube at an angle of swirling $\alpha=61°$ and for tube No. 06 [10]. The comparison of the results has yielded a satisfactory agreement; the disagreement did not exceed 5%.

The analysis of the experimental results (Fig. 3) on tubes with internal helical finning revealed that the hydraulic resistance coefficients increased with increasing angle of swirling $\alpha=(14\text{-}87)°$. It should be noted that the values of hydraulic resistance coefficient for tubes with $\alpha=14°$ and $\alpha=32°$ were close. A maximal increase in hydraulic resistance coefficient in the turbulent Reynolds number range was $\xi/\xi_0=11.7$ times in comparison with a smooth channel.

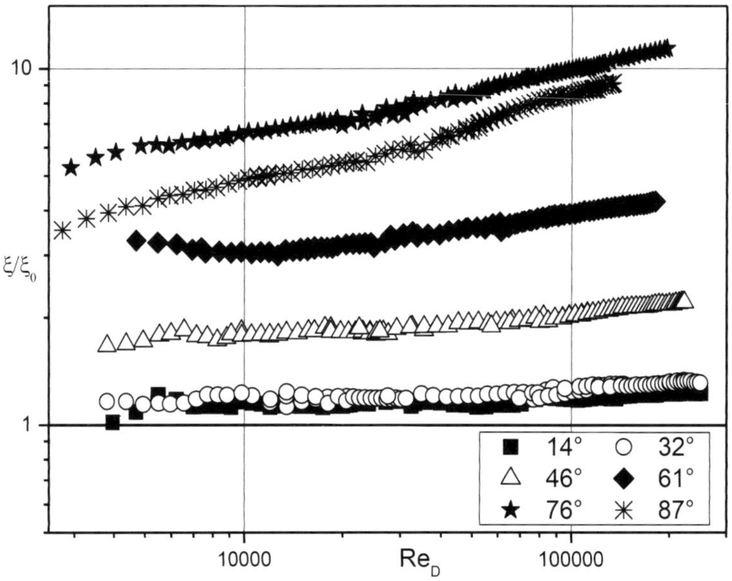

Figure 4. Comparison of hydraulic resistance coefficients of tubes at different angles of swirling

It should be noted (Fig. 4) that at $Re_D=4500$, the ratio of hydraulic resistance coefficient of tubes with internal helical finning ξ to hydraulic resistance coefficient of tubes without finning ξ_0 is $\xi/\xi_0=1.05\text{-}1.1$ for a tube with an angle of fin twisting $\alpha=14°$; $\xi/\xi_0=1.1\text{-}1.15$ at $\alpha=32°$; $\xi/\xi_0=1.7\text{-}1.8$ at $\alpha=46°$; $\xi/\xi_0=3.0\text{-}3.1$ at $\alpha=61°$; $\xi/\xi_0=5.9\text{-}6.1$ at $\alpha=76°$; $\xi/\xi_0=3.95\text{-}4.05$ at $\alpha=87°$. Attention should be paid to the fact that the tubes with swirling angles $\alpha=76°$ and 87° (Fig. 3) are characterized by the similar behavior of hydraulic resistance over the Reynolds number range. At that, the hydraulic resistance coefficient ξ for a tube at the angle of swirling $\alpha=76°$ is 1.25 time larger than for a tube at $\alpha=87°$. To explain this phenomenon, the flow structure should be visualized. It may be assumed that, as the angle of swirling is increased and the axial pitch is decreased, the circumferential velocity component becomes much smaller than the axial one.

Analysis of numerical simulation results

Because of the difficulties associated both with flow structure visualization in the tube with internal helical finning and with effect explanations, particularly the similarity effect of hydraulic resistance in tubes with $\alpha=14°$, $\alpha=32°$, $\alpha=46°$ and $\alpha=61°$, as well as tubes with $\alpha=76°$ and $\alpha=87°$, numerical simulation results were adopted. The computational algorithm was preliminarily tested for flow in the tube without internal finning; the numerical simulation results were compared with the

Blasius law $\xi_0 = 0.3164\mathrm{Re}^{-0.25}$. Over the Reynolds number range 3×10^3–2.5×10^5 the numerical simulation results differed from the Blasius' law by 8-12%.

At small angles of helical swirling (Fig. 5, a) finning slightly influences the flow structure in the tube, as well as in the case of the tube without finning the following flow structure is formed: the velocity on the flow axis has the highest magnitude and sharply changes near the wall [15]. Increasing the angle of helical swirling (Fig. 5, *b*, c) disturbs the boundary layer. The finning height exceeds the boundary layer thickness, and this results in a hydraulic resistance growth.

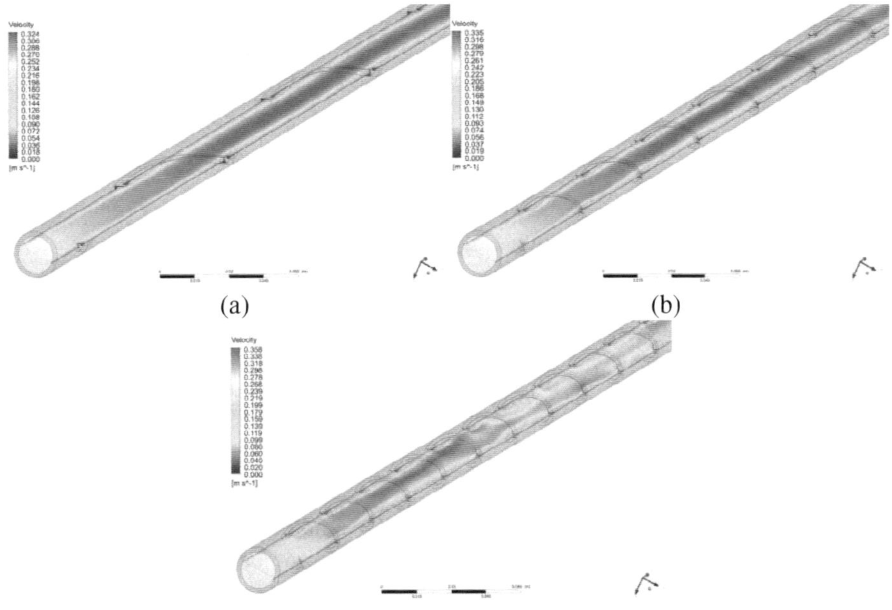

(a) (b)

(c)

Figure 5. Velocity distribution over the tube middle cross section: (a) – α=14°; (b) – α=32°; (c) – α=46°

It was earlier mentioned that as the angle of helical finning (Fig. 6) is increased, in the near-wall layers the share of the circumferential velocity component increases and the share of the longitudinal velocity component decreases. Thus, tubes with a large angle of swirling and a small axial pitch can be considered as tubes with transverse rolling.

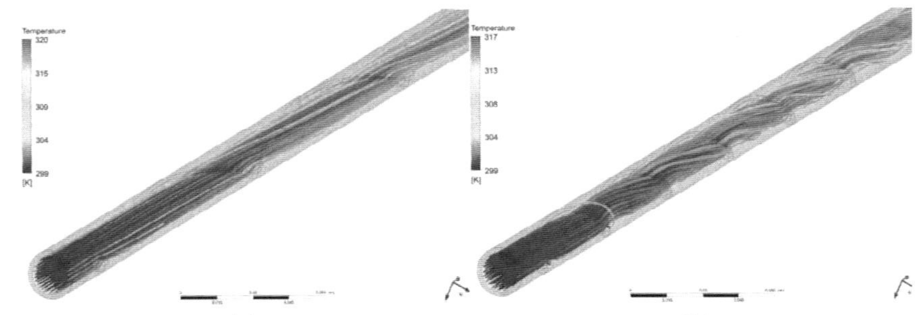

(a) (b)

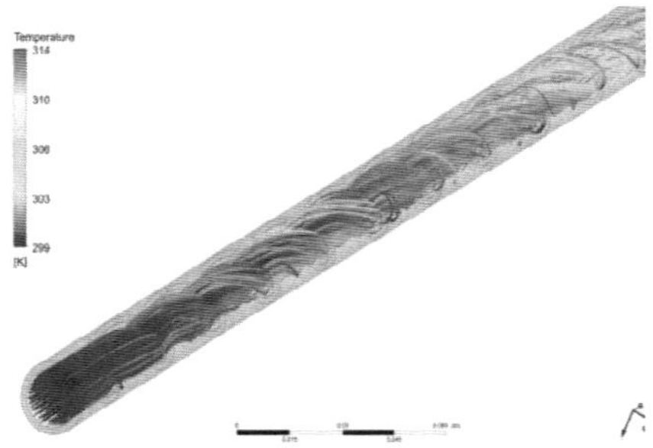

(c)

Figure 6. Flow structure in the tube with internal helical finning: (a) – α=14°; (b) – α=32°; (c) – α=46°

Conclusion

1. Hydraulic resistance of tubes with internal helical finning has been studied over a wide range of operating and geometric parameters: $Re_D=2\cdot10^3...2.5\cdot10^5$, under the variation of the angle of swirling $\alpha=14^\circ$-87°, the relative height of a protrusion $h/d=(25\text{-}87{,}5)\cdot10^{-3}$, the relative axial pitch p/d=0.16-12.73. It has been revealed that the hydraulic resistance of tubes with helical finning increases by a factor of 1.1 to 11.7.
2. The numerical simulation results have shown that, as the angle of helical swirling is increased, in the near-wall layers the share of the circumferential velocity component increases and the share of the longitudinal component decreases. And since the finning height exceeds the boundary layer thickness, the hydraulic resistance grows.

Acknowledgement

The authors are grateful to the Russian Foundation for Basic Research (grant 16-38-00066 mol_a, 15-38-20444 mol_a_ved), and the Belarusian Republican Foundation for Fundamental Research (grant T17PM-081) and to Elvira Zharkova for preparation and translation of the manuscript.

References

[1] Rabas T J, Bergles A E and Moen D L 1988 Heat transfer and pressure drop correlations for spirally grooved (rope) tubes used in surface condensers and multistage flash evaporators *New York: Proc. ASME Symposium* pp 693–704
[2] Mehta M H and Raja Rao M 1979 Investigations on heat transfer and frictional characteristic of enhanced tubes for condensers *New York Proc. ASME Symposium* pp 11–21
[3] Gupta R K and Raja Rao M 1979 Heat transfer and friction characteristics of Newtonian and power law type of non-Newtonian fluids in smooth and spirally corrugated tubes *New York: Proc. ASME Symposium* pp 103–113
[4] Withers J G Tubeside 1980 Heat transfer and pressure drop for tubes having helical internal ridging with turbulent/transitional flow of single phase fluid. Part I –Single-helix ridging *Heat Transfer Eng. J.* **1** 48–58
[5] Li H M, Ye K S, Tan Y K and Deng S J 1998 Investigation of tube-side flow visualisation. Friction factor and heat transfer characteristics of helical-ridging tubes *Washington: Proc. 7th Heat Transfer Conference* pp 75–80
[6] Ganeshan S and Raja Rao M 1982 Studies on thermohydraulics of single and multistart spirally corrugated tubes for water and time-dependent power law fluids *Int. J. Heat Mass Transfer* **25** 1013–1022
[7] Sethumadhavan R and Raja Rao M 1982 Turbulent flow friction and heat transfer characteristics of single and multistart spirally enhanced tubes *J. Heat Transfer* **108** 55–61
[8] Zimparov V D, Vulchanov N L and Delov L B 1991 Heat transfer and friction characteristics of spirally corrugated tubes for power plant condensers. 1. Experimental investigation and performance evaluation *Int. J. Heat Mass Transfer* **34** 2187–2197

[9] Ravigururajan T S and Bergles A E 1995 Prandtl number influence on heat transfer enhancement in turbulent flow of water at low temperatures *Trans. ASME* **117** 276–282

[10] Vicente P G, Garc'ıa A and Viedma A 2004 Experimental investigation on heat transfer and frictional characteristics of spirally corrugated tubes in turbulent flow at different Prandtl numbers *IJHMT* **47** 671–681

[11] Shchelchkov A 2016 Thermohydraulic characteristics of discretely rough tubes for the transitional flow regime *Heat Transfer Research* **47**(6) 545-557

[12] Zubkov N N, Ovchnnikov A I, Kononov O V 1993 Manufacture of new-class heat transfer surfaces by deforming cutting *Vestnik MSTU* **4** 79–82

[13] Zubkov N N 2005 Finning of heat exchanger tubes by deforming and cutting of surface layers *Heating News* **4** 51–53

[14] Ansys Fluent 2011 Available at: http//www.ansys.com

[15] Mikheev M A and Mikheeva IM 1977 *Fundamentals of heat transfer* (Moscow: Energiya) p 344

IOP Conf. Series: Journal of Physics: Conf. Series **980** (2018) 012005 doi:10.1088/1742-6596/980/1/012005

Dynamics of the aircraft in a vortex wake

A M Gaifullin[1] and Yu N Sviridenko[1]

[1] Central Aerohydrodynamic Institute, 1 Zhukovsky Street, TsAGI, Zhukovsky, Moscow Region, 140180, Russian Federation

Abstract. This paper considers the aerodynamics and the dynamics of an aircraft on various modes when the aircraft enters a strongly swirling flow. This is the case when an aircraft purposefully enters the jet-vortex wake of another aircraft in the course of in-flight refuelling, when an aircraft is flying in the trail of an aircraft carrier during landing, or when an aircraft accidentally enters other aircrafts' vortex wakes. These situations, according to pilots' evaluation, are the most dangerous and the most difficult modes for piloting. That is why their real time modelling on flight simulators has taken on particular importance. This article provides the algorithms and methodology of mathematical modelling of aerodynamic forces and moments which act upon an aircraft in vortex wakes.

1. Introduction

An aircraft sometimes happens to enter a swirling air flow during the flight. This work contains numerical studies of how a vortex wake behind an aircraft and a disturbed air flow from an aircraft carrier affects the aircraft's aerodynamic characteristics. This article also contains the algorithms and methodology for creating mathematical models of aerodynamic forces and moments which act upon an aircraft in a vortex wake, which enable modelling an aircraft's dynamics on a flight simulator on-line.

In order to determine additional aerodynamic forces and moments acting on an aircraft which are induced by a vortex wake, it is necessary to find the distribution of fluid dynamics parameters across the whole area of the flow. At the same time, a dangerous wake can be up to 20-30 kilometers (13-20 miles) long for the cases of wakes behind heavy aircrafts, so the solution domain has to be divided into the near wake and the far wake areas.

In the near wake, the parameters of flows in vortex wakes are calculated on the basis of solving 3D steady Reynolds-averaged Navier-Stokes equations. The distribution of fluid-dynamic parameters thus obtained is used for determining the vorticity distribution, the inflow temperature and the velocity on the boundary for further calculation of the far wake. In this domain, the unsteady-state two-dimensional Reynolds-averaged Navier-Stokes equations are solved.

After the flow fields have been determined, it is necessary to create real time mathematical models of the forces and moments which act upon an aircraft given its space and angular position in the vortex wake.

Determining additional forces and moments acting upon an aircraft in a jet-vortex wake is based on calculating the configuration flow using the panel method with symmetrizing the singularities.

In real time modelling of an aircraft's dynamics it is necessary to reduce the computing time to 0.01 seconds, which is determined by the integration step of the equations describing the aircraft's flight on

Content from this work may be used under the terms of the Creative Commons Attribution 3.0 licence. Any further distribution of this work must maintain attribution to the author(s) and the title of the work, journal citation and DOI.
Published under licence by IOP Publishing Ltd

a flight simulator. An approach based on approximating the obtained aerodynamic vaulting with the help of artificial neural networks was used to solve the task. Meanwhile, preliminary calculations of additional forces and moments for a large number of options for the space and angular position of an aircraft in a vortex wake were conducted. Then 6 neural networks whose vector contained the information about the space and angular position of an aircraft, and whose output was the increment of aerodynamic forces and moments caused by the vortex wake were trained.

2. Modelling the effect of vortex wake disturbances

A strongly swirling flow which an aircraft enters is caused by another aircraft. The solution to this problem consists of solving several sub-tasks: determining the characteristics of the flow near the aircraft generating the vortex wake, the wake's parameters and, finally, calculating the forces and moments which act upon an aircraft entering the area of a vortex wake from a tanker aircraft.

Several computational methods from aerodynamics have been used to determine the flow fields in the wake following the generator aircraft. In that case the flow range has been divided into three sub-ranges. The representation of dividing the area behind the tanker aircraft IL-78 into sub-areas of a near field, a near wake and a far wake is given in fig. 1. The border between the near field and the near wake is the initial cross-section for calculating the near wake, then, the far wake.

The velocity field in the near field behind the aircraft was computed using steady Reynolds-averaged Navier-Stokes equations (RANS) for the case of gas, compressed at the temperature which corresponds to the altitude of a cruise or refuelling. The expected angle of attack of an aircraft is determined basing on the necessity to perform a horizontal flight with a given mass and a given velocity at a given altitude. The aircraft is supposed to be flying with zero sideslip.

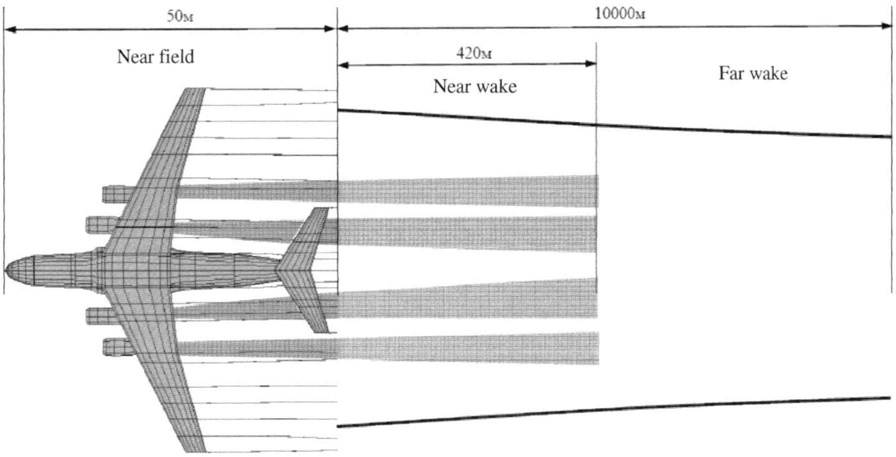

Figure 1. The representation of dividing solution domain.

The flow fields and temperatures in the near wake have been calculated using the JVWAKE [2] code, which is based on solving two-dimensional unsteady-state Reynolds-averaged Navier-Stokes equations. In this software, the diffusion of an aircraft's vortex wake was calculated with the help of invariant modelling method developed at Princeton University [3-4] and modified by the authors for calculating the characteristics of a wake [2]. The modification involves the models of closure for Reynold's equations, which, as has been proved several times, lead to a stronger diffusion in the wakes' cores than in the experimental investigations.

Starting from the output cross-section of the computational domain and until the wake dies out, the flow fields are calculated using a simplified procedure. All the vorticity behind an aircraft is considered to be concentrated in two vortices, whose coordinates in the output cross-section coincide with the coordinates of the mass centres of the right and the left halves of the vortex system. The vortices descend according to a non-standard analogy:

$$\frac{dy}{dx}=\frac{\Gamma}{2\pi b u_\infty},$$

here b is the distance between the vortices, u_∞ - is the velocity and Γ is the circulation of vortices, which fades according to the following empirical law:

$$\Gamma=\Gamma_0\exp\left(-\frac{0.8qx}{bu_\infty}\right).$$

Here q is the atmospheric turbulence.

The vortices do not descend uniformly because they fade. The larger the distance from the wing, the lower the velocity with which the vortices descend. That is why the vortex wake is distorted. Curved vortex lines are replaced with a system of broken wake segments from which the velocities are calculated according to Bio-Savara formulas. These are also calculated in the point for which the velocity to be calculated was outside the computational domain of a near wake.

3. Modelling the effect of disturbances caused by an aircraft carrier wake

The aircraft's motion in a vortex wake following an aircraft carrier is considered. This section contains the solutions to a number of problems which are concerned with modelling a turbulent wind above the water surface, with determining the characteristics of the flow near and behind the ship, as well as with determining the forces and moments which act on a landing or taking off aircraft. The field of velocities near the ship is determined by the velocity and the direction of a gradient wind, which depends on the state of the atmosphere and the ship's velocity. The air flow fields near the ship have been calculated using CFD methods. The solutions to the boundary-value problem for Reynolds-averaged Navier-Stokes equations have been used. The picture of the flow thus obtained at zero angle of the apparent wind with the velocity of 15m/sec is shown in fig. 2. The left part of the figure shows the space pattern of vorticity isolines behind the ship. The streamlines of the flow behind the antenna tower are shown in the lower right corner of the figure.

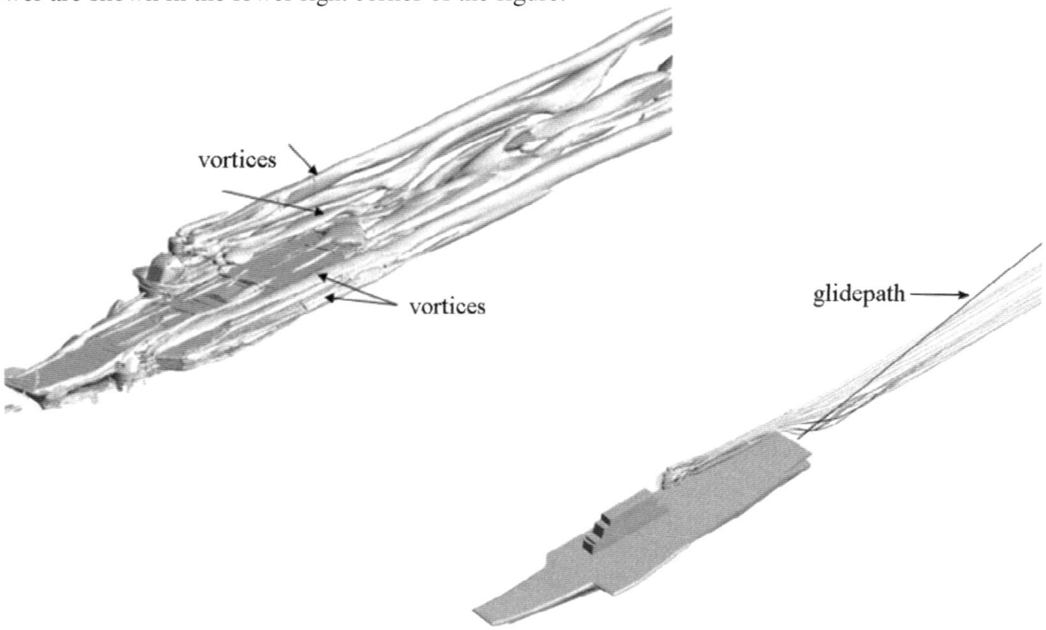

Figure 2. The flow in wake following an aircraft carrier.

4. Determining the additional forces and moments caused by a vortex wake

Determining the additional forces and moments which act upon an aircraft in a vortex wake is based on calculating the airflow using a panel method and is performed according to the following algorithm:

1 - aerodynamic forces and moments which act on an aircraft in a body axis coordinate system are singled out from a uniform flow with the help of the panel method;

2 - using the flow fields obtained through calculations in the wake following a generator aircraft or an aircraft carrier velocities in the control points of the aircraft's panels are determined using interpolation;

3 - the airflow is calculated considering additional velocities, and aerodynamic forces and moments are calculated;

4 - additional forces and moments are calculated through subtracting the values obtained at step 3 from the values obtained at step 1;

This work uses the modification of a panel code PANSYM [5] where hydrodynamic characteristics on the lifting surfaces (wings, aircraft's empennage and the flap systems) are symmetrized in order to calculate the full configuration of the aircraft to be refuelled. The software has proven useful for solving a wide range of applied problems. In this method the aircraft's surface is modelled using a series of tetragonal panels with the sources and vortices distributed along them. Non-structural elements like the body are modelled using the panels with a piecewise-constant distribution of the sources and the outflows. The surface of the wings' lifting surfaces, the empennage and the flap systems are modelled using panels with piecewise-linear distribution of the vorticity along a chord and a piecewise-constant distribution of the sources and the outflows. The principle of symmetrizing the characteristics of an element's upper and lower surfaces is used for modelling the lifting elements. In this case the characteristics on the respective panels of the upper and the lower surfaces are equally intensive. A typical configuration of an aircraft in a panel representation is shown in fig. 3.

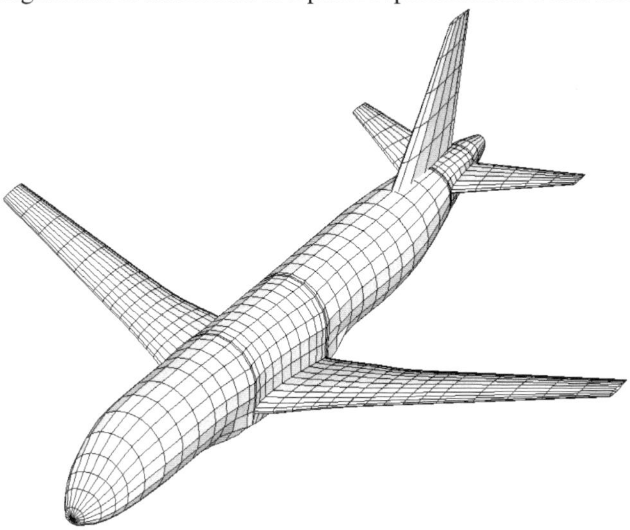

Figure 3. Computational model for the panel method.

5. Creating on-line mathematical models on the basis of artificial neural networks

In order to create a system modelling the impact on an aircraft and the vortex wake from a generator aircraft or an aircraft carrier on-line, it is necessary to:

1 - conduct computational research of the airflow of an object generating the vortex wake and form a numerical array of gas-dynamic parameters of flow fields in the wake;

2 - determine the additional forces and moments acting on an aircraft in a vortex wake;

3 - create real time software modules which can determine the increment of forces and moments during flight simulator modelling.

Computing the airflow and obtaining numerical arrays of gas-dynamic parameters of the flows require considerable amounts of time and modern industrial CFD packages. In order to create on-line software modules on the basis of the data obtained it is necessary to employ fast approximation methods.

While using this information to model an aircraft dynamics in a vortex wake on a flight simulator in real time it is necessary to reduce the time required to determine aerodynamic characteristics to 0.01 seconds, which is determined by the time integration step of the equations describing the aircraft's motion. To solve this task, an approach based on approximating the obtained array of aerodynamic characteristics with the help of artificial neural networks has been used. Under this approach a considerable number of options of the angular and space positions of an aircraft relative to the wake generator object were calculated in advance. Following that, 6 neural networks with the input vector containing the information about the aircraft's position (three coordinates and three angles), and with the output being the increment of aerodynamic forces and moments caused by the wake were trained. A comparison between the approximation accuracy and the computational result has shown that the accuracy of determining aerodynamic characteristics is adequate. Artificial neural networks that have been trained beforehand with the help of computational data serve as aerodynamic software modules for the simulator's mathematical support under this approach. This procedure has enabled to reduce the time required to calculate an aircraft's aerodynamic characteristics in one point dramatically (to 0.0001 - 0.0002 sec) with only a slight decrease in accuracy. Multilayer perceptron-type neural networks with two hidden layers have been used. There were 11 neurons on the first layer and 5 on the second one. In total, 6 neural networks have been trained for approximating the increments of three coefficients of aerodynamic forces and three coefficients of aerodynamic moments. In order to form a set of patterns which were used for teaching and testing the neural networks, about 200000 calculations for each aircraft type, using random values of its position in the wake, have been carried out. The ranges of the computational parameters of the given set of patterns correspond to the area where an aircraft could be located.

After training neural networks the accuracy of determining the increments to forces and moments caused by an aircraft's position in a vortex wake as compared to the computational characteristics obtained with the help of the panel software was evaluated.

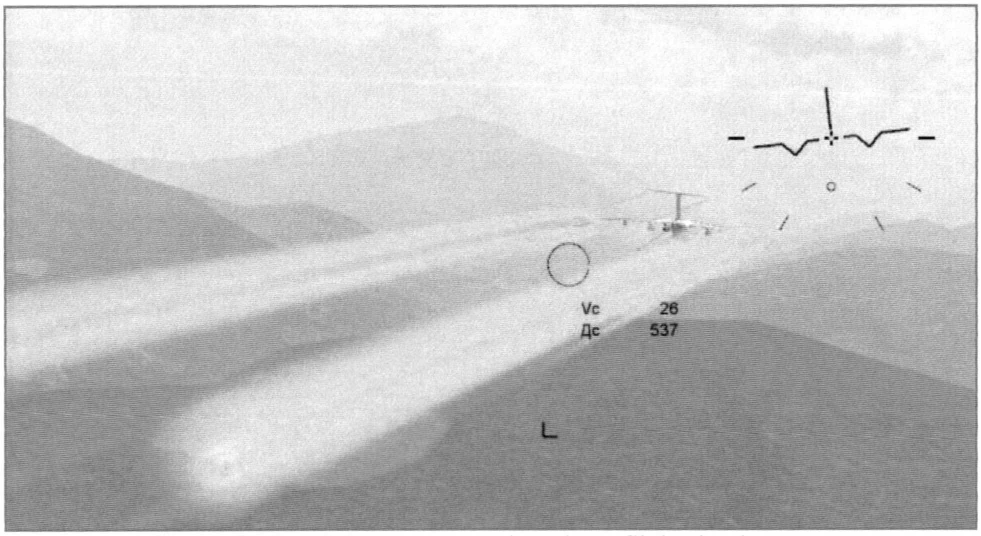

Figure 4. Visualising a vortex wake using a flight simulator.

Then basing on the array of weighting coefficients of neural networks the functions were formed in an algorithmic language C++ with the help of special software. The execution speed of the functions was evaluated on a PC with an Intel (R) Core (TM) processor 17-3820 CPU, 3.60 GHz, with an installed memory (RAM) of 8.00 GB in a 64-bit operating system Windows. It took about 0.0001 sec CPU to calculate the increments of three coefficients of forces and three coefficients of moments in one point on the flight trajectory. The developed mathematical software enables one to model the dynamics of an aircraft when it enters vortex wakes on a flight simulator and to visualize dangerous areas, which is an important part in training air staff. A flight simulator visualization of the calculated distribution of vorticity behind an IL-78 aircraft is shown in fig. 4.

6. Conclusion

The work presents the results of theoretical research of vortex wakes in a turbulent atmosphere. Mathematical models of an aircraft flying in the area of a strong vortex wake influence, which are applicable to modelling an aircraft's dynamics in real time have been created. The technique of visualizing dangerous vortex wake areas on a flight simulator has been improved. The software based on the suggested mathematical models has been implemented on flight simulators in TsAGI, the Department of Aerodynamics and Flight Engineering at Moscow Institute of Physics and Technology (State University), Russian Aircraft Corporation «MiG» and Beriev Aircraft Company, [1].

Acknowledgments

This work has been carried out with funding from Russian Ministry of Education and Science into the subject "Applying artificial neural networks to promote flight safety", project 14.624.21.0046.

7. References

[1] Gaifullin A M 2016 Modelling an aircraft flying in severe environment *Trudy TsAGI* **2752** 88
[2] Voyevodin A V, Vyshinsky V V, Gaifullin A M and Sviridenko Yu N 2003 Evolution of civil aircraft jet-vortex wake *Aeromech Gas Dyn* vol **4** 23–31
[3] Donaldson C 1972 Calculation of turbulent shear flows for atmospheric and vortex motions *AIAA J* **10**(1) 4–12
[4] Frost W and Moulden T 1980 *Turbulence Principles and applications*. (Moscow: Mir Press)
[5] Sviridenko Yu N and Ineshin Yu L 1996 Application of panel method with symmetrization of singularities for calculating flow around aircraft with regard to engine jets influence *Trudy TsAGI* **2622** 41–53

IOP Conf. Series: Journal of Physics: Conf. Series **980** (2018) 012006 doi:10.1088/1742-6596/980/1/012006

Mathematical modelling of thermolabile solutions concentration in vortex chamber

V V Kharkov

Assistant Professor, Department of Food Production Equipment,
Kazan National Research Technological University, Kazan, Russia

E-mail: v.v.kharkov@gmail.com

Abstract. The study represents a computation scheme for a fixed cylindrical chamber with an air tangential swirler, used for the thermal concentration of liquids. During the evaporation, the drops are not in direct contact with hot surfaces, which is critically important for thermosensitive substances. This is achieved by adjusting droplets motion in the form of a suspended rotatory drop layer. For this purpose, a detailed mathematical model was constructed, and a numerical study of heat-and-mass transfer was performed in terms of the real flow mechanics. It has been found that the time needed to change the drop volume during the process of concentration to the desired final volume uniquely depends on the construction of the chamber and operating regime in the workspace. So, numerical calculation of evaporating drops motion in a swirling gas flow has been performed over a wide range of different parameters such as heating gas temperature, inlet gas velocity, liquid-gas mass-flow ratio, inclination angle of the swirler blades and circulation ratio.

1. Introduction and mathematical model description

Thermal concentration processes are of widest use for producing commercial and intermediate products in chemical, pharmaceutical, food and other industries. Traditional techniques and equipment for concentrating and drying thermolabile liquids are bounded by their strongly marked temperature-sensitive properties, i.e. depend critically on the physicochemical properties of the starting materials. Even under low heat, irreversible changes in the chemical structure of the product occur. For instance, it is the decrease in the concentration of the desired substance because of thermal decomposition processes. The problem has been explored by using a vortex chamber which design and technological operating regime completely account for material thermosensitivity to achieve required quality of the finished products.

Despite the principles of designing vortex apparatuses have been studied for a long time, wide industrial use is complicated by the absence of valid and well-grounded methods for calculating the processes in the chamber [1–2]. So, the purpose of this paper is to present a mathematical model and perform a numerical study of heat-and-mass transfer in terms of using a vortex chamber for concentrating thermolabile liquids. In order to achieve this, the real flow mechanics principles in the workspace of the chamber should be considered.

The vortex chamber operates as follows (figure 1). The heating agent, in particular, hot air, is inserted through the gas distribution device into the vortex chamber. Gas in passing through the tangential blade swirler generates a rotary motion, and swirling gas flow is formed in the workspace of

Content from this work may be used under the terms of the Creative Commons Attribution 3.0 licence. Any further distribution of this work must maintain attribution to the author(s) and the title of the work, journal citation and DOI.
Published under licence by IOP Publishing Ltd

IOP Conf. Series: Journal of Physics: Conf. Series **980** (2018) 012006 doi:10.1088/1742-6596/980/1/012006

the chamber. Liquid enters the chamber through holes in the tubular distributing device and instantly breaks up into discrete droplets to produce a suspended rotatory gas-liquid layer between the swirler blades and the discharge radius. The entry of the droplets into the inner surface of the chamber with a temperature close to that of the heating gas may be dangerous, because it causes the decomposition of thermosensitive substances and burning.

Moving droplets gradually come out and revolve in the equilibrium radii which match associated diameters and operating parameters of the chamber. This forms suspended droplets which have no direct contact with hot surfaces. Drops move down under gravity. Droplet size and equilibrium radius decrease as evaporation proceeds; that is because the radial position of the drop layer tends to the axis. The radius of gas discharge is governed by the equilibrium radius of drops with the desired content of solids. The cooling-water jacket is mounted in the lower part of the bottom, where drops tend to settle, in order to prevent the chemical decomposition of the desired product. Concentrated liquid crosses the separation gap device and flows out from the chamber. Gas is removed through the lower axial discharge.

When the vortex chamber operates without recirculation, the entire liquid phase interacts with heating gas once. By contrast, in operation using a circulation circuit, the liquid phase outflow is returned to the chamber after adding fresh flow. Therefore, the solids content of the incoming fluid is higher in case of the circulation scheme than the single-entry one under comparable conditions. Liquid recirculation can help to augment the concentration and increase the final solids content of the product.

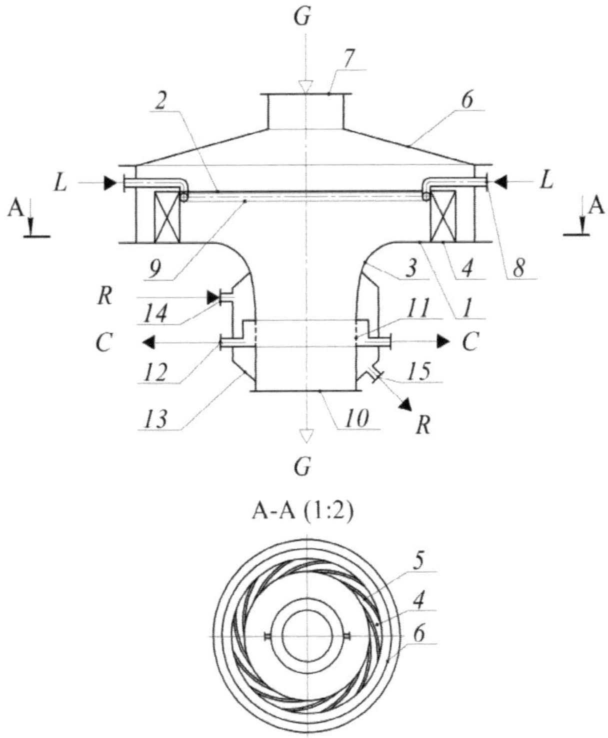

Figure 1. A schematic of the vortex chamber:
1 — body; 2 — cover; 3 — bottom; 4 — tangential swirler; 5 — blade of the swirler; 6 — gas distributing device; 7 — heating gas intake; 8 — liquid intake; 9 — tubular liquid distributing device; 10 — gas discharge; 11 — separation gap device; 12 — concentrated liquid discharge; 13 — cooling-water jacket; 14, 15 — input-output technological pipes.

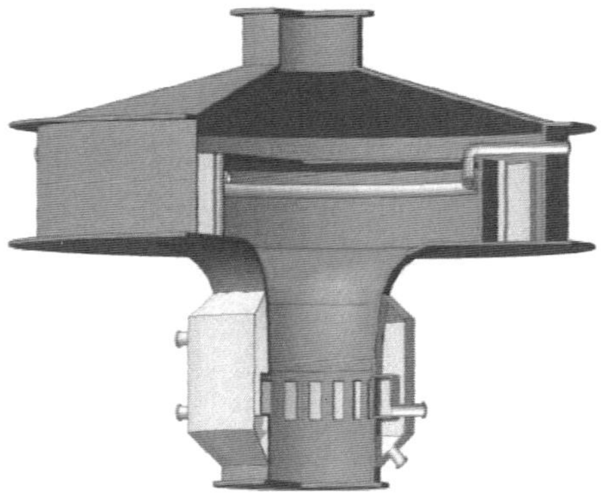

The objectives in the calculation of the vortex chamber are to determine the design parameters and operating conditions based on the mathematical model in terms of the real flow mechanics and heat-and-mass transfer principles in the workspace of the chamber. This takes into account the following assumptions:

- polydisperse liquid spray is replaced with monodisperse one. The Sauter mean diameter is taken as an averaging parameter;
- potential flow has been used to describe the velocity distribution of gas swirling flow;
- heat-and-mass transfer is quasistationary;

- all physical properties of gas, namely, air-vapor mixture flowing around a drop, must be estimated at average temperature in the thermal boundary layer and at average concentration of vapor in the diffusion boundary layer;
- it is assumed to be the ideal stirred model in the bulk of the chamber;
- liquid phase has wet-bulb temperature at initial time.

The system of equations for the unit drop motion in terms of the cylindrical fixed coordinate system (axis of the chamber extends in an z-axis direction) can be written as follows [3]:

$$\begin{cases} \dfrac{\mathrm{d}V_{\mathrm{r}}}{\mathrm{d}\tau} = \dfrac{V_{\varphi}^2}{r} + \dfrac{3}{4}\dfrac{\rho_{\mathrm{g}}}{\rho_{\mathrm{liq}}} c_{\mathrm{a}} \dfrac{V_{\mathrm{rel}}}{a}\left(W_{\mathrm{r}} - V_{\mathrm{r}}\right) \\ \dfrac{\mathrm{d}V_{\varphi}}{\mathrm{d}\tau} = -\dfrac{V_{\mathrm{r}}V_{\varphi}}{r} + \dfrac{3}{4}\dfrac{\rho_{\mathrm{g}}}{\rho_{\mathrm{liq}}} c_{\mathrm{a}} \dfrac{V_{\mathrm{rel}}}{a}\left(W_{\varphi} - V_{\varphi}\right) \\ \dfrac{\mathrm{d}V_{\mathrm{z}}}{\mathrm{d}\tau} = g + \dfrac{3}{4}\dfrac{\rho_{\mathrm{g}}}{\rho_{\mathrm{liq}}} c_{\mathrm{a}} \dfrac{V_{\mathrm{rel}}}{a}\left(W_{\mathrm{z}} - V_{\mathrm{z}}\right) \\ \dfrac{\mathrm{d}r}{\mathrm{d}\tau} = V_{\mathrm{r}}, \quad \dfrac{\mathrm{d}\varphi}{\mathrm{d}\tau} = V_{\varphi}, \quad \dfrac{\mathrm{d}z}{\mathrm{d}\tau} = V_{\mathrm{z}} \\ \dfrac{\mathrm{d}a}{\mathrm{d}\tau} = -2\dfrac{\beta_y}{\rho_{\mathrm{liq}}}\left(y_{\mathrm{s}} - y\right) \end{cases} \tag{1}$$

The numerical solution of system (1) using four-order Runge-Kutta method with initial conditions is as follows:

$$\tau = 0, \quad r = r_1, \quad \varphi = 0, \quad z = z_1, \quad V_{\mathrm{r}} = V_{\mathrm{r1}}, \quad V_{\varphi} = V_{\varphi 1}, \quad V_{\mathrm{z}} = 0, \quad a = a_1, \tag{2}$$

where r_1 is the radius of liquid input; z_1 is the distance from chamber cover; and V_1 is the feeder velocity.

The radial component of the drop velocity can be found from the constant gas mass flow condition in the chamber in terms of its geometry, whereas the tangential component — from the law of circulation conservation. The presence of liquid is accounted for an empirical coefficient of effective velocity reduction in the chamber [4, 5].

The formed calculation algorithm has the following features:

1) The temperature at the drop surface t_{s} is determined by bisection method from a balance between the rate of heat and mass transfer at interface — $\beta_y r_{\mathrm{s}}\left(y_{\mathrm{s}} - y\right) = \alpha\left(t - t_{\mathrm{s}}\right)$. Experimental data on mass-transfer rates for spheres may be correlated by the following empirical equation of the form used by Froessling [6]:

$$\mathrm{Nu} = 2 + 0{,}6\mathrm{Re}^{1/2}\mathrm{Pr}^{1/3}, \quad \mathrm{Sh} = 2 + 0{,}6\mathrm{Re}^{1/2}\mathrm{Sc}^{1/3}; \tag{3}$$

2) The moisture content of the gas mixture X should be corrected at each time step by the expression derived from mass balance:

$$X = X_1 + \Lambda\left(1 + X_1\right)\left[1 - \frac{\rho_{\mathrm{liq2}}}{\rho_{\mathrm{liq1}}}\left(\frac{a_2}{a_1}\right)^3\right], \tag{4}$$

where Λ is the input liquid-gas mass-flow ratio, $L_{\mathrm{m1}} / G_{\mathrm{m}}$;

3) Using the same procedure, temperature of the gas mixture t should be corrected at each time step by the expression derived from heat balance:

$$t=\frac{t_1\left(\bar{c}_p^{'}+X_1\bar{c}_p^{''}\right)-\Lambda\left(1+X_1\right)\left[1-\frac{\rho_{\text{liq}2}}{\rho_{\text{liq}1}}\left(\frac{a_2}{a_1}\right)^3\right]\left(r_s-\bar{c}_p t_s\right)}{\Lambda\left(1+X_1\right)\left[1-\frac{\rho_{\text{liq}2}}{\rho_{\text{liq}1}}\left(\frac{a_2}{a_1}\right)^3\right]\bar{c}_p+\bar{c}_p^{'}+X_1\bar{c}_p^{''}}. \quad (5)$$

Computations have been made by using an air-vapor mixture, thermolabile solution with known physical properties, as an example of liquid-gas system. The mean heat capacity of gas was estimated using the rule of mass additivity, and the mean density was estimated according to volume additivity. The heat capacity at constant pressure for heating gas components and the vaporization heat of pure water were taken as an approximation by Beard from reference data [7].

The dynamic viscosity and thermal conductivity of the gas mixture were calculated from the Wilke equation [8] as functions of the molecular weights and appropriate characteristics of the pure components determined from Sutherland type equation [9]. The calculation of the diffusion coefficient for binary gas mixture followed the method given by Fuller [10].

2. Results and discussion

The system of equations (1) under initial conditions (2) was used to establish the behavior of the evaporating droplets in time, depending on drop diameters, and circulation ratio of liquid and gas temperature for the algorithm given above. A necessary condition of the concentration completion is to achieve droplet diameters a_2 that are in proportion to the desired concentration.

A numerical calculation of the drop trajectory in the vortex chamber with part-load recirculation indicated that both the cylindrical height *H* and the radius of the discharge R_0 were determined by the circulation ratio K_{liq}, closely associated with the final diameter of the drop with coordinates (*h*, *r*), moreover $h = H$, $r = R_0$.

Figure 2 presents data for the lifetime of the drops evaporated to the final concentration of 70 % in the vortex chamber under changes in the liquid circulation ratio. In the absence of the circulation circuit (K_{liq} = 0) the lifetime of drops with a diameter of 0.25 mm in the chamber with 0.35 m radius is 3.56 s, but when the circulation ratio K_{liq} lies in the range from 0 to 5, the duration of stay decreases 7.91 times. For drops with a diameter of 0.30 mm in the chamber with 0.35 m radius the decrease is 9.07 times, and for drops with a diameter of 0.37 mm in the chamber with a diameter of 0.65 m it is 9.37 times.

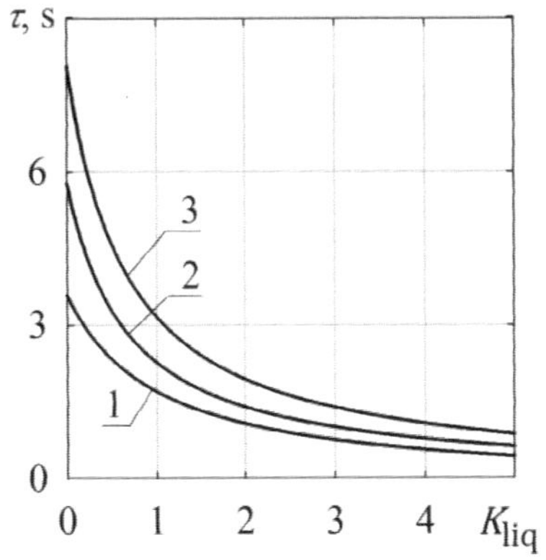

Figure 2. Effect of circulation ratio on lifetime for evaporating droplets with different diameters a_1, mm: 1 — 0.25 (R = 0.35 m, r_1 = 0.34 m); 2 — 0.30 (R = 0.45 m, r_1 = 0.44 m); 3 — 0.37 (R = 0.65 m, r_1 = 0.63 m):

$\alpha = 35°$; n = 12 pcs; W_{in} = 13 m/s; $L_{\text{m}} / G_{\text{m}}$ = 0.2; X_1 = 0.03 kg/kg; t_1 = 110 °C, z_1 = 0.05 m; V_{r1} = 0; $V_{\varphi 1}$ = 9 m/s.

IOP Conf. Series: Journal of Physics: Conf. Series **980** (2018) 012006 doi:10.1088/1742-6596/980/1/012006

There are several parameters that have a direct and substantial impact on drop trajectory during evaporation in the vortex chamber. The system (1) was computed under a wide range of variables, such as averaged gas velocity in the effective cross-section of the swirler W_{in} — from 10 to 20 m/s (figure 3a), liquid-gas mass-flow ratio L_m / G_m — from 0.1 to 0.5 (figure 3b) and inclination angle of the swirler blades □ — from 25 to 45° (figure 3c) for drops with initial diameter a_1 = 0.25, 0.30, 0.37 mm (at concentration degree of 70 % and K_{liq} = 3 the values of the final droplet diameters are a_2 = 0.225, 0.270 and 0.333 mm respectively).

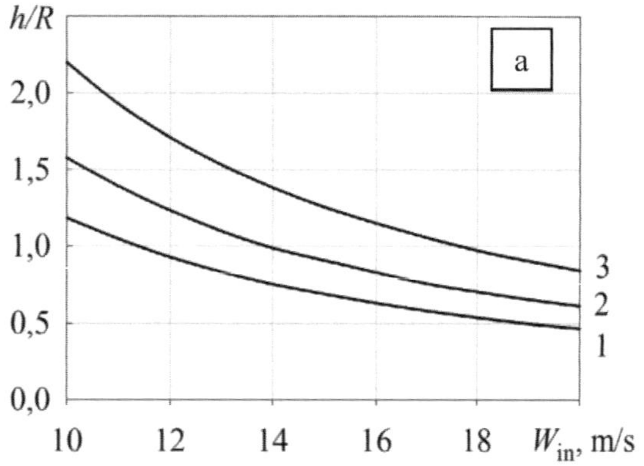

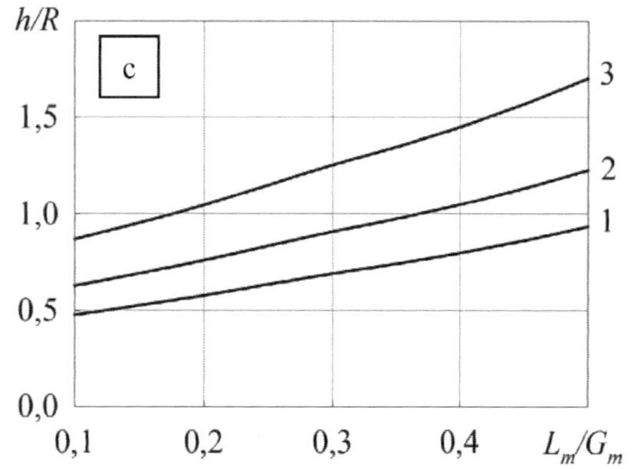

Figure 3. Dependence of the relative displacement of drops in z-direction on:
a) averaged gas velocity in the effective cross-section of the swirler (α = 35°; n = 12 pcs; L_m / G_m = 0,2);
b) inclination angle of the swirler blades (n = 12 pcs; L_m / G_m = 0,2; W_{in} = 15 m/s);
c) liquid-gas mass-flow ratio (α = 35°; n = 12 pcs.; W_{in} = 15 m/s);
a_1, mm: 1 — 0,25; 2 — 0,30; 3 — 0,37.

Figure 4 shows results of computations for the relative displacement of drops in z-direction under changes of gas temperature for different liquid circulation ratio.

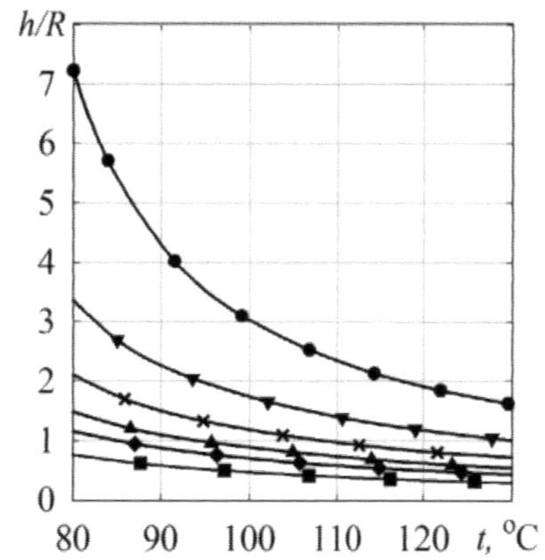

Figure 4. Dependence of the relative displacement of drops in z-direction on heating gas temperature at varying circulation ratio K_{liq} : ● — 0; ▼ — 1; × — 2; ▲ — 3; ♦ — 4; ■ — 5:

R = 0.35 m; a_1 = 0.25 mm; α = 35°; n = 12 pcs; W_{in} = 13 m/s; L_m / G_m = 0.2; X_1 = 0.03 kg/kg; t_1 = 110 °C, z_1 = 0.05 m; r_1 = 0.34 m; V_{r1} = 0; $V_{\varphi 1}$ = 8.5 m/s.

IOP Conf. Series: Journal of Physics: Conf. Series **980** (2018) 012006 doi:10.1088/1742-6596/980/1/012006

The value of h/R decreases with a rise in temperature from 80 to 130 °C for drops of 0.25 mm in diameter. Furthermore, there is more substantial impact on the relative height of the chamber at $K_{\text{liq}} = 0$ and $K_{\text{liq}} = 1$ than in the other cases. For example, temperature on the drop surface $t_s = 52.12$ °C ($t_1 = 105$ °C, $K_{\text{liq}} = 3$) is sufficient to prevent chemical decomposition processes of thermolabile substances.

As a result of the numerical experiment the following conclusions can be made:

1) the lifetime of the evaporating drops with different diameters tend to decrease on average of 8.8 times when the circulation ratio of the concentrated liquid increases from 0 to 5;
2) the reduction of the relative displacement of drops in z-direction has been established under an increase in the averaged gas velocity in the effective cross-section of the swirler and a decrease of inclination angle of the swirler blades and liquid-gas mass-flow ratio for drops with different initial diameters;
3) the gas temperature within 105 ± 25 °C has no significant effect on chamber geometry at the circulation ratio $K_{\text{liq}} \geq 2$.

Notation

a = diameter of drop; c_a = aerodynamic drag coefficient; $\bar{c}_{\text{p}}, \bar{c}'_{\text{p}}, \bar{c}''_{\text{p}}$ = averaged isobaric heat capacity of the gas mixture, dry air and water vapor at constant pressure, respectively; g = acceleration of gravity; K_{liq} = liquid circulation ratio; Nu = Nusselt number, $\alpha a/\lambda_{\text{g}}$; Pr = Prandtl number, $c_{\text{p}}\mu_{\text{g}}/\lambda_{\text{g}}$; r, □, z = cylindrical coordinates; Re = Reynolds number, $aV_{\text{rel}}/\nu_{\text{g}}$; r_{s} = specific heat of vaporization on the drop surface; Sc = Schmidt number, $\nu_{\text{g}}/D_{\text{g}}$; Sh = Sherwood number, $\beta_y a/D_{\text{g}}$; t, t_{s} = gas temperature in main stream and on the drop surface, respectively; V_{r}, V_{φ}, V_{z} = components of vector fields for the drop velocity; V_{rel} = droplet velocity related to the gas; W_{r}, W_{φ}, W_{z} = components of vector fields for the gas flow velocity; X = moisture content of the gas mixture; y_{s}, y = water vapor concentration on the drop surface and outside the boundary layer, respectively; α = heat transfer coefficient; β_y = mass transfer coefficient; λ_{g} = thermal conductivity of gas; μ_{g}, ν_{g}, = dynamic and kinematic viscosity of gas; ρ_{g} = density of gas; ρ_{liq} = density of liquid in drop;□□ = time; subscripts: g = gas; l = liquid; m = mass; p = at constant pressure; r = radial component; s = on the drop surface; z = axial component;□□□□□angular component; 1 = initial conditions; 2 = final conditions; superscripts: ' = dry air; "= vapor.

References

[1] Leontev A I, Kuzma-Kichta Yu A and Popov I A 2017 *Thermal Engineering* vol 64 (Luxembourg: Springer) pp 111–126

[2] Jawarneh A M, Sakaris P and Vatistas G H 2006 *J. Fluids Eng.* vol 129 (New York: ASME Technical Publishing Office) pp 100–105

[3] Kharkov V 2016 *Scientific and Technical Volga region Bulletin* vol 1 (Kazan: NTVP) pp 37–44

[4] Kharkov V and Nikolaev A 2017 *Chemical Industry today* vol 1 (Moscow: Himprom segodnya) pp 16–21

[5] Ovchinnikov A 2005 *Dynamics of two–phase swirling turbulent flows in the vortex separators* (Kazan: Novoe znanie) p 288

[6] Frössling N 1938 *Gerlands Beitr. Geophys.* vol 52 (Stuttgart: AGU) pp 170–216

[7] Thompson C W and White H W 1983 *Am. J. Phys.* vol 51 (College Park: AAPT) pp 362–364

[8] Wilke C R 1950 *J. Chem. Phys.* vol 18 (New York: AIP) pp 517–519

[9] Schetz J A and Fuhs A 1999 *Fundamentals of Fluid Mechanics* (New York: John Wiley & Sons) p 935

[10] Fuller E N, Schettler P D and Giddings J C 1966 *Ind. Eng. Chem.* vol 58(5) (Easton: ACS) pp 18–27

IOP Conf. Series: Journal of Physics: Conf. Series **980** (2018) 012007 doi:10.1088/1742-6596/980/1/012007

Heat transfer simulation of unsteady swirling flow in a vortex tube

S V Veretennikov[1], Sh A Piralishvili[1], O A Evdokimov[1], A I Guryanov[1]

[1] Department of General and Technical Physics, P.A. Soloviev Rybinsk State Aviation Technical University University, 173a Pushkina st., Rybinsk, 152934, Russian Federation

e-mail: serveret@yandex.ru

Abstract. Effectiveness of not-adiabatic vortex tube application in the cooling systems of gas turbine blades depends on characteristics of swirling flows formed in the energy separation chamber. An analysis of the flow structure in the vortex tube channels has shown a presence of a complex three-dimensional spiral vortex, formed under relatively high turbulence intensity and vortex core precession. This indicates the presence of a significant unsteady flow in the energy separation chamber of the vortex tube that has a great influence on convective heat transfer of the swirling flow to the inner surface of tube. The paper contains the results of investigation of gas dynamics and heat transfer in the vortex tube taking into account the flow unsteadiness.

1. Introduction

Application of swirling flows in technology (flow rotary motion by means of various swirling devices) leads to a large-scale effect on flow parameters and, consequently, on the heat transfer [1]. Three-dimensionality of the velocity field and proportionality of tangential and axial velocity components (that is specific for swirling flows) determine a formation of three-dimensional pressure field with values of radial gradient comparable to longitudinal gradient. A presence of transverse velocity components (tangential and axial) enhances the convective transport of momentum, energy and mass allowing change in the swirling flow structure. It defines some specific properties of swirling flows in the context of technical application and their ability to intensify heat and mass transfer processes, to align the local temperature inhomogeneities due to convective mixing, to inhibit or to enhance the random perturbations, to stabilize the flow in a complex heat transfer with chemical reactions and combustion, to provide recirculating flow in the combustion chambers with high efficiency of fuel burn-out, and etc [1].
The necessity of studying the heat transfer of swirling flows in the axisymmetric channels of vortex energy separator is connected with requirements to improve the cooling systems of turbine blades [2]. Swirling flow application allows intensifying the heat transfer and leads to an increase of heat transfer coefficient across the channel length. Energy separation allows increasing the cooling efficiency of the most heat-stressed areas of the blade [2, 3].

2. Problem Statement

The analysis of flow structure in the vortex energy separator and the estimation of heat transfer intensity have been performed using the method of 3D numerical simulation. The research object was the vortex tube with a diameter of the energy separation chamber equal to d = 10 mm. An untwist plate

Content from this work may be used under the terms of the Creative Commons Attribution 3.0 licence. Any further distribution of this work must maintain attribution to the author(s) and the title of the work, journal citation and DOI.
Published under licence by IOP Publishing Ltd

was set at the "hot" end of the tube. The area of the throttle disclosure was chosen to obtain the required value of the relative proportion of cooled stream μ. The flow in the channels of the vortex energy separator could be described by the system of differential equations of continuity, momentum, energy and state, and the heat transfer in solid body could be described by Fourier equation. There were specified conditions of continuity of temperature and heat flow equation for conjugation at the boundary "gas – solid body". The 3D numerical simulation was performed for the steady, viscous, turbulent flow, taking into account compressibility, nonisothermality, and absence of chemical reactions and phase transformations. The described system of equations was closed by the two-parameter turbulence model k-ω SST.

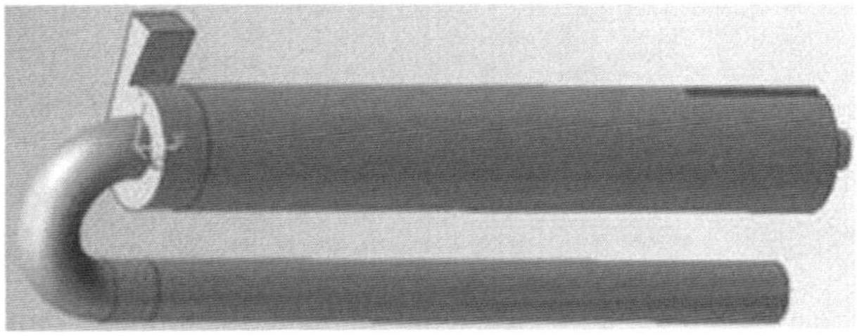

Figure 1. Vortex tube geometry

3. Numerical modeling of heat transfer under stationary conditions

The conjugate problem of gas dynamics and heat transfer in the vortex tube under stationary conditions was numerically simulated to estimate the heat transfer intensity on the energy separation chamber. The calculations were performed for air and superheated water steam flows. The results have shown that the heat transfer coefficient increases with the growth of expansion ratio in a vortex tube and decreases along the length of the energy separation chamber. Such dependences of α are fair for all the studied regimes for air and water steam.

Properties of steam as a cooler are noticeably better than those of air. It could be explained by higher heat capacity of steam. Graphical dependences are shown in fig. 2. It can be seen that these dependences are equidistant with the aspect ratio $\alpha_{st}/\alpha_{air}=1.65$.

The influence of swirling on the interaction between the flow and the channel walls may be specified by the critical angle of flow swirling on the channel surface that equals:

$$\tan \varepsilon_{\varphi}=\tau_{\varphi\omega}/\tau_{x\omega}, \quad (1)$$

$\tau_{\varphi\omega}$, $\tau_{x\omega}$ – are the projections of surface friction stress on the coordinate axes. The swirling angle in the wall region changes little along the radius of the channel. This is why:

$$\tan \varepsilon_{\varphi}=\omega/u, \quad (2)$$

where u, ω – are the flow velocity components (axial and peripheral) in the wall region.

The value of $\tan \varepsilon_{\varphi}$ characterizes the degree of deviation of the limit streamline from the axial direction, i.e. the curvature of the streamline in the wall region.

The degree of flow swirling is gradually reduced along the energy separation chamber, from the swirling device to the middle of the tube. There is some increase of the degree of flow swirling in the region of the unswirling device that is caused by significant reduction of axial velocity component. Formation of recirculation zone and near-axial reverse flow are shown in fig. 3.

Post-processing of the calculation results allowed obtaining an equation of swirling flow damping in the initial section of the energy separation chamber of the vortex tube

for air:

$$\varepsilon_{\varphi} = 1{,}184\,\bar{L}^{\,-0{,}014}; \quad (3)$$

IOP Conf. Series: Journal of Physics: Conf. Series **980** (2018) 012007 doi:10.1088/1742-6596/980/1/012007

for water steam:

$$\varepsilon_\varphi = 1{,}272\,\bar{L}^{\,-0{,}024}. \quad (4)$$

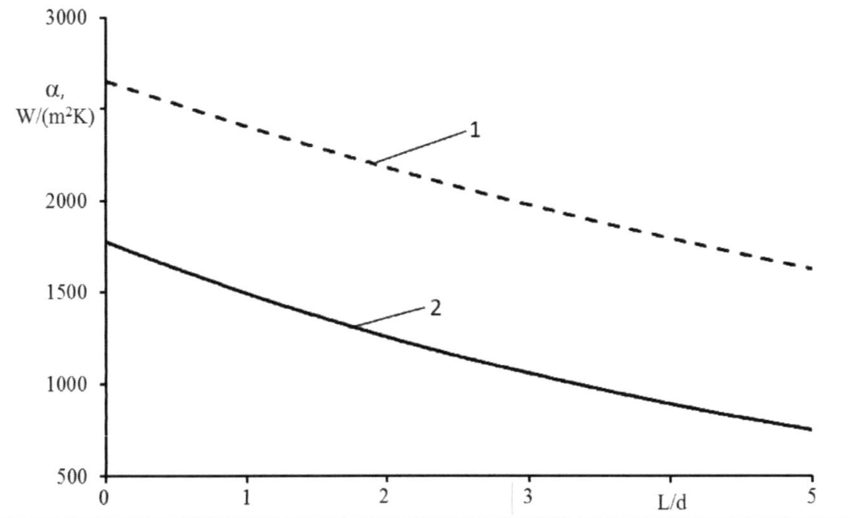

Figure 2. Comparison of heat transfer coefficients distribution along the energy separation chamber for air (1) and water steam (2)

Post-processing of the calculation results allowed to get a criteria equation of heat transfer:

$$Nu_{d,fl}=0{,}419\cdot Re^{0{,}646}{}_{d,fl}\cdot Re^{0{,}646}{}_{d,fl}\cdot \varepsilon_\varphi^{\ 10}. \quad (5)$$

These dependences allow determining the distribution of heat transfer coefficients along the energy separation chamber, taking into account the flow swirling, and may be applied for calculating the regime and geometrical parameters of non-adiabatic vortex tubes.

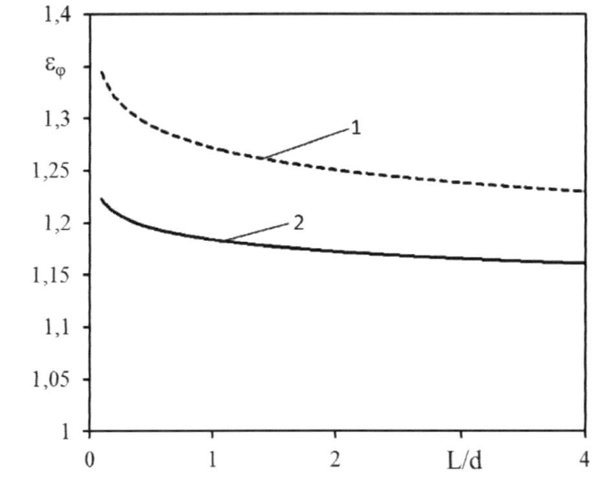

Figure 3. Distribution of the degree of flow swirling along the vortex tube: 1 – air, 2 – water steam

4. Numerical modeling of heat transfer under stationary conditions

An unsteady calculation of heat transfer allowed revealing sight mass flow oscillations through the vortex energy separator. Oscillation frequency is 660 Hz. The results of calculation were fixed in three points in time: for maximal mass flow $t=t_0$, for minimal mass flow $t=1/2t_0$ and for medium mass flow value $t=3/4t_0$. The results of calculation are shown in figures 4-6.

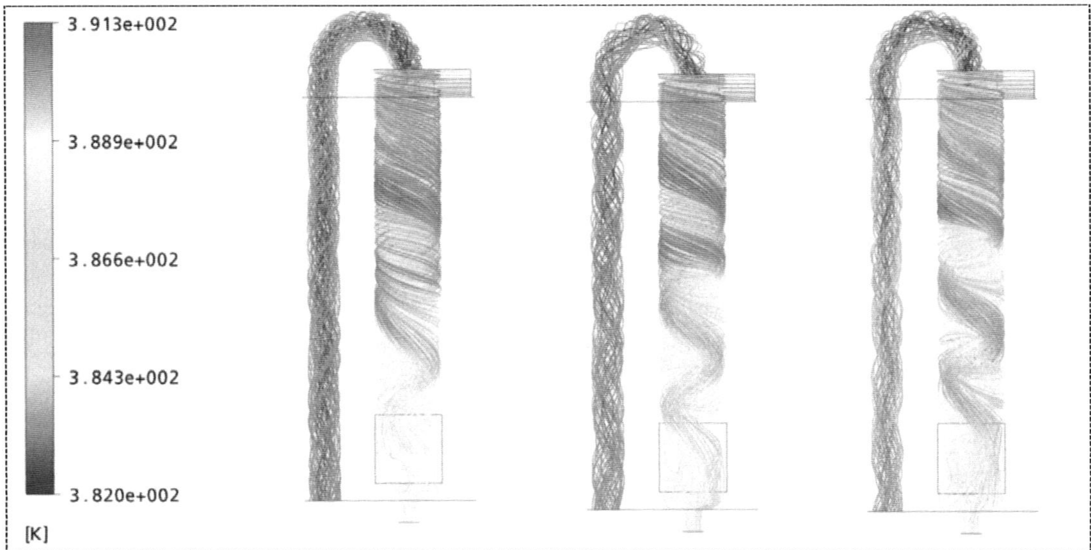

a) b) c)

Figure 4. Streamlines (a color matches the total temperature): a) $t=1/2t_0$; b) $t=3/4t_0$; c) $t= t_0$

Temperature distributions in the channel (fig. 4) have revealed the energy separation effect; there is a difference of average temperatures of peripheral and axial flows (fig. 6).

The centrifugal forces act on the swirling gas flow and forming of a radial static pressure gradient (fig. 5). The radial pressure gradient decreases along the tube from nozzle to throttle because of dissipation. Besides, it causes setting pressure gradient in the axial region energy of the separation chamber. This pressure gradient is directed from throttle to diaphragm and forms the reverse flow shown in fig. 5. It is shown that the size of the axial region of this reverse flow constantly changes over time.

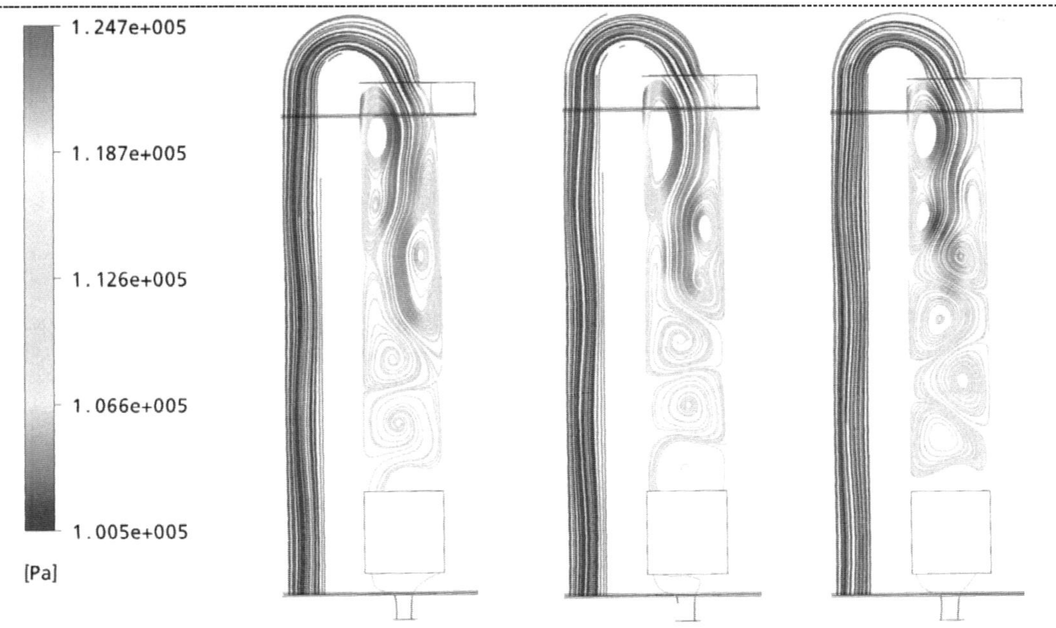

a) b) c)

Figure 5. Streamlines in axial section (a color matches the total pressure of the flow): a) $t=1/2t_0$; b) $t=3/4t_0$; c) $t= t_0$

The temperature distribution (fig. 4) and form of the streamlines (fig. 6) allow revealing the presence of unsteady interaction of gas flows, rotating at different distances from the axis. There is an influence of large and small scale turbulence that defines a formation of secondary vortex structures and precessing vortex core of the flow.

Vortex precession is clearly visible in the main flow sections (fig. 6) that are located at 0.01 m from the center of the tangential nozzle. The core of the stable swirling flow is displaced from the axis of the channel starting from the third section, and, as a result, the flow becomes a pulsating precessing vortex flow. It is obvious that mass flow oscillations at the channel outlet are the result of these pulsations.

Flow swirling intensifies heat transfer from the walls of the channel. Heat transfer coefficient significantly increases in the region of tangential nozzle and reaches value α=1380 W/(m^2·K), and then it decreases along the channel by 40%. The region of the most effective regimes (for air) relative to energy separation efficiency and heat transfer coefficients is in ranges μ=0,5÷0,7 and π=2÷3,3.

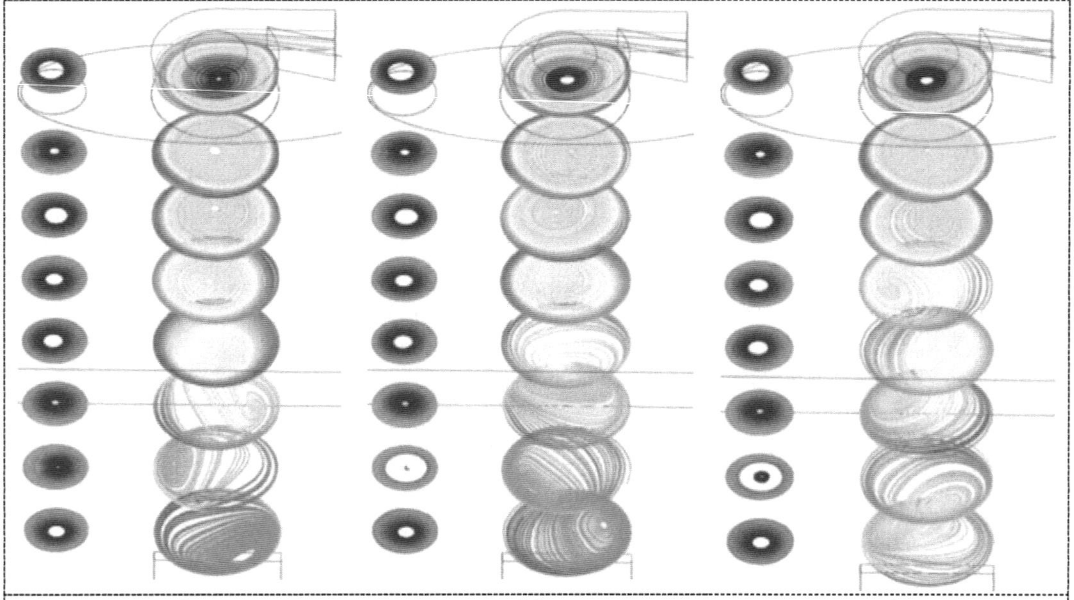

a) **b)** **c)**

Figure 6. Streamlines in the cross sections (a color matches the total temperature of the flow): a) $t=1/2t_0$; b) $t=3/4t_0$; c) $t= t_0$

5. Conclusion

Numerical modeling of conjugate gas dynamics and heat transfer allowed analyzing the flow structure in the energy separation chamber of the vortex tube for steady and unsteady regimes. The modeling of the flow structure that forms in the vortex tube channels has shown the presence of the complex three-dimension vortex system under the conditions of significant flow unsteadiness. Forms of the streamlines are determinant for the fact that unsteady interaction of flows rotating at different distances from the axis with the influence of significant large and small scale turbulence leads to the formation of secondary vortex structures and precessing vortex core of the flow.

The results of numerical calculation allowed defining the heat transfer coefficient at the inner surface of blade cooling channel α=1380 W/(m2·K). This agrees well with the experimental data obtained earlier. Unsteady calculation of heat transfer allowed revealing mass flow oscillations with frequency 660 Hz.

The research of heat transfer in the vortex tube channels has been carried out using air and water steam as coolers. There effect of damping swirling flow on the heat transfer rate has been estimated.

The flow swirling degree gradually decreases along the energy separation chamber from swirling device to the middle of the tube. Significant decrease of axial velocity component in the region of

IOP Conf. Series: Journal of Physics: Conf. Series **980** (2018) 012007 doi:10.1088/1742-6596/980/1/012007

unswirling plate causes small growth of the swirling degree. Post-processing of the calculation results allowed obtaining the equation of swirling flow damping in the initial section of the energy separation chamber of the vortex tube.

Post-processing of the calculation results allowed obtaining the equations to define the distribution of heat transfer coefficients along the energy separation chamber for calculating the regime and geometrical parameters of non-adiabatic vortex tubes.

The results obtained are not exhaustive in the context of heat transfer from swirling flows in the channels. This serves a basis for the research continue.

6. Acknowledgments

This work was financially supported by RFBR (Russian Foundation for Basic Research) grant No. 16-08-00974 "Development of a model description of heat and mass exchange processes within swirling limited flows and its application in the development of advanced technology devices."

7. References

[1] Biruk V V, Veretennikov S V, Guryanov A I and Piralishvili Sh A 2014 *Vortex effect. Technical applications* Vol. 2 Part 2 (Moscow: Nauchtechlitizdat)

[2] Piralishvili Sh A and Zhornik I V 1988 *Proc. of Scientific and Technical Conference «Vortex effect and its technical application»* (Kuibyshev: KuAI Press) 87-91

[3] Piralishvili Sh A and Veretennikov S V 2011 *Journal of Samara State Aerospace University* (Samara: SGAU Press) vol 3 241-247

IOP Conf. Series: Journal of Physics: Conf. Series **980** (2018) 012008 doi:10.1088/1742-6596/980/1/012008

Column formation and hysteresis in a two-fluid tornado

B R Sharifullin[1], I V Naumov[1], M A Herrada[2], V N Shtern[3]

[1]S.S. Kutateladze Institute of Thermophysics SB RAS, 630090, Russia, Novosibirsk, Lavrentyeva str., 1
[2]E.S.I, Universidad de Sevilla, Camino de los Descubrimientos s/n 41092, Spain
[3]Shtern Research and Consulting, Houston, Texas 77096, USA

E-mail: naumov@itp.nsc.ru

Abstract. This experimental and numerical study addresses a flow of water and sunflower oil. This flow is driven by the rotating lid in a sealed vertical cylinder. The experiments were performed in a glass container with a radius of 45 mm and a height of 45 mm with the water volume fraction of 20%. Different densities and immiscibility of liquids provide the stable and sharp interface. At the rest, the interface is flat and horizontal. As the rotation speeds up, a new water-flow cell emerges near the bottom center. This cell expands and occupies almost the entire water domain while the initial water circulation shrinks into a thin layer adjacent to the interface. The water, rising near the container axis, strongly deforms the interface (upward near the axis and downward near the sidewall). A new oil-flow cell emerges above the interface near the axis. This cell disappears as the interface approaches the lid. The water separates from the sidewall, reaches the lid, and forms a column. As the rotation is decreased, the scenario reverses, but the flow states differ from those for the increasing rotation, i.e., a hysteresis is observed. The numerical simulations agree with the experiment and help explain the flow metamorphoses.

1. Introduction

Swirling two-fluid flows attract the attention of researchers due to applications in aerial vortex bioreactors and are of fundamental interest due to striking features including their nontrivial topology [1-5]. Fujimoto & Takeda [4] experimentally studied interface patterns of a flow of water and silicone oil driven by the rotating lid in a vertical sealed cylindrical container. As the rotation intensifies, the interface becomes strongly and unusually deformed. The experiment of Tsai et al. [5] revealed vortex breakdown in the oil flow, and the numerical simulations by Carrion et al. [3] predicted that vortex breakdown also occurs in the water flow.

The current experimental and numerical study addresses a similar flow of water and sunflower oil in a sealed vertical cylindrical container driven by the lid rotating with angular velocity ω. As ω increases, the oil-water interface deforms upward near the axis and downward near the sidewall as typical of water spouts. At $\omega = \omega_1$, the flow separates from the sidewall. At $\omega = \omega_{up}$, the interface touches the rotating lid and water spreads over the near-axis lid part. A bell-shape water column forms near the axis, separated from the sidewall and extended from the bottom to the lid. The resulting density stratification with water (oil) being located near the axis (at the periphery) is unstable for fast rotation. Accordingly as ω further increases, the interface breaks and water mixes with oil. When ω is decreases before the interface breaks, the water separates from the lid at $\omega = \omega_{down}$, which is slightly smaller than ω_{up}, and touches the sidewall at $\omega = \omega_2 < \omega_1$.

Content from this work may be used under the terms of the Creative Commons Attribution 3.0 licence. Any further distribution of this work must maintain attribution to the author(s) and the title of the work, journal citation and DOI.
Published under licence by IOP Publishing Ltd

Our numerical and experimental results agree for the dependence of the water height at the axis on the rotation strength. The numerical study allows for detailed investigations of topological metamorphoses occurring in this flow. As the rotation intensifies, first the water flow becomes two-cellular, and then a new cell emerges and disappears in the oil flow. Thin circulation water layer adjacent to the interface emerges at a weak rotation when the interface is almost horizontal, has the bubble-ring shape in some ranges of ω, and still exists as the interface becomes nearly vertical.

2. Experimental setup and numerical technique

Figure 1(a) is a schematic of the experimental setup, showing the cylindrical container (1), CCD camera (2), stepping motor (3), LED 250W lamp (4) and reflecting white screen (5) [6,7]. The apparatus was housed in a darkened room, with lighting, provided by a LED 250W lamp reflected off a white sheet of paper. The photos of interface were shot by CCD 6-megapixel camera CASIO EX-F1. The cylindrical container is made of transparent optical glass with an inner radius of 45.75 mm and a height of 126 mm. The lid of radius R = 45 mm can be positioned at a varying height. The container is filled with tap water ($0 < z < h_w$, dark in figure 1) and sunflower oil ($h_w < z < h$, light in figure 1); z is the distance from the container bottom. Here we use h = R and $h_w = 0.2R$; these values are relevant for studying the hysteresis and the column formation in this flow.

At the room temperature of 22.6°C, the fluid densities and kinematic viscosities are $\rho_w = 1000$ kg/m^3, $\rho_o = 920$ kg/m^3, $\nu_w = 1$ cSt, and $\nu_o = 55$ cSt. Hereafter subscripts "w" and "o" denote "water" and "oil". The oil-water interface tension is $\sigma_{ow} = 26$ mN/m. The container is still except for the lid, which can rotate with angular velocity ω driven by the stepping motor. The rotation strength is characterized by the Reynolds number $Re = \omega R^2/\nu_o$; Re = 100 corresponds to approximately 0.5 RPS. For the Re range considered here, the variation in ω during a revolution is less than 0.3%.

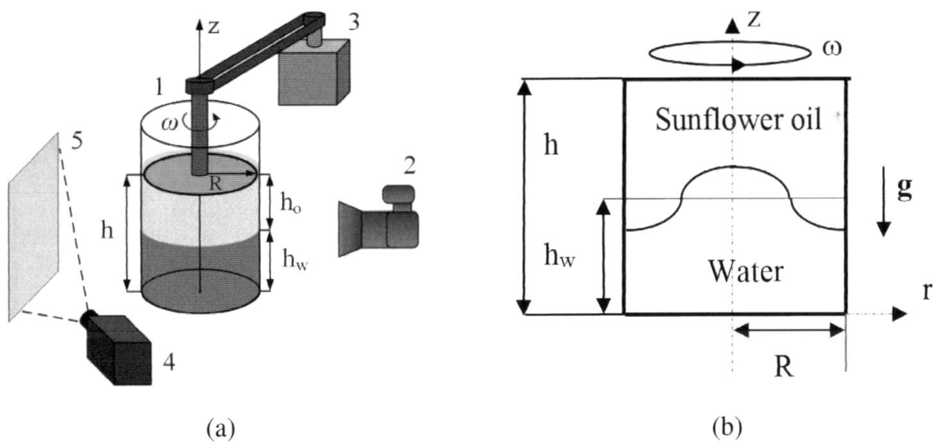

Figure 1. Schematic of experimental set up (a) and the numerical problem (b).

Figure 1(b) is a schematic of the numerical problem. One control parameter is aspect ratio, H = h/R. Another control parameter is the dimensionless water height, $H_w = h_w/R$. These values are fixed here: H = 1 and $H_w = 0.2$ to be consistent with the experiment. For the numerical procedure, it is convenient to introduce the Reynolds number based on the water viscosity, $Re = \omega R^2/\nu_w$.

To simulate the nonlinear problem for the basic flow, we use a numerical technique which is a variation of that described in detail by Herrada & Montanero [8]. First, the heavy-fluid (water) and light-fluid (oil) regions are mapped onto the standard square domain. Then, each variable (velocities, pressure and the interface shape) and all its spatial and temporal derivatives, which appear in the transformed equations, are composed as a single symbolic vector. The next step is to use a symbolic toolbox to calculate the analytical Jacobians of all the equations with respect to all the symbolic vectors. Using these analytical Jacobians we generate functions which then are evaluated point by point in the square domains. Then, we carry out the spatial and temporal discretization of the problem. To summarize, the numeric procedure includes the mapping of water and oil regions, the proper spatial and temporal discretization creating the discrete Jacobian matrix of the Newton procedure. For the established flow, we get the final steady solution through an unsteady process. The final step is to set up the numerical matrices allowing us to solve the problem by using a Newton procedure for the basic steady flow and by solving the generalized eigenvalue problem for disturbances [1-3].

3. Topological metamorphoses of flows

It is instructive to start with a creeping flow and to gradually increase the Reynolds number. Figure 2 depicts streamlines at Re_o = 1 (a), 100 (b) and 300 (c). The rotating lid pushes the upper fluid to periphery near the lid, as the upper arrow indicates in figure 2(a), and thus drives the clockwise circulation in CRO1; CR is an abbreviation for circulation region and O denotes the oil.

Near the interface, the oil moves toward the container axis, r = 0, and tends to drive the anticlockwise circulation of water. On the other hand, the oil flow transports the angular momentum from the lid to the interface. This angular momentum rotates the water. The centrifugal force pushes the water toward the sidewall near the interface and tends to drive the clockwise circulation of water.

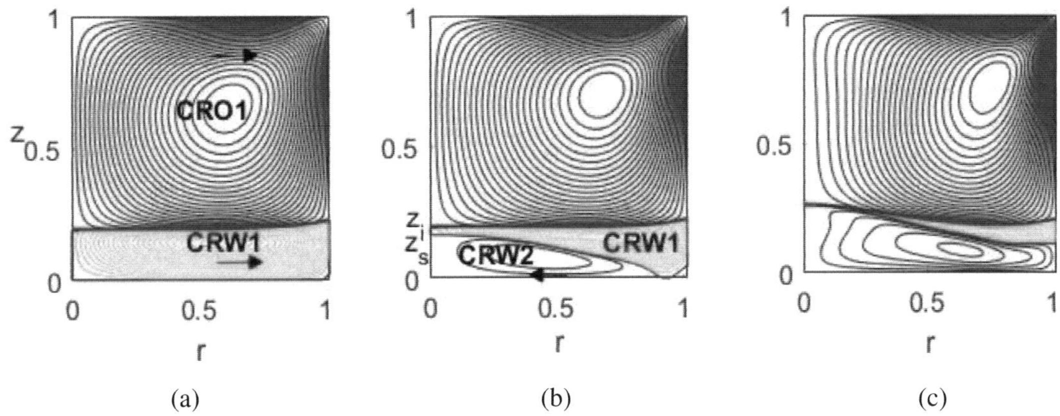

Figure 2. Clockwise circulation emerges in water as Re increases: Re_o = 1 (a), 100 (b) and 300 (c).

Thus, the two factors, (*i*) swirl and (*ii*) meridional velocities at the interface, push the water in opposite directions. At Re_o = 1, the swirl velocity at the interface is smaller than ωR by three orders of magnitude [3]. This yields that factor (*ii*) dominates over (*i*) and drives the bulk anticlockwise circulation, marked as CRW1 in figure 2(a), where W denotes the water.

In figure 2(a), the water moves upward in the bulk flow near the sidewall and thus blocks the downward transfer of the angular momentum from the interface. The angular momentum is transported toward the axis along the interface and then downward near the axis in the water domain. Therefore, the swirl effect is the strongest near the axis-bottom intersection where a cell of clockwise circulation emerges as Re_o increases, and factor (*i*) strengthens; this cell is denoted as CRW2 in figure 2(b). CRW1 separates from the bottom and becomes thin in figure 2(c).

Figure 3 shows photos of the oil-water interface (left) and streamline pattern (right) as Re increases. Figure 3(a) illustrate the development of the interface hump shape. At Re_o = 687 (figure 3(b)), the water separates from the sidewall. At Re_o = 840, the interface has the "bell" shape with its top being close to the lid as figure 3(c) illustrates. At Re_o = 865, the rising interface reaches the rotating lid. Then, the water spreads over a near-axis area of the lid as figure 3(d) shows at Re_o = 935. The centrifugal force pushes water, located near the lid, away from the axis, thus increasing the water radius at the lid and developing a "neck" in the interface upper part.

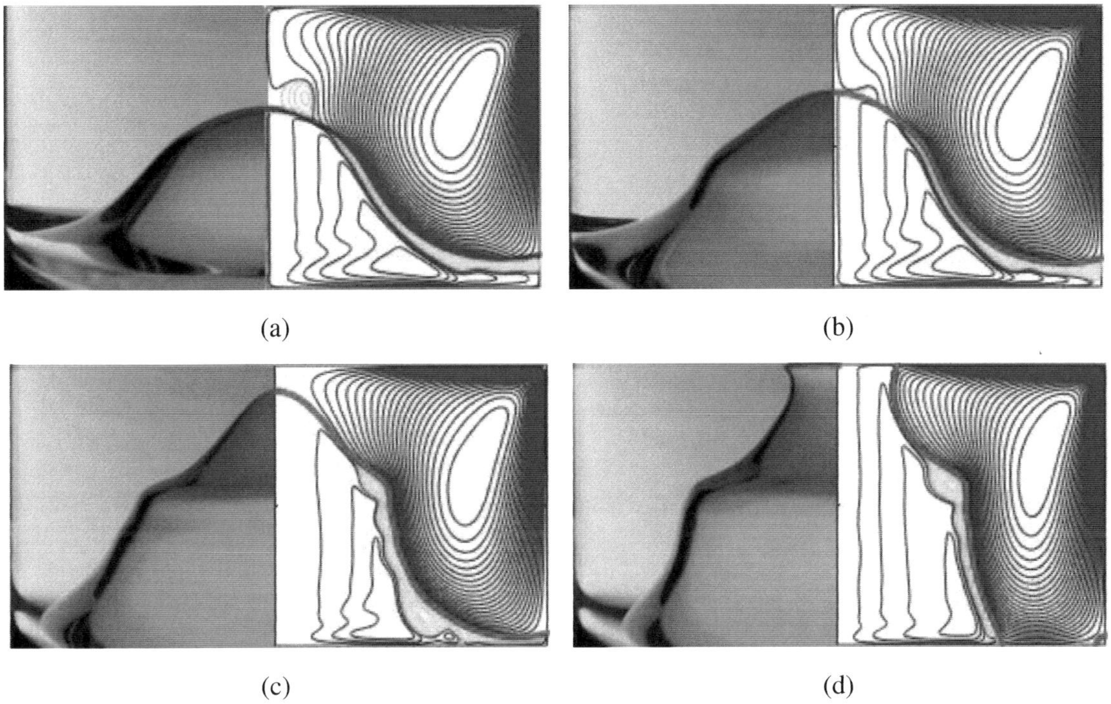

Figure 3. Interface shapes (left) and streamline pattern (right) at Re_o=650 (a), 687 (b), 840 (c), 935 (d).

Figure 4 characterizes the hysteresis by presenting the dependence of the interface height at the axis, h_i, on the rotation speed, Re. To measure the interface height at the axis, we use the experimental technique described in detail by Naumov et al [6, 7]. The interface height at the axis was measured by a laser beam and a precise scale. A semiconductor laser was fixed on the coordinate device. The laser beam was shifted up starting from the bottom along z-axis with a step of 0.1 mm until the reflection of laser beam from the interface terminates and the beam becomes tangent to the interface, touching the interface top.

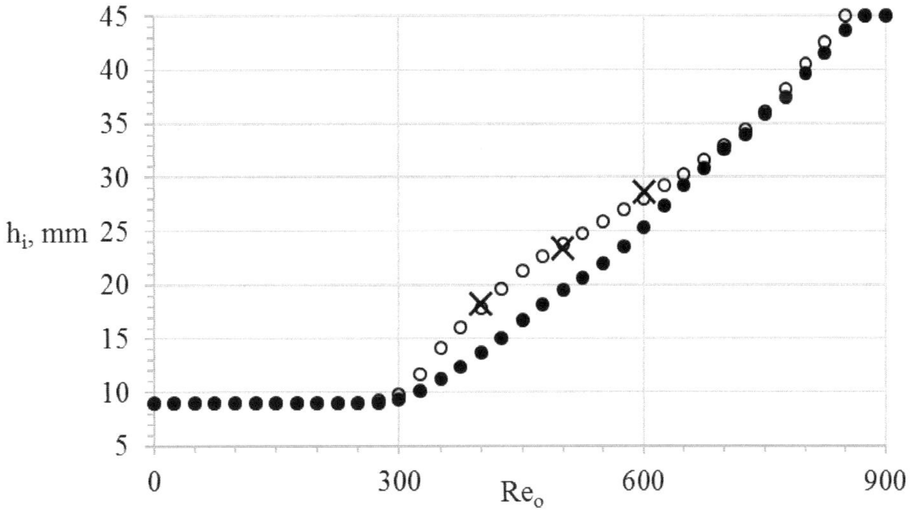

Figure 4. Dependence of interface height on the rotation speed as Re_o increases (filled circles) and decreases (experimental (empty circles) and numerical (crosses) results).

First, Re_o was increased from 0 with an increment ≤ 25. The results are presented by filled circles in figure 4. The interface height looks unchanged up to Re_o, being around 300, and then linearly grows with Re_o. The slope slightly increases for $Re_o > 687$. It seems that this occurs due the interface becomes separation from the sidewall. The horizontal line is for the interface attached to the lid for $Re_o > Re_{ol1} = 865$. Next, Re_o was decreases with the step down being ≤ 25. The results are presented by empty circles in figure 4. Interpolation estimates that the interface separates from the lid at $Re_o = Re_{ol2} = 854$. While the difference, $Re_{ol1}–Re_{ol2}$, is small and close to the measurement accuracy, the difference of the interface height at the axis, Δh_i, for the filled and empty circles remarkably increases for $Re_o < 650$ (figure 5). This increase is due to the fact that the interface is detached from the sidewall for the emplty circles while it is attached to the sidewall for the filled circles. The difference diminishes as the interface touches the sidewall at $Re_o = 343$ for decreasing Re_o (figure 5).

Figure 4 also shows the numerical (crosses) results describing the dependence of the interface height at the axis on decreasing Re_o. The numerical results correspond to $Re_o = 600$, 500, and 400 where the water is separated from the sidewall. Since our software is not applicable for such flow patterns (because of the applied mapping), these simulations were performed with the help of the commercial software FLUENT. The interface-sidewall contact angle is taken as 120° at $Re_o = 600$, 100° at $Re_o = 500$, and 80° at $Re_o = 400$.

It is known [9], that the contact angle is a monotonically increasing function of the interface velocity at a wall, U, and has a jump at U = 0. Therefore, the angle depends on the flow prehistory even for the established steady state. The jump value depends among others on the wall-material features such as roughness and wetting. Since it is problematic to experimentally determine this dependence, here the angle is chosen to better fit the experimental data.

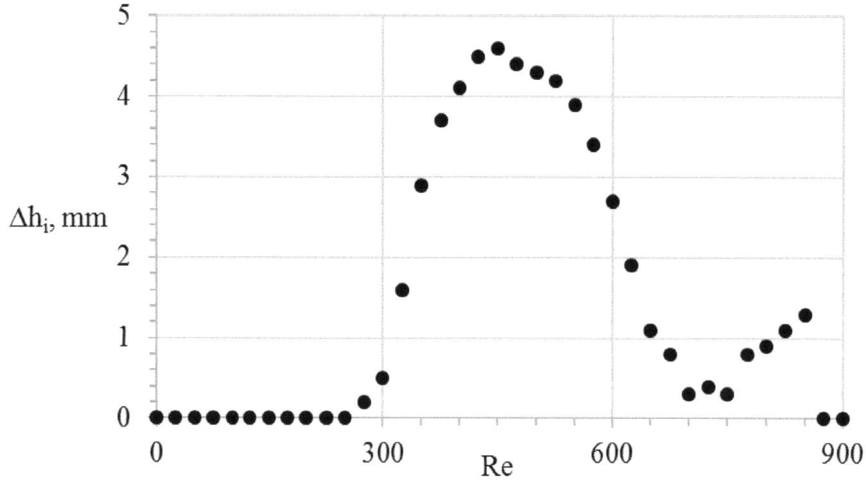

Figure 5. Dependence of the interface height difference at the axis on Re_o.

The flow states differ for increasing and decreasing Re_o, that is well observed in figure 6 at $Re_o = 450$ where Δh_i have maximal value of 4.6 mm. Figure 6(a) corresponds to increasing Re_o, and figure 6(b) corresponds to decreasing Re_o. The water is well attached to the sidewall in figure 6(a) while the water is still detached from the sidewall in figure 6(b). Accordingly, the water height at the axis is significantly smaller in figure 6(a) than that in figure 6(b). As Re_o further decreases, the water attaches the sidewall at $Re_o = 343$.

To sum up, the experimental observations reveal a hysteresis related to the interface separation from and reattachment to the sidewall with the hysteresis range being $343<Re_o<687$.

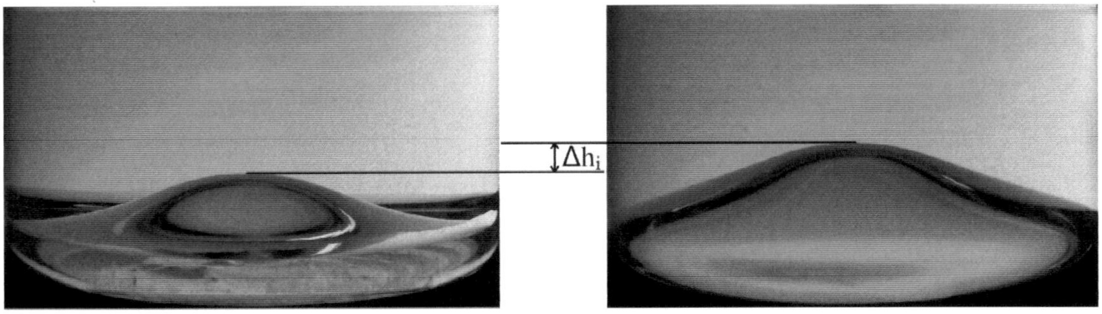

(a) (b)

Figure 6. Two different flow states at Re_0 =450 (a) as Re_0 increases and (b) as Re_0 decreases, where the difference of the interface height have maximal value 4.6 mm.

4. Conclusion

This pater explores two intriguing fluid-mechanics phenomena: (a) the formation of water column and (b) hysteresis in a water-spout flow. To observe large interface deformations in a laboratory, we study a water-sunflower-oil flow. The close densities and very different viscosities of water and oil allow analyzing the phenomena in a compact set-up: a cylindrical container of 45 mm in both radius and height, filled with 20% (80%) volume fraction of water (oil). The rotating lid of container drives the meridional circulation and swirl of both liquids.

The numerical calculations and the visual experimental observations have revealed that as the rotation speeds up, the interface rises near the axis, descends near the sidewall, separates from the sidewall, touches the lid, and then destroys. As rotation decreases, the interface pattern differs from that as the rotation increases. The hysteresis is quantitatively explored by measuring the dependence of the interface height at the axis, h_i, on the rotation strength, Re_0. The numerical and experimental results agree concerning the interface shape and height at the axis.

To investigate flow features (especially topology) in more detail, the numerical simulations were performed with the help of the original software [8]. The simulations have revealed that as Re_0 increases, the water flow becomes two-cellular, ascending (descending) along the axis near the bottom (interface). The ascending circulation expands and touches the interface while the opposite water circulation shrinks into a thin layer adjacent to the interface. A new cell appears and disappears in the oil flow. The agreement of the numerical and experimental results verifies the paper findings.

5. Acknowledgments

Research was supported by the Russian Science Foundation (Project № 14-29-00093).

References

[1] Balci A, Brøns M, Herrada M A and Shtern V N 2015 *Fluid Dyn. Res.* **47** 065503
[2] Herrada M A and Shtern V N 2014 *J. Fluid Mech.* **744** 65-88
[3] Carrión L, Herrada M A and Shtern V N 2017 *Phys. Fluids* **27** 032109
[4] Fujimoto S and Takeda Ya 2009 *Phys. Rev.* E **80** 015304(R)
[5] Tsai J C, Tao C Y, Sun Y C, Lai C Y, Huang K H, Juan W T and Huang J R 2015 *Phys. Rev.* E **92** 031002(R)
[6] Naumov I V, Sharifullin B R and Shtern V N 2017 *J. Eng. Thermophys.* **26** 391-8
[7] Naumov I V, Sharifullin B R and Shtern V N 2017 *J. Phys. Conf. Ser.* **899** 032015
[8] Herrada M A and Montanero J M 2016 *J. of Comput. Phys.* **306** 137-47
[9] Eral H B, 't Mannetje D J C M and Oh J M 2013 *Colloid & Polymer Sci.* **291** 247-60

Interaction of laser beams with magnetized substance in a strong magnetic field

V V Kuzenov [1,2]

[1] Bauman Moscow State Technical University, 2 Baumanskaya Street 5, 1, Moscow 105005, Russia
[2] N.L. Dukhov All-Russian Research Institute of Automatics (State Corporation «ROSATOM» company), Suschevskaya Street 22, Moscow 127055, Russia

E-mail: vik.kuzenov@gmail.com, svryzhkov@bmstu.ru

Abstract. Laser-driven magneto-inertial fusion assumed plasma and magnetic flux compression by quasisymmetric laser-driven implosion of magnetized target. We develop a 2D radiation magnetohydrodynamic code and a formulation for the one-fluid two-temperature equations for simulating compressible non-equilibrium magnetized target plasma. Laser system with pulse radiation with 10 ns duration is considered for numerical experiments. A numerical study of a scheme of magnetized laser-driven implosion in the external magnetic field is carried out.

1. Introduction

Development of new powerful sources of implosion with external magnetic field (magnetized target fusion or magneto-inertial fusion [1-10]) makes demands of radiation hydrodynamic simulations. Magneto-inertial fusion (MIF) is an alternative path to the magnetic and inertial fusion, where magnetic field suppresses the thermal conductivity in a magnetized plasma (the target), thereby reducing the heat losses, increasing the lifetime of the configuration, and hence the thermophysical properties and energy parameters of the reactor. The mathematical model is based on the calculation of the transport (diffusion, thermal, radiative) processes of atoms, ions, electrons, molecules, photons in a strong magnetic field.

This work is devoted to the simulation of laser beams interaction with an axisymmetric dense magnetized plasma, e.g. such burning plasma parameters can be produced in magneto-inertial fusion. For this reason the numerical model of the process is based on the multi-component radiation gas dynamics with turbulence model. The radiation transfer is taken into account in multi-group diffusive approach. The main distinguish of this proposed model is the presence of the externally applied magnetic field.

2. Model and System

A model of laser-driven imploding target in an externally applied magnetic field is created in [11-13]. We consider the mathematical model of fusion burn in targets under compressed magnetic fields for magneto-inertial fusion in centrally symmetric coordinate system. This model is based on the one-fluid, two-temperature equations of magnetic radiation gas dynamics. They take into account the equation of state and the absorption coefficients for laser radiation, resonance absorption, the

Content from this work may be used under the terms of the Creative Commons Attribution 3.0 licence. Any further distribution of this work must maintain attribution to the author(s) and the title of the work, journal citation and DOI.
Published under licence by IOP Publishing Ltd

generation of overthermal electrons and transfer of internal plasma energy, ion heating by laser radiation (direct ion heating), determine conditions of achieving self-sustaining fusion. The electromagnetic processes in the fusion target and a surrounding plasma region described by Maxwell's equations and Ohm's law with finite plasma conductivity. The radiation transport is considered in a multi-group approximation framework "back and forth".

A numerical technique using dynamic adaptive grid and computer code for the nonstationary radiation magneto-gasdynamics models for describing the plasmodynamic processes occurring in different MIF schemes is developed. A computer code MATTINA (MAgnetized inerTial Target In Numerical Analysis) is developed. Modified alternatively triangular three-layered iterative scheme is applied for the solution of radiation transport equations, where the time step was selected via conjugate directions method.

It has been proposed to use laser and plasma beams to implode a magnetized target [14-19]. The device consists of: 1) the system of target irradiation involves the lasers or hohlraum; 2) the "dynamic" system to compress the initial seed magnetic flux directed along the geometric axis of symmetry. The magnetic system is located outside the target area. The target is located in the center of the dynamic magnetic trap and can be irradiated with laser energy in the direct drive configuration or indirect drive system (conventional hohlraum).

The laser system does not participate in the coils compression. They represent a separate system designed to compress the target through radiative effects. I.e. laser system and "shrink coil" are not connected, there two different (decoupled) technical systems. Thus, the target is compressed in two ways: by external radiation (laser energy) and by means of the dynamic magnetic field of the trap, e.g. probkotron.

Hohlraum means the system of lasers that generates hot temperature plasma, that is the source of broadband radiation. One can create hohlraum by applying high current on the surface of the capsule (i.e. not laser-generated plasma) in an experimental stand. The external magnetic field will penetrate inside the hohlraum.

3. Calculation results

Modeling calculations based on the proposed numerical method for solving the system of gas dynamics equations are performed. A comparison of the numerical and modeling solutions was carried out using the values of decision errors in the standards of various kinds.

Laser system with pulse radiation with 5÷10 ns duration is considered. Numerical simulation of the magnetized target compression is done for $q_{las} \leq 10^{14}$ W/cm^2. Initial values of the intensity of a seed magnetic field in the rarefied medium reach a fraction of Tesla. A computational domain and target of magneto-inertial fusion consist of the central part and one coaxial layer. They have a cylindrical shape with the following range of values of initial parameters of target and medium. The central part of the target (inner radius R = 0.05 cm) is filled with deuterium-tritium (D-T) gas mixture with density $\rho = 5\cdot10^{-2}$ g/cm^3 and temperature of T = 297 K. It was surrounded by the coaxial layer (outer radius R_c = 0.1 cm) consisting of metal (Al) with density ρ = 2.7 g/cm^3 and temperature of T = 297 K. The computational domain has outer radius r = 0.2 cm. Thermodynamic parameters of external rarefied medium (Ar) are defined: T = 297 K, $\rho = 1.29\cdot10^{-3}$ g/cm^3.

The dynamics of the magnetized plasma expansion into an ambient gas is considered. The spatial distribution of pressure P (atm) and temperature T (kK) is shown in figures 1-2. Plasma velocity V and density ρ versus coordinate r at time t = 8.5 ns are given in figures 3-4. Plots of contact boundary (1st and 2nd) versus r are presented in figures 5-6. The distribution of the laser radiation W_{las} along the radial coordinate is shown in figure 7.

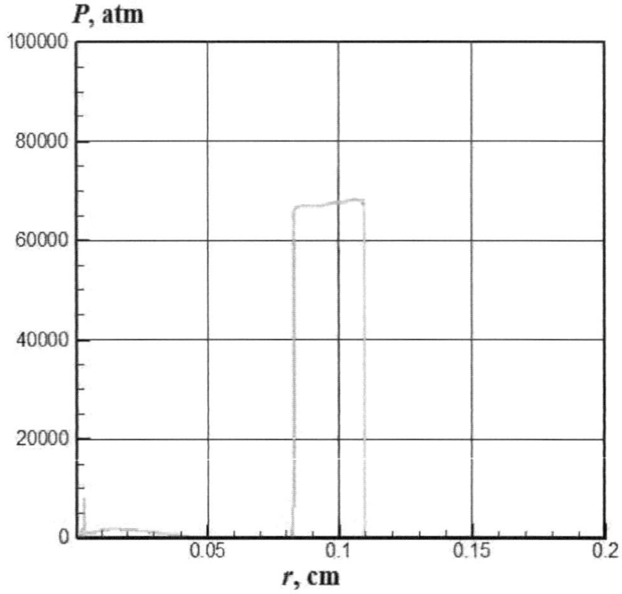

Figure 1. Spatial distribution of the plasma pressure (P) at 8.5 ns (t_{Las} = 10 ns).

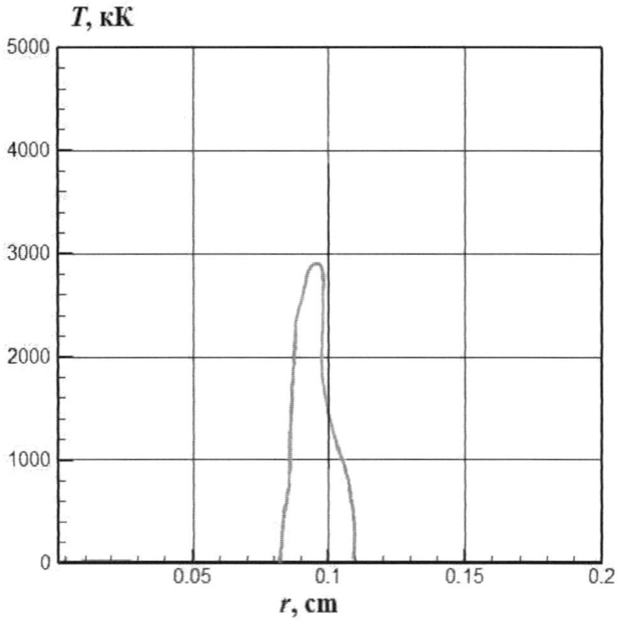

Figure 2. Spatial distribution of the plasma temperature (T) at t = 8.5 ns (t_{Las} = 10 ns).

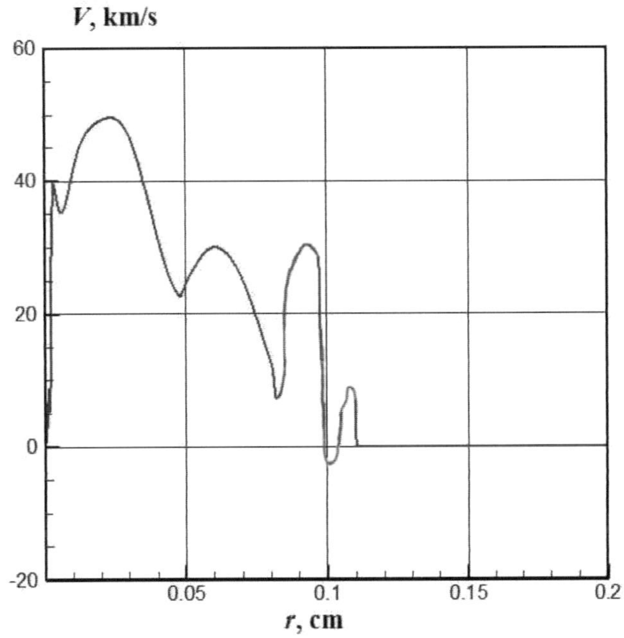

Figure 3. Distribution of the plasma velocity (V) at t = 8.5 ns along the radial coordinate.

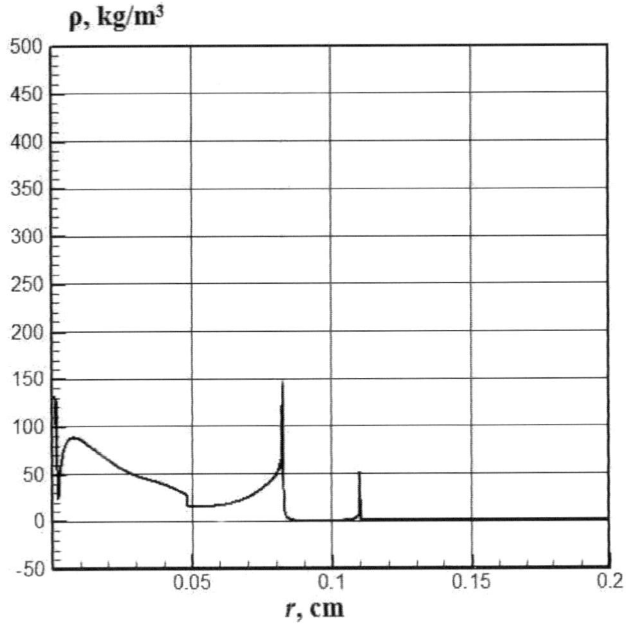

Figure 4. Distribution of the plasma density (ρ) at t = 8.5 ns (t_{Las} = 10 ns).

The key parameters of the laser-target system are calculated. Calculations carried out for the Nd laser radiation for forming rectangular impulse shape (duration = 10 ns). The radiant flux value is q_{las} = $2{\cdot}10^{14}$ W/cm^2. A thin metallic cylindrical shell material is Al. The target radius is 0.1 cm.

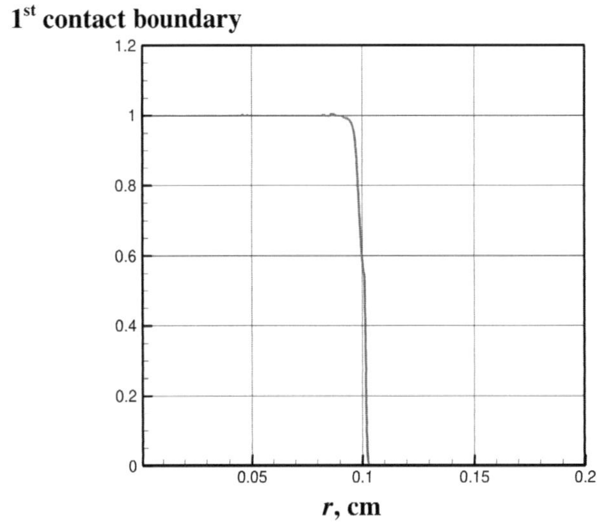

Figure 5. Spatial distribution of the first contact boundary (a) at time t = 8.5 ns.

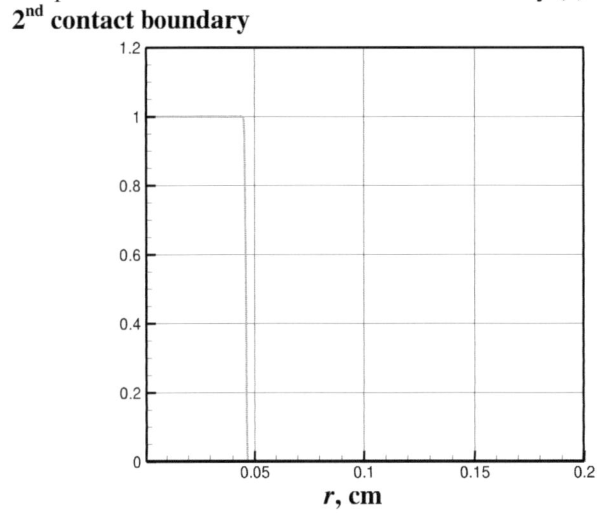

Figure 6. Spatial distribution of the second contact boundary (b) at time t = 8.5 ns.

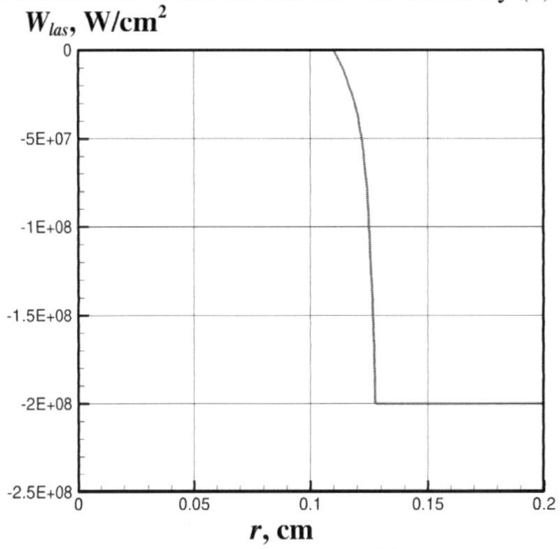

Figure 7. Distribution of the laser radiation (W_{las}) at t = 8.5 ns (t_{Las} = 10 ns).

4. Conclusion

We have developed a model for laser-driven magneto-inertial fusion - laser-driven target implosion in an externally applied magnetic field. The main plasma parameters (pressure P, temperature T, velocity V and density ρ), and laser energy W_{las} along the radial coordinate are calculated. Numerical modeling of the laser target compression process showed the following. The central part of a target is optically transparent for both laser and own broadband plasma radiation during compression. However at the same time heat flow density on the first wall of the reactor chamber at a certain moment of time can reach 10^8 W/cm^2. The magnetic pressure P_{mag} during compression of a target changes in time and is comparable to the static pressure P, reaching values 10^6 atm. The maximum values of plasma pressure and temperature are observed after reflection of a shock wave from a geometrical axis of symmetry at $t \approx 5$ ns.

5. Acknowledgment

This work was supported by the Russian Ministry of Education and Science, Project 13.5240.2017/8.9.

References

[1] Velikhov E P, Vedenov A A, Bogdanets A D, Golubev V S, Kasharskii E G, Kiselev A A, Rutberg F G and Chernukha V V 1973 *Sov. J. Tech. Phys.* **18** 274–283
[2] Gasilov V A, Zakharov S V and Smirnov V P 1991 *JETP Lett.* **53** 85–88
[3] Ryzhkov S V 2010 *Problems of Atomic Science and Technology* **4** 105–110
[4] Ryzhkov S V 2011 *Plasma Physics Reports* **37** 1075
[5] Ryzhkov S V 2014 *Bulletin of the Russian Academy of Sciences. Physics* **78** 456–461
[6] Ryzhkov S V 2015 *Sustainable Cities and Society* **14** 313–315
[7] Aleksandrov V V, Bolkhovitinov E A, Volkov G S, Grabovski E V, Gritsuk A N, Medovshchikov S F, Oleinik G M, Rupasov A A and Frolov I N 2016 *Plasma Phys. Rep.* **42** 1091–1100
[8] Honrubia J J, Murakami M, Mima K, Johzaki T, Sunahara A, Nagatomo H, Fujioka S, Shiraga H and Azechi H 2016 *J. Phys.: Conf. Ser.* **688** 012033
[9] Ryzhkov S V 2005 *Fusion Science and Technology* **47** (1T) 342
[10] Thio Y C F 2008 *J. Phys.: Conf. Ser.* **112** 042084
[11] Kuzenov V V and Ryzhkov S V 2013 *Problems of Atomic Science and Technology* **4** (86) 103–107
[12] Ryzhkov S V, Chirkov A Y and Ivanov A A 2013 *Fusion Sci. Technol.* 63 135–138
[13] Kuzenov V V and Ryzhkov S V 2013 *Problems of Atomic Science and Technology* **1** (83) 12–14
[14] Kostyukov I Y and Ryzhkov S V 2010 *Plasma Physics Reports* **37** 1092–1098
[15] Kuzenov V V and Ryzhkov S V 2016 *Bulletin of the Russian Academy of Sciences. Physics* **80** 598–602
[16] Loktionov E Yu, Protasov Yu S, Protasov Yu Yu and Telekh V D 2015 *Opt. Spectrosc.* **118** 300
[17] Sivkov A, Ivashutenko A, Shanenkova Y and Shanenkov I *Advanced Powder Technology* 2016 **27** 1506–1513
[18] Kuzenov V V and Ryzhkov S V 2018 *Mathematical Models and Computer Simulations* **10**
[19] Kuzenov V V and Ryzhkov S V 2017 *J. Phys.: Conf. Ser.* **830** 012124

IOP Conf. Series: Journal of Physics: Conf. Series **980** (2018) 012010 doi:10.1088/1742-6596/980/1/012010

Numerical simulation of an argon swirling flow in the presence of a DC discharge

P N Kazanskii[1;2], A I Klimov[1;2], N E Molevich[1;3], D P Porfiriev[1;3] and I P Zavershinskii[1]

[1] Samara National Research University, Samara, 443086, Russia

[2] Joint Institute for High Temperatures RAS, Moscow, 125412, Russia

[3] Lebedev Physical Institute, Samara, 443011, Russia

E-mail: ipzav63@mail.ru

Abstract. Numerical simulations of the non-stationary three-dimensional swirling Ar flow are presented for plasma vortex reactor (PVR) with a paraxial heat source at various positions of the heat source and electrode forms. Flow and duct parameters correspond to the experimental conditions. Flow velocity and thermal fields have been obtained.

1. Introduction

Swirling flows are known to be important for devices with heat release such as plasma dynamic systems [1], aero-engines [2,3], hydrogen plasma generators [4,5], etc. Recently, the problem of creating efficient compact energy sources, including alternative ones, has become increasingly noticeable [4,5]. One of the most promising technologies is the use of plasma vortex reactor (PVR), which can be both heat and hydrogen generator.

Intensive studies of a plasma chemical reactor started a few years ago [5-7]. Peculiar properties of gas-plasma and heat fluxes in the presence of longitudinal heterogeneous electric discharge in the swirl flow were studied in [6,7]. The test gases used in this setup are the following: argon, water vapor, aluminium nanoparticles, and their mixtures. It is shown that in the heterogeneous plasmoid formed in the argon-water vapor-nanoparticle mixture, UV-radiation and soft X-ray radiation are generated, whereas in the diffuse plasma produced during the discharge in pure argon or in the argon-nanoparticle mixture, this radiation is absent.

In addition, in a heterogeneous plasmoid there is a significant release of thermal energy ~ 1-2 keV per metal atom. It is shown that the coefficient of performance (COP) depends on the parameters of the vortex flow itself, the electric discharge and the used material. The energy cost of the hydrogen molecule production in the PVR has been estimated. It turns out to be much lower than the market price, due to the release of energy during the interaction of water vapor with excited nanoclusters of aluminium.

The efficiency of hydrogen and heat production obviously increases in the field of vortex flows, which transfer heat predominantly downstream. At the same time, the structure of the flow is strongly influenced by a number of factors: the organization of the input and output of the mixture components, the shape and arrangement of the discharge electrodes, swirl number and swirl rate, and the mass flow of components.

Content from this work may be used under the terms of the Creative Commons Attribution 3.0 licence. Any further distribution of this work must maintain attribution to the author(s) and the title of the work, journal citation and DOI.
Published under licence by IOP Publishing Ltd

The self-consistent simulation of the flow structure is very complicated and requires joint modeling of the gas-dynamic and discharge subsystems. In this paper, as a starting step, a simulation of the structure of the gasdynamic and thermal fields formed in a three-dimensional unsteady viscous turbulent vortex flow of pure argon in a PVR duct under experimental conditions [6,7] is carried out for different configurations of electrodes system. The ultimate goal of investigation was to ascertain geometry and positions of electrodes to minimize the upstream heat flux which leads to energy losses through dissipation by heating parts of the apparatus. The spatial distributions of the source power were chosen phenomenologically.

2. Mathematical modelling

2.1. Geometrical configurations, governing equations and boundary conditions

The schematic swirler and duct geometry is sketched in figure 1. The swirler has four tangential inlets with independent gas supply. The system of electrodes is only outlined. Its varying parameters are the shape and the distance from the swirler z_0. The heat source is located between two electrodes. The outlet of the duct has the form of a nozzle.

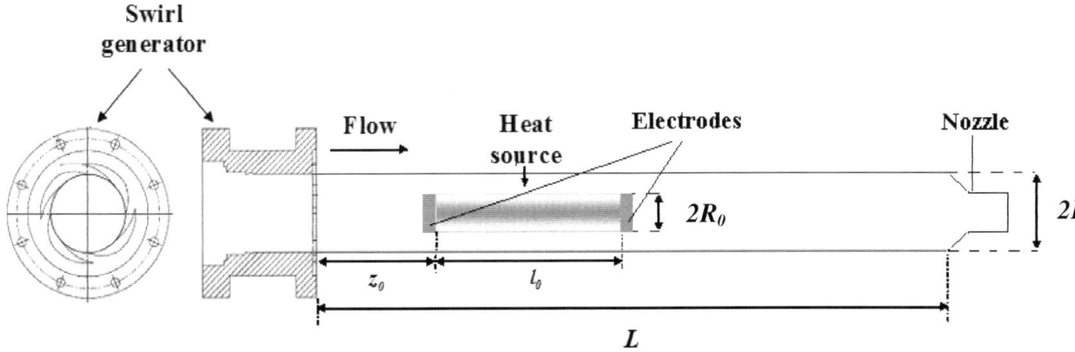

Figure 1. A schematic diagram of the geometry of swirl generator and duct used for numerical simulations, L= 70 cm, R = 2.5 cm, l_0 = 12.0 cm. Here R_0 is the electrode radius.

Standard unsteady Reynolds averaged Navier-Stokes (URANS) equations were used to describe the flow:

$$\frac{\partial \rho}{\partial t}+\frac{\partial(\rho v_i)}{\partial x_i}=0,$$

$$\frac{\partial(\rho v_i)}{\partial t}+\frac{\partial(\rho v_i v_j)}{\partial x_i}=-\frac{\partial P}{\partial x_i}+\frac{\partial}{\partial x_i}\left[\mu\left(\frac{\partial v_j}{\partial x_i}+\frac{\partial v_i}{\partial x_j}-\frac{2}{3}\delta_{ij}\frac{\partial v_k}{\partial x_k}\right)\right]+\frac{\partial}{\partial x_i}\left[-\rho\overline{v'_i v'_j}\right] \tag{1}$$

$$\frac{\partial(\rho E)}{\partial t}+\frac{\partial[v_i(\rho E+P)]}{\partial x_i}=\frac{\partial}{\partial x_j}\left[\left(\kappa+\frac{c_P\mu_t}{Pr_t}\right)\frac{\partial T}{\partial x_j}+v_i\left(\tau_{ij}\right)_{eff}\right]+\Im(\vec{x}),\ E=h-\frac{P}{\rho}+\frac{v^2}{2},\ P=\frac{\rho T}{M},$$

where $\left(\tau_{ij}\right)_{eff}=\mu_{eff}\left(\frac{\partial v_j}{\partial x_i}+\frac{\partial v_i}{\partial x_j}\right)-\frac{2}{3}\mu_{eff}\frac{\partial v_k}{\partial x_k}\delta_{ij}$ is the deviatoric stress tensor; $\left[-\rho\overline{v'_i v'_j}\right]$ are the Reynolds stresses which must be modeled to close the set of equations, v_i, v_i', ρ, T P, E, and h are the mean and fluctuating velocity components, density, temperature, pressure, total energy, and enthalpy, respectively; $\Im$ is the energy source, μ, μ_t, μ_{eff} are the molecular, turbulent, and effective viscosity coefficients, respectively; c_p is the molar specific heat capacity at constant pressure; κ is the thermal conductivity coefficient; and Pr_t is the turbulent Prandtl number.

The no-slip velocity and fixed temperature conditions were imposed along the tube and electrode surfaces. At the duct inlet and outlet, fixed mass flux conditions were used. The inlet turbulent intensity set to moderate values of 1-5% did not affect the flow structure discernibly. At the outlet, we imposed the boundary conditions with the static pressure equal to the atmosphere pressure.

2.2. Flow, heat source and electrodes parameters

In what follows, we investigate the problem for Ar close to the experimental conditions [6,7] corresponding to the total mass flux rate m_t = 2 g/sec, the homogeneous heat source power N =50 W and the gas pressure P = 1 Bar.

Three different types of electrodes are considered in the simulation: a sharp or plain ended thick electrode with a diameter $D_0 = 2R_0$ of 20 mm and a length of 40 mm, a sharp or plain ended thin electrode with a diameter D_0 of 5 mm and a length of 40 mm, and a pipe-like electrode with an external diameter of 20 mm and a thickness of 1 mm. The interelectrode distance l_0 = 12 cm remains constant, whereas the position of the input electrode z_0 varies from 52 to 60 cm from the inlet. The heat capacity of the swirler, electrodes and nozzle elements, as well as the thermal conductivity of the walls were neglected in this simulation.

2.3. Numerical procedure

The numerical simulation of the non-stationary 3D turbulent swirling flow was made using the ANSYS FLUENT 15.0 program package, which solves governing equations using the finite volumes method. In our simulations, the Spalart-Allmaras model was used. A detailed analysis of the use of the turbulence models in similar problems is given in [1].

For spatial discretization of density, momentum, energy and turbulent quantities, a second-order upwind scheme is applied. The use of higher-order schemes has not changed the flow parameters considerably. The diffusion terms are central-differenced and second-order accurate. We interpolate the pressure values at the faces using the PRESTO! scheme because of the strong swirl nature of the flow.

For transient terms, we used the fully implicit scheme of the second-order accuracy. Different pressure-velocity coupling schemes were tested and gave equal results. So, we used the SIMPLE scheme as the least resource intensive. The convergence was obtained when the residual reached 10^{-6} for the energy equation and 10^{-4} for the continuity equation, the momentum equation, and the equations for turbulent quantities.

The computational grids consisted of about $4{\cdot}10^6$ (from $3.7{\cdot}10^6$ to $4.3{\cdot}10^6$) hexahedral cells. The grid refinement did not considerably alter the gasdynamic parameters. Calculations with different values of time step from $1{\cdot}10^{-4}$ sec to $1{\cdot}10^{-6}$ sec were carried out. Finally, the time step equal to $5{\cdot}10^{-5}$ sec was set since its lower values led only to an increase in the computational time.

3. Results of numerical simulation and discussion

The calculations have showed the following features of the formation of the flow velocity field and the thermal field.

To a great extent, the structure of the flow is determined by the paraxial counterflow zone, typical for strongly swirled flows [1]. Essentially, there are two types of such phenomena: bounded by the thick electrode (figure 2a, c) and faded away due to a gradual decrease of angular momentum of the fluid and a consequent fall of the swirl number below the critical value (figure 2b).

Working towards a design system with desirable behavior, we analyzed several cases. The first one is two thick cylindrical electrodes (figure 3 a-c). Such a pattern has a fundamental drawback: between the electrodes, a dead zone is formed where the pressure (figure 3a) and inertia forces almost prevent from the mass exchange with the outer flow (figure 3b). Eventually, it leads to the zone overheating (figure 3c) and makes the employed modelling approach inapplicable. The second case of thick and thin electrodes (figure 3 d-e) is roughly the same, with only slight progress. Structures of counterflow (figure 2a) and dead zones (figure 3a, b, d, e) in both setups have several common features. First and foremost, the thick electrode placed in the upstream position virtually stops the counterflow. No less important is the fact that the pressure between electrodes is lower than in peripheral areas, nevertheless it is higher than pressure downstream from the second electrode. Taking into account those facts, we proposed two configurations (figure 4). In the case of thick and pipe-like electrodes, the pressure gradient along the symmetry axis draws out the hot gas from the interelectrode area. In the fourth version, electrodes set position is downstream from the end of the counterflow zone.

IOP Conf. Series: Journal of Physics: Conf. Series **980** (2018) 012010 doi:10.1088/1742-6596/980/1/012010

Still, it is not possible to replace a pipe-like electrode with solid one because, in that case, the counterflow zone prolongs (figure 2c), and heat is partly carried away upstream.

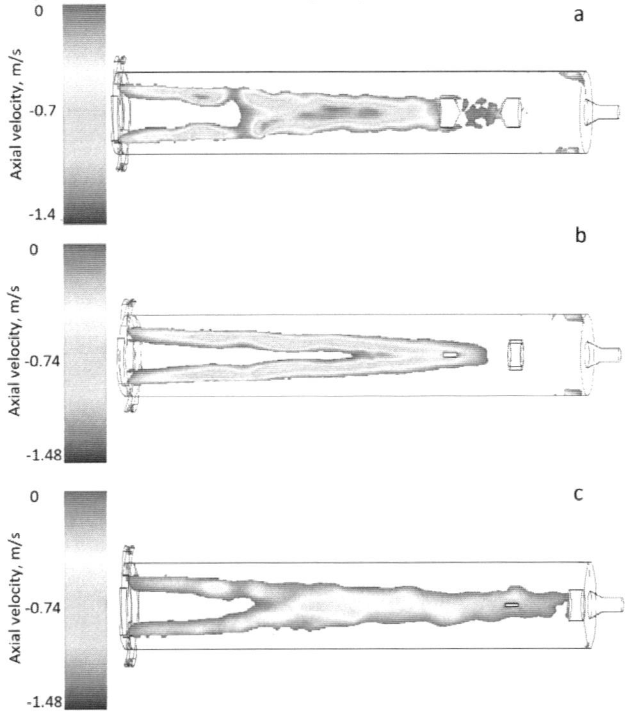

Figure 2. Typical structures of the counterflow zone. Only areas of negative velocity are shown.

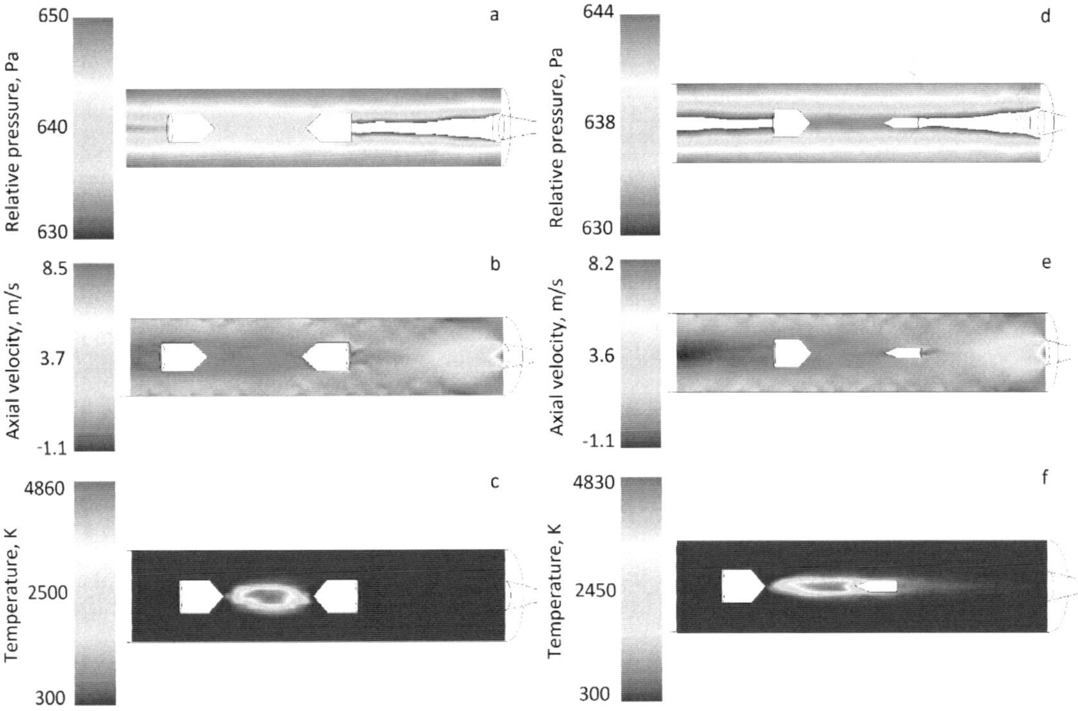

Figure 3. Longitudinal fields of relative (to atmospheric) static pressure (a, d) (only values higher than 630 Pa are shown for clear visualization), axial velocity (b, e) and temperature (c, f). Thick electrodes

are of radius $R_0 = 1$ cm, thin electrode is of radius $R_0 = 0.25$ cm, upstream coordinate of electrodes set $z_0 = 48$ cm.

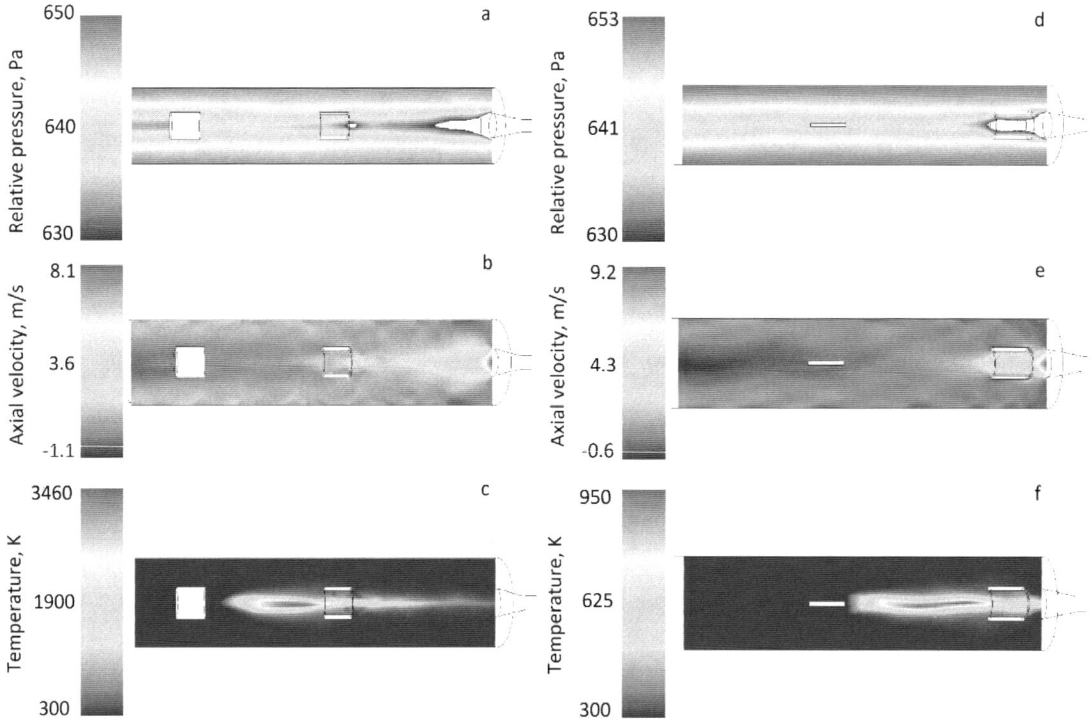

Figure 4. Longitudinal fields of relative (to atmospheric) static pressure (a, d) (only values higher than 630 Pa are shown for clear visualization), axial velocity (b, e) and temperature (c, f). Thick electrodes are of radius $R_0 = 1$ cm, thin electrode is of radius $R_0 = 0.1$ cm, upstream coordinate of electrodes set $z_0 = 48$ cm (a, b, c) and $z_0 = 57$ cm (d, e, f).

4. Concluding remarks

The numerical modelling has demonstrated that it is possible to configure electrodes for the energy to be transferred mainly downstream from the interelectrode zone. It may be accomplished by the pipe-like electrode through which, in one case (figure 4a-c), areas of different pressures are connected. Pressure forces, created in this way, draw out the hot gas towards the exit nozzle. The second set (figure 4d-f) ensures the desirable behavior by placing the electrodes downstream from the counterflow zone. The pipe-like electrode is still crucial since the solid one creates an obstacle producing the negative axial pressure gradient and prolonging the counterflow zone.

A qualitative agreement between the results of the calculations and the experimental data for pure argon [6,7] has been obtained. The next step in the simulation of the PVR is the calculation for real mixtures of Al, H_2O, Ar, taking into account the plasma-chemical kinetics of the discharge mixture components as well as thermal characteristics of the experimental set up.

Acknowledgments

The study was supported in part by the Ministry of Education and Science of Russia under the public contract with educational and research institutions within the project 3.1158.2017 and by RFBR under grant 16-41-630591.

References

[1] Gorbunova A, Klimov A, Molevich N, Moralev I, Porfiriev D, Sugak S and Zavershinskii I 2016 Precessing vortex core in a swirling wake with heat release *International J. Heat and Fluid Flow.* **59** 100-108

[2] Hassa C, Voigt B, Lehmann B, Schodl C and Carl M 2002 Flow field mixing characteristics of an aero-engine combustor - part I: Experimental results *AIAA Paper* (2002)3709
[3] Yiguang J and Wenting S 2015 Plasma assisted combustion: Dynamics and chemistry. *Progress in Energy and Combustion Science*. **48** 21-26
[4] Jasiński M, Dors M, Nowakowska H and Mizeraczyk J 2008 Hydrogen production via methane reforming using various microwave plasma sources. *Chem. Listy* **102** 1332-1339
[5] Klimov A, Bitiurin V, Tolkunov B, Chinnov V, Godin S, Efimov A, Kutuzov D and Polyakov L 2011 Hydrogen Plasma Flow Creation for MHD Power Generation. *AIAA Paper* (2011) 3285
[6] Bitiurin V, Klimov A, Korshunov O and Chinnov V 2014 Kinetic Model of Aluminum Oxidation by Water Vapor in Heterogeneous Plasma: Gas-Phase *Kinetics. High Temperature*. **52** 657-663
[7] Bitiurin V, Klimov A, Korshunov O and Chinnov V 2015 Kinetic Model of Al Oxidation by Water Vapor in Heterogeneous Plasma: Heterophase *Kinetics. High Temperature* **53** 21-26

Hydrodynamic characteristics of viscous fluid flow in screw channels formed by two ribs

A I Kadyirov[1], B R Abaydullin[1], E K Vachagina[1]

[1]FIC Kazan SC of RAS, 2 Lovachevsky, Kazan, 420111Russia

Email: aidarik@rambler.ru

Abstract. The mathematical model of laminar viscous flows in screw channels, formed by two ribs, is developed using the helical coordinates. The numerical results of the flow with helical symmetry are presented.

1. Introduction

Helical flows are widely used to enhance heat transfer in channels of heat exchange equipment. The most common types of twisted tapes are long and short screw inserts. There are a lot of works dedicated this type of inserts [1-9]. Long screw inserts are known to intensify the heat transfer under high viscosity fluid flow. Whereas short inserts are not effective for such fluids. However the long screw inserts give high values of friction factor. To reduce the negative influence it is possible to use screw channels formed by one, two or more ribs. The hydrodynamic characteristics of viscous fluid flow in channels formed by two screw ribs are analyzed.

2. Mathematical model

The mathematical model is developed under the following assumption: the stationary viscous flow is laminar and isothermal; the velocity profile is fully developed; and thermophysical properties of fluid are constant during the process. Due to helical symmetry in screw channels it is more convenient to use helical coordinates as follows [10]:

$$\begin{cases} x = R\left(\xi^1 \cos\left(\xi^3\right) - \xi^2 \sin\left(\xi^3\right)\right) \\ y = R\left(\xi^1 \sin\left(\xi^3\right) + \xi^2 \cos\left(\xi^3\right)\right) \\ z = Rw\xi^3 \end{cases} \quad (1)$$

where R is the radius of the tube with screw ribs; and S is the pitch of the screw channel; $w = S/\left(2\pi R\right)$.

In this case, a simpler form of helical coordinates [3] is not applicable. There are some features in the area at the tube axis, where velocity profile is not symmetrical.

The presented helical coordinates (1) have the following parameters:

There are non-zero Christoffel symbols of the first kind, related to R^2

$\Gamma_{33,1} = \Gamma_{33,1} = -\xi^1$, $\Gamma_{23,1} = \Gamma_{32,1} = -1$, $\Gamma_{33,2} = \Gamma_{33,2} = -\xi^2$, $\Gamma_{13,2} = \Gamma_{31,2} = 1$, $\Gamma_{13,3} = \Gamma_{31,3} = \xi^1$, $\Gamma_{23,3} = \Gamma_{32,3} = \xi^2$;

There are non-zero Christoffel symbols of the second kind, related to R^2

$\Gamma^1_{23} = \Gamma^1_{32} = -1$, $\Gamma^2_{13} = \Gamma^2_{31} = 1$, $\Gamma^1_{33} = -\xi^1$, $\Gamma^2_{33} = -\xi^2$;

The covariant components of the metric tensor are related to R^2

Content from this work may be used under the terms of the Creative Commons Attribution 3.0 licence. Any further distribution of this work must maintain attribution to the author(s) and the title of the work, journal citation and DOI.
Published under licence by IOP Publishing Ltd

$$g_{11}=1,\ g_{22}=1,\ g_{33}=\left(\left(\xi^1\right)^2+\left(\xi^2\right)^2+w^2\right),\ g_{12}=0,\ g_{13}=-\xi^2,\ g_{23}=\xi^1;$$

The contravariant components of the metric tensor are related to $1/R^2$

$$g^{11}=\left(w^2+\left(\xi^2\right)^2\right)/w^2,\ g^{22}=\left(w^2+\left(\xi^1\right)^2\right)/w^2,\ g^{33}=1/w^2,\ g^{12}=g^{21}=-\xi^1\xi^2/w^2,$$
$$g^{13}=g^{31}=\xi^2/w^2,\ g^{23}=g^{32}=-\xi^1/w^2.$$

Thus the system of equations in the projections onto the directions of natural basis vectors can be represented in the form:

$$\mathrm{Re}_*\left(V^1\frac{\partial V^1}{\partial\xi^1}+V^2\frac{\partial V^1}{\partial\xi^2}-\xi^2V^1\frac{\partial V^3}{\partial\xi^1}-\xi^2V^2\frac{\partial V^3}{\partial\xi^2}-2V^2V^3-\xi^1V^3V^3\right)=-\frac{\partial p}{\partial\xi^1}+$$
$$+\left(\frac{\partial(\nabla\mathbf{V})_1^{\ 1}}{\partial\xi^1}+\frac{\partial(\nabla\mathbf{V})_1^{\ 2}}{\partial\xi^2}+(\nabla\mathbf{V})_2^{\ 3}\right),\tag{2}$$

$$\mathrm{Re}_*\left(V^1\frac{\partial V^2}{\partial\xi^1}+V^2\frac{\partial V^2}{\partial\xi^2}+\xi^1V^1\frac{\partial V^3}{\partial\xi^1}+\xi^1V^2\frac{\partial V^3}{\partial\xi^2}+2V^1V^3-\xi^2V^3V^3\right)=-\frac{\partial p}{\partial\xi^2}+$$
$$+\left(\frac{\partial(\nabla\mathbf{V})_2^{\ 1}}{\partial\xi^1}+\frac{\partial(\nabla\mathbf{V})_2^{\ 2}}{\partial\xi^2}+(\nabla\mathbf{V})_1^{\ 3}\right),\tag{3}$$

$$\mathrm{Re}_*\left(-\xi^2V^1\frac{\partial V^1}{\partial\xi^1}-\xi^2V^2\frac{\partial V^1}{\partial\xi^2}+\xi^1V^1\frac{\partial V^2}{\partial\xi^1}+\xi^1V^2\frac{\partial V^2}{\partial\xi^2}+\right.$$
$$\left.+\left(\left(\xi^1\right)^2+\left(\xi^2\right)^2+w^2\right)\left(V^1\frac{\partial V^3}{\partial\xi^1}+V^2\frac{\partial V^3}{\partial\xi^2}\right)+2\xi^1V^1V^3+2\xi^2V^2V^3\right)=$$
$$=-\frac{\partial p}{\partial\xi^3}+\left(\frac{\partial(\nabla\mathbf{V})_3^{\ 1}}{\partial\xi^1}+\frac{\partial(\nabla\mathbf{V})_3^{\ 2}}{\partial\xi^2}+(\nabla\mathbf{V})_1^{\ 2}+\xi^1(\nabla\mathbf{V})_1^{\ 3}-(\nabla\mathbf{V})_2^{\ 1}+\xi^2(\nabla\mathbf{V})_2^{\ 3}\right),\tag{4}$$

$$\frac{\partial V^1}{\partial\xi^1}+\frac{\partial V^2}{\partial\xi^2}=0,\tag{5}$$

where $\mathrm{Re}_*=\left(V_aR\rho\right)/\mu$ is the modified Reynolds number; $p=\left(RP\right)/\mu V_a$ is the dimensionless pressure; P is the pressure; V_a is the – some specific velocity; ρ is the density; μ is the dynamic viscosity; and $V^i\left(i=1,2,3\right)$ are the contravariant components of velocity vector.

The boundary conditions: the no-slip conditions are taken at the walls; and the flow rate is predefined.

Due to the form of helical coordinates and type of screw channels, formed by ribs, the additional condition appears for the function $p\left(\xi^1,\xi^2,\xi^3\right)$: $\partial p/\partial\xi^3=C_0=const$, $p=C_0\xi^3+f\left(\xi^1,\xi^2\right)$. This leads to a simple form of the system (2)-(5). The sought function $f\left(\xi^1,\xi^2\right)$ is non zero only in the first two equations, which greatly facilitates the final solution. The finite element method is used to solve the problem.

3. Numerical results

It is known that secondary flows in the tubes' cross sections lead to heat transfer enhancement. Figure 1 shows the lines of the axial velocity component for different Reynolds numbers ($l = 0.36D$ is the length of the rib). It is observed that the area of maximum axial component is in the periphery for small Reynolds numbers $\mathrm{Re} \le 500$, in contrast to the case of large Reynolds numbers. This is due to a higher pressure drop required for large flow rate, therefore the maximum velocity appears on the tube axis.

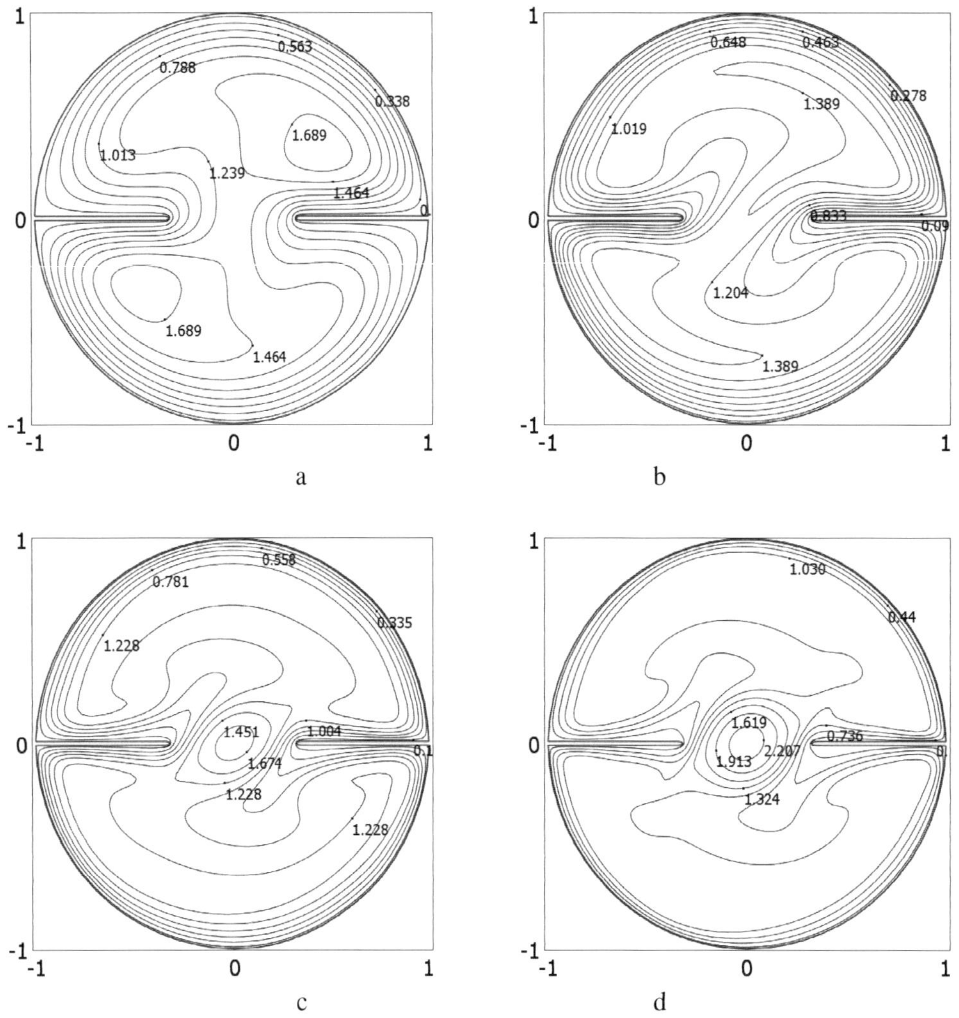

Figure 1. Level lines of axial velocity component for Reynolds numbers: a - $\mathrm{Re} = 200$; b - $\mathrm{Re} = 500$; c - $\mathrm{Re} = 900$; d - $\mathrm{Re} = 1300$ ($l = 0.36D$).

4. Summary

The screw ribs lead to the appearance of secondary flows in the near-wall regions. For some range of the ribs' length, the axial velocity components are found to prevail in the central part of the channel. Similar to the case of tape inserts, such fluid flow structure leads to heat transfer enhancement due to the occurrence of secondary flows. However, screw ribs do not completely overlap the channel cross section and provide a lesser friction factor, compared to tape inserts.

Acknowledgments

The reported study was funded by RFBR № 16-31-00425 mol_a

References

[1] Gortyshov Yu F, Popov I A, Olimpiev V V, Schelkov A V and Kas`kov S I 2009 Thermohydraulic efficiency of perspective methods of intensification of heat transfer in the channels of heat-exchange equipment. *Heat transfer enhancement: monograph (Kazan: KNRTU-KAI)* 531

[2] Mitrofanova O V 2003 Hydrodynamics and heat exchange under swirl flows in channels with twisted tapes (Review) *High temperature* **41**(4) 587-633

[3] Vachagina E K and Kadyirov A I 2014 The use of helical coordinate systems *Quarterly journal Mechanics Applied Mathematics* **67**(4) 553-566

[4] Kalinin E K, Dreycer G A and Yarcho S A 1990 Intensification of heat transfer in channels *Mechanical engineering* 208

[5] Bergles A E, Nirmalan V, Junkhan G H and Webb R L 1983 *Bibliography of Augmentation of Convective Heat and Mass Transfer (part II) Report HTL-31, ISU-ERI-Ames-84221, Heat Transfer Laboratory* (Ames: Iowa State University, IA)

[6] Manglik R M and You L 2002 *Computational Modeling of Laminar Swirl Flows and Heat Transfer in Circular Tubes with Twisted-Tape Inserts Report TFTPL-7, Thermal-Fluids and Thermal Processing Laboratory* (Cincinnati: University of Cincinnati, OH)

[7] Schukin V K and Chalatov A A 1982 *Heat and mass transfer and hydrodynamics of swirling flows in axisymmetric channels* (Moscow: Mashinostroenie) p 200

[8] Kadyirov A I and Vachagina E K 2016 Hydrodynamic structures of viscous flow in a coaxial channel with screw inserts *Transactions of Academenergo* **3** 58-74

Heat transfer and structure of flow at boiling of refrigerant R134a in channels with inserts in the form of finned twisted tape

A V Shishkin, S E Tarasevich

Kazan national research technical university named after A N Tupolev, Kazan, Russia

e-mail: shishkin-andrei@mail.ru

Abstract. The paper presents the results of experimental study of heat transfer for the refrigerant R134a flow in the channels with finned twisted tape inserts at bubble boiling. The flow regimes implemented under the given conditions are shown. The stable cord-like flows appear at boiling in channels with twisted tape inserts and high vapor content when the liquid phase moves in the form of a stream (cord) along the central part of the tape, which is not an active heat exchange surface. At boiling this can lead to an increase in the length required for complete liquid evaporation. Existing geometric modifications of twisted tapes are used in the heat-exchange equipment at forced convection of the coolant and do not solve the problem of cord-like flows elimination. The present work discusses the experimental study of heat transfer at boiling of refrigerant R134a in the channels with twisted tape inserts that have fins on its surface.

1. Heat transfer and structure of the flow at boiling of refrigerant R134a in the channels with inserts in the form of finned twisted tapes

Heat transfer at boiling of refrigerant R134a in the channels with inserted finned tapes having a relative pitch of twist $y=s/d=3$; 4; 6, pitch of fin installation $t=60$; 40; 20 mm, at fixed values of the fin height $h=1$ mm and installation angle $\varphi=45^\circ$ (the fins are installed in the opposite direction to the tape twisting) was experimentally studied in the range of the regime parameters: Reynolds number depending on fluid circulation velocity $Re_0=3\cdot10^4\div6.5\cdot10^4$, heat flow density $q=99\div326$ kW/m^2, boiling parameter $Bo=0.00052\div0.0045$, and the estimated mass flow vapor content according to the supplied heat flow at the outlet of the working section reached $X=0.07\div0.74$. The diameter of the working section channel $d=0.01$ m, and the relative tube length $L/d=50$ [1, 2].

The evaluation of the regime parameters influence on the heat transfer at boiling of the refrigerant R134a in the channel with finned twisted tape inserts has shown that the influence of Reynolds number Re_0, the heat flow density q and relative pitch of twist $y=s/d$ in channels with twisted tape inserts is similar to that detected in channels with smooth twisted tapes [1].

With a decrease of the dimensionless pitch the heat transfer increases insignificantly in the area from $t/s=1$ to $t/s>1$ and decreases with further reduction of the dimensionless pitch at $t/s<1$. The decrease in heat transfer at the tightest pitch of the fin installation $t=20$ mm is related to the change in the flow structure, namely, to the formation of stagnant zones on the channel surface beyond the

Content from this work may be used under the terms of the Creative Commons Attribution 3.0 licence. Any further distribution of this work must maintain attribution to the author(s) and the title of the work, journal citation and DOI.
Published under licence by IOP Publishing Ltd

contact point of the fin butt end with the channel wall. With a decrease of the Reynolds number below $4.5 \cdot 10^4$ the formation of a stable dry spot in this area was visually recorded (Figure 1). With further increase in the heat flux density above q>222.9 kW/m^2 a heat exchange crisis occurred along with the sharp increase of the channel wall temperature followed by the channel destruction.

Figure 1 shows a visually recorded flow regime in the channel with finned twisted tape y=4, with pitch of the fin installation t=20 mm at the boundary of the heat transfer crisis beginning. Local values of the heat transfer along the working section length at the crisis boundary are shown in figure 2. The heat exchange crisis is in section No. 5, as indicated by the decrease of the heat transfer coefficient even before the crisis began, with the crisis coming the wall temperature rises sharply by more than 200 °C in this section with subsequent growth in the remaining sections.

The experimental data on heat transfer at boiling of the refrigerant R134a in the channel with finned twisted tape inserts are shown in figure 3. As it can be seen from the graph, the influence of the pitch of fin installation on the twisted tape surface on the heat transfer at various relative pitches does not qualitatively differ from the conclusions drawn earlier, that in the area from t/s=1 to t/s>1 the heat transfer increases and at pitch t/s<1 the heat transfer decreases to the level of the smooth twisted tapes.

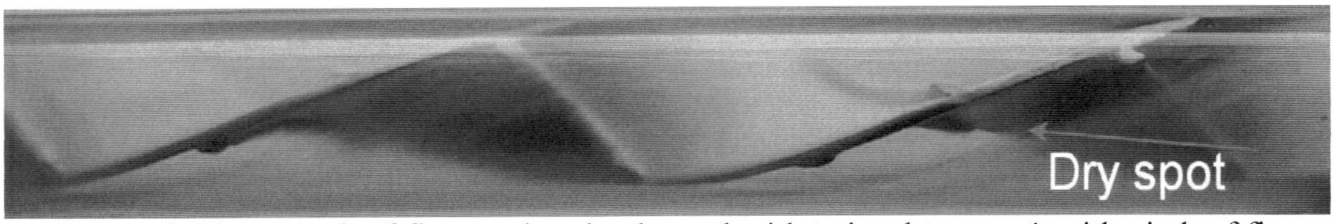

Figure 1. Photograph of flow regime in channel with twisted tape y=4, with pitch of fin installation t=20 mm at: Re_0=31459.8; q=222.9 kW/m^2; p=627.9 kPa; t_{inlet}=23.6°C; X=0.488

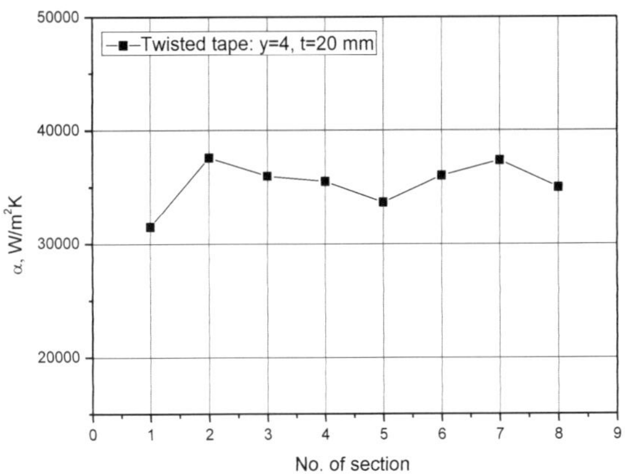

Figure 2. Local values of heat transfer coefficient at boiling of refrigerant R134a in channel with twisted tape y=4, with pitch of fin installation t=20 mm at: Re_0=31459.8; q=222.9 kW/m^2; p=627.9 kPa; t_{inlet}=23.6°C; X=0.488

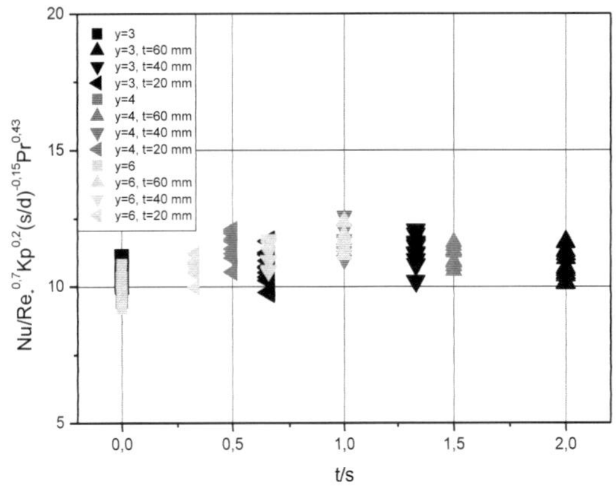

Figure 3. Heat transfer at boiling of refrigerant R134a in channel with finned twisted tapes y=3; 4; 6 with pitches of fin installation t=60; 40; 20 mm

The presence of fins affects the heat transfer at boiling insignificantly but fin installation changes the flow structure. The fin installation on the surface leads to an increase of the vapor amount inside the channel due to the formation of vortex structures behind the fin, in which active vaporization takes place, and also because of the decrease of fluid saturation temperature as a result of the increase of hydraulic resistance. The increase of vapor content with other regime parameters being equal in the channels with finned twisted tapes leads to an earlier coming of the heat exchange crisis, as compared to channels with smooth twisted tape inserts, due to the appearance of dry spots on the heat-release surface.

As noted above at boiling of refrigerant in the channels with twisted tape insert y=4 with pitch of fin installation t=20 mm, the formation of stagnant zones with the decrease of the liquid film thickness and the formation of a stable dry spot behind the contact point of the fin butt end with the channel wall at a decrease of the Reynolds number $Re_0<4.5\cdot10^4$ was recorded. In the channel with twisted tape insert y=6 with pitch of fin installation t=20 mm, a similar picture of flow regime with the formation of a stable dry spot in this area was observed. With an increase of Reynolds number $Re_0>4.5\cdot10^4$ the liquid inleakage with a short-term wetting of the surface by a thin liquid film and the formation of ruptures of the liquid film in the contact zone of the fin butt end and the channel wall occurred. The flow of liquid film on the wall surface in the channel with finned twisted tape y=6 and t=20 mm was less uniform than in the channel with the tape y=4 and t=20 mm.

The boundary of the beginning of heat transfer crisis in the channel with finned tape with the relative pitch of twist y=6 and pitch of fin installation t=20 mm is lower than for y=4 and t=20 mm and corresponds to the heat flow density $q\geq161.8$ kW/m^2. Local values of heat transfer along the working section length at the heat transfer crisis boundary in the channel with twisted tape y=6 and pitch of fin installation t=20 mm are shown in figure 4.

We failed to achieve the critical thermal loads in the channel with finned twisted tape y=3 with pitch of fin installation t=20 mm even at the maximum possible power of the transformer.

It was observed that due to the fins installed on the surface the increase of vaporization occurs inside the channel because of the formation of vortex structures behind the fin in which active vaporization takes place. It is noted that with a decrease of the relative pitch of tape twisting a more uniform distribution of the liquid on the heat exchange surfaces occurs due to the increase of mass forces as a result of which the critical heat flows increase.

It is important that in this channel there are neither dry spots in the entire studied range, nor discontinuities of the liquid film on the heat-release surface and in the contact zone of the fin butt end with the channel wall. The liquid distribution on the heat-release surface in the channel with finned tape y=3 and t=20 mm is more uniform than with the tape y=4 and t=20 mm as indicated by higher heat transfer coefficients. The local values of heat transfer along the working section length at the maximum possible heat flow density q=326.5 kW/m^2 in the channel with twisted tape y=6 and pitch of fin installation t=20 mm are shown in figure 5.

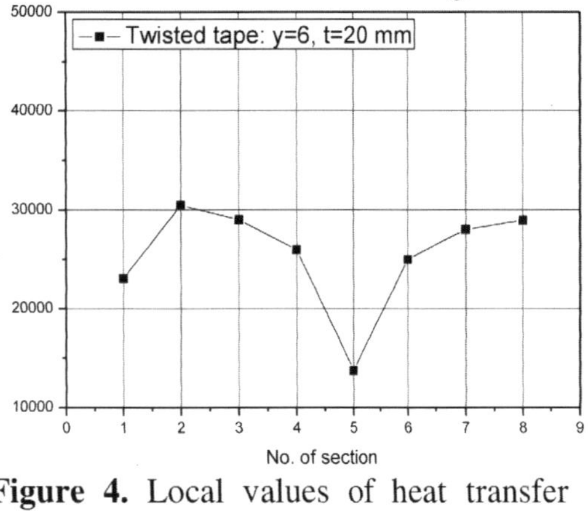

Figure 4. Local values of heat transfer coefficient at boiling of refrigerant R134a in channel with twisted tape y=6, with pitch of fin installation t=20 mm at: Re_0=32462.5; q=161.8 kW/m^2; p=618.4 kPa; t_{inlet}=23.0°C; X=0.330

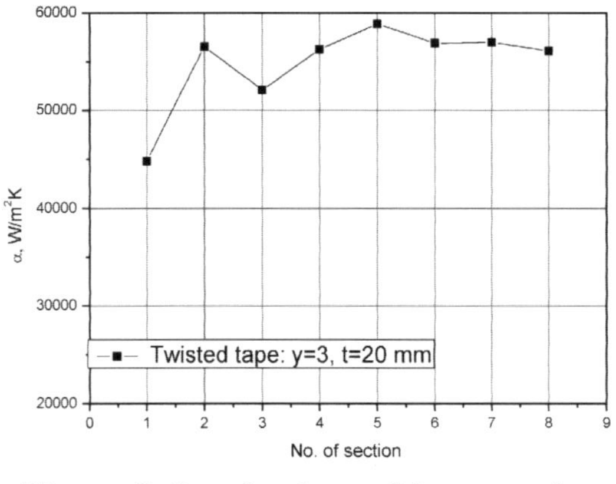

Figure 5. Local values of heat transfer coefficient at boiling of refrigerant R134a in channel with twisted tape y=3, with pitch of fin installation t=20 mm at: Re_0=31433.1; q=326.5 kW/m^2; p=661.4 kPa; t_{inlet}=24.3°C; X=0.740

On the basis of the presented data on the flow regimes in the channels with finned twisted tapes at boiling of refrigerant R134a we can conclude that fin installation affects the heat transfer at boiling insignificantly. The fin installation with the smallest pitch t=20 mm at $Re_0<4.5\cdot10^4$ in channels with relative pitch of twist $y>4$ leads to the formation of stagnant zones that promote the formation of dry spots resulting in an earlier heat transfer crisis than in the channels with smooth twisted tapes. This again confirms the data of the V.I. Subbotin [3] that the maximum effect of twisting is observed when s/d is less than 4.

The increase of the heat transfer coefficient in the channels with finned twisted tape inserts in the studied range of geometric parameters compared to the straight channel at boiling is on average from 1.52 to 1.75 times.

The experimental data generalization allowed obtaining the dependence for calculation of the heat transfer coefficient with finned twisted tape inserts in the form (1):

$$Nu = 10.33\, Re_*^{0.7}\, K_p^{0.2}\, (s/d)^{-0.15} Pr^{0.43} \tag{1}$$

Dependence (1) with an error of ±22% at a confidence probability of 0.95 generalizes the experimental data, and it is valid for calculation of the heat transfer at boiling of refrigerant R134a in the channels with finned twisted tape inserts with relative pitches of twist $y=s/d$=3; 4; 6, pitch of fin installation t=60; 40; 20 mm, at fixed values of the fin height h=1 mm and installation angle $\varphi=45^\circ$ (the fins are installed in the opposite direction to the tape twisting) in the range of the regime parameters: Reynolds number depending on the fluid circulation velocity $Re_0=3\cdot10^4\div6.5\cdot10^4$, heat flux density $q=99\div326$ kW/m^2, boiling parameter $Bo=0.00052\div0.0045$ and the estimated mass flow vapor content according to the supplied heat flow at the outlet of the working section reached $X=0.07\div0.74$. The diameter of the working section channel d=0.01 m, and the relative tube length L/d=50.

2. Conclusion

It is found that the influence of regime parameters on the heat transfer at boiling of refrigerant R134a in the channel with finned twisted tape inserts is the same as in the channels with smooth twisted tapes. Because of the fins installed on the surface the increase of vaporization occurs inside the channel as a result of the formation of vortex structures behind the fin in which active vaporization takes place. The fin installation affects the heat transfer at boiling insignificantly, however, the fin installation changes the flow structure, increases the amount of vapor by additional generation of vapor inside the channel at the decrease of the liquid saturation temperature as a result of the increases of hydraulic resistance. The increase of vapor content with other regime parameters being equal in the channels with finned twisted tapes leads to an earlier coming of the heat exchange crisis than in channels with smooth twisted tape inserts due to the appearance of dry spots on the heat-release surface. The increase of the heat transfer coefficient in the channels with finned twisted tape inserts in the studied range of geometric parameters compared to the straight channel at boiling is on average from 1.52 to 1.75 times.

It is noted that with a decrease of the relative pitch of tape twisting a more uniform distribution of the liquid on the heat exchange surfaces occurs under the action of mass forces, as a result of which the critical heat flows increase. The maximum effect of twisting is observed when s/d is less than 4.

The work was carried out within the framework of the project No.14.Z50.31.0003 dated March 04 2014"Establishment of multidisciplinary diversified laboratory for modeling of the physical and technical processes when solving the adjoint problems of aeromechanics, thermal physics, acoustics and vibration resistance, ventilation and microclimate, ecology and monitoring of operation of trucks and their aggregates" (headed by Dr. Phys.-Math.Sci., prof. S.A. Isaev)

References

[1] Shishkin A V and Tarasevich S E Thermohydraulic Characteristics of Flow at Boiling of Refrigerant R134a in Tubes with Inserts in the Form of Twisted Tapes *Proc. of the X school-seminar of young scientists and specialists of the V.E. Alemasov academician of RAS "Problems of heat and mass transfer and hydrodynamics in mechanical engineering",* Kazan, Russia, 13-15 of September 2016 pp. 132-135 (in Russian)

[2] Tarasevich S E, Shishkin A V, and Yakovlev A B 2013 Stand for Experimental Research of Freon Boiling in Various Channels *Vestnik KSTU named after A.N. Tupolev,* **2**(2) 60-62 (in Russian)

[3] Ibragimov M Kh, Nomofilov E V and Subbotin V I 1961 Heat Transfer and Hydraulic Resistance for Fluid Helical Motion in the Tube *Teploenergetika* **7** 57-60 (in Russian)

IOP Conf. Series: Journal of Physics: Conf. Series **980** (2018) 012013 doi:10.1088/1742-6596/980/1/012013

Determining the parameters of vortex structures in a hydrodynamic vortex chamber

V S Dremov[1,2], S I Shtork[1,2] and I K Kabardin[1]

[1] Kutateladze Institute of Thermophysics, SB RAS, Novosibirsk, 630090, Russia

[2] Novosibirsk State University, Novosibirsk, 630090, Russia

E-mail: supersonny@mail.ru

Abstract. This study focused on the large-scale vortex structures. The working section was a tangential vortex chamber with a cylindrical cross-section. The flow was swirled by tangential supply of fluid into the chamber, using 12 direct-flow rectangular nozzles. The design swirl parameter was equal to 6.6 and was determined based on the vortex chamber geometry. The Reynolds number was maintained equal to 30000. Quantitative measurements were realized with the use of the optical technique of laser Doppler anemometry. The results of optical measurements allowed determining main parameters of the vortex core for two basic modes of the vortex flow, including the central direct-flow vortex and the spiral vortex.

1. Introduction

Studies of large-scale vortex structures are of great importance and interest, as these structures regularly occur in nature in a wide range of scales, from quantum vortices to astrophysical objects, such as rotating galaxies [1]. Vortex structures are run across many technical applications, for example, in vortex furnaces, combustion chambers, cyclonic separators and various chemical and biological reactors [2]. Special attention should be paid to the vortex structures, generated behind turbine impellers. Despite numerous studies of swirling flows, there are many effects, for example, the vortex core precession, which require further research. The same is also relevant for new effects and new flow regimes arising with time. In addition to the practical interest, investigation of vortex structures is also of great scientific interest.

The aim of this work is an experimental study of large-scale vortex structures with different helical symmetry in the tangential vortex chamber.

2. Experimental installation

The experiment was conducted in a closed hydrodynamic circuit with fixed values of input parameters. The working fluid (distilled water) was supplied from the tank into the vortex chamber, using a centrifugal pump. The liquid flow rate was maintained in the range from 3 to 25 m^3/h and measured by ultrasound flowmeter with a relative error of less than 1.5%. The chamber, where a vortex structure formed, was a cylindrical channel with a diameter D =190 mm and a height of 600 mm. The working section was made of transparent plexiglass, which allowed using the contactless

Content from this work may be used under the terms of the Creative Commons Attribution 3.0 licence. Any further distribution of this work must maintain attribution to the author(s) and the title of the work, journal citation and DOI.
Published under licence by IOP Publishing Ltd

optical measurement technique to investigate the swirling flow. For creating different types of vortex structures, special diaphragms with orifices were placed at the outlet of the vortex chamber. The diameter of the orifice in the diaphragm was 70 mm. In the first case, the hole was placed at the center of the channel, and the second orifice was shifted by 70 mm from the channel axis. The working fluid was delivered into the chamber, using 12 tangential direct-flow rectangular nozzles, combined into the corner blocks (Fig. 1.a).

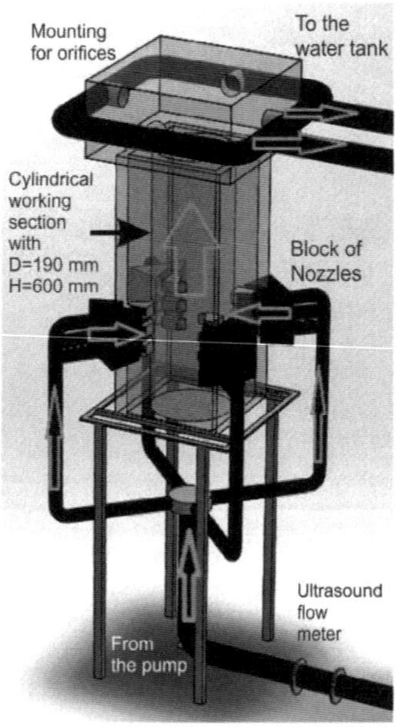

Figure 1.a Vortex chamber geometry.

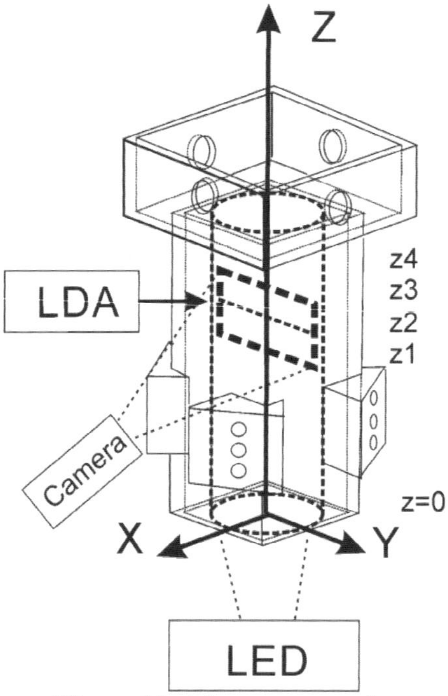

Figure 1.b. Measurement scheme.

The main parameters in our work were the design swirl parameter S and Reynolds number Re. The design swirl parameter was determined from the geometry of the vortex chamber: $S = D^2 \sin(\gamma) / N\sigma_n$, where D = 190 mm was the chamber diameter, N = 12 was the number of nozzles, $\sigma_n = 14x23 = 322$ mm^2 was the area of the nozzle, and $\gamma = 45$ was the angle of nozzle rotation relative to the center of the chamber. Thus, the swirl parameter remained equal to 6.6. The Reynolds number was determined as $Re = DU_0 / \nu$, where D = 190 mm was the chamber diameter, $U_0 = Q / \sigma$ was the average velocity, calculated using the flow rate Q = 15 m^3/ h and the cross section of the vortex chamber $\sigma = \pi D^2 / 4$, and ν was the kinematic viscosity of water. Accordingly, the Reynolds number equaled to 30000.

3. The experiment and results.

The first stage of the experiment was flow visualization. The used light source was LED projector with an aperture ratio of 7000 lm. The tracers for visualization were air bubbles with a diameter of about 1 mm, which were sucked out of the atmosphere into the flow. The flow structure image was obtained using a digital camera CanonEOS 7Dc with a resolution of 18Mpx and video recording frequency of 30 Hz. The visualization has revealed the presence of two major vortex structures: the central stationary direct-flow vortex and the stationary swirl vortex, appearing in the form of a thin elongated concentrated vortex filament. In addition, the researchers used experimental data on precession double vortex [3].

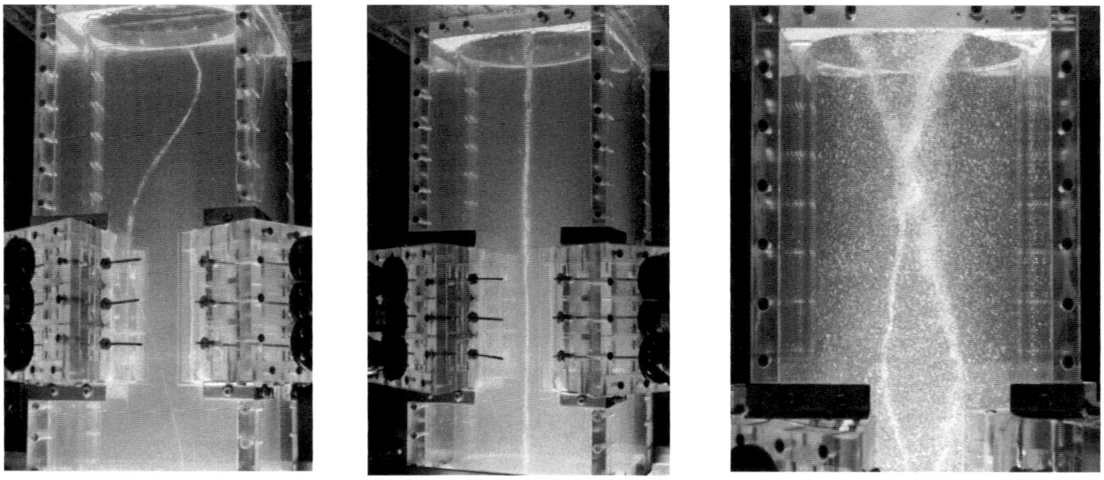

Figure 2. Visualization of different vortex flow regimes. Spiral vortex (left). Axial vortex (in the center). Two spiral vortex structures (right) [3].

Velocity profiles were measured using laser Doppler anemometry (LDA). Anemometer LAD-05 was manufactured at the Kutateladze Institute of Thermophysics SB RAS, in Novosibirsk. The measurements were carried out at the height of cross section of the vortex chamber z = 350, 400, 450, 500 mm for the central vortex, and z = 413, 428, 438, 453 mm for the spiral vortex. Coordinate x took

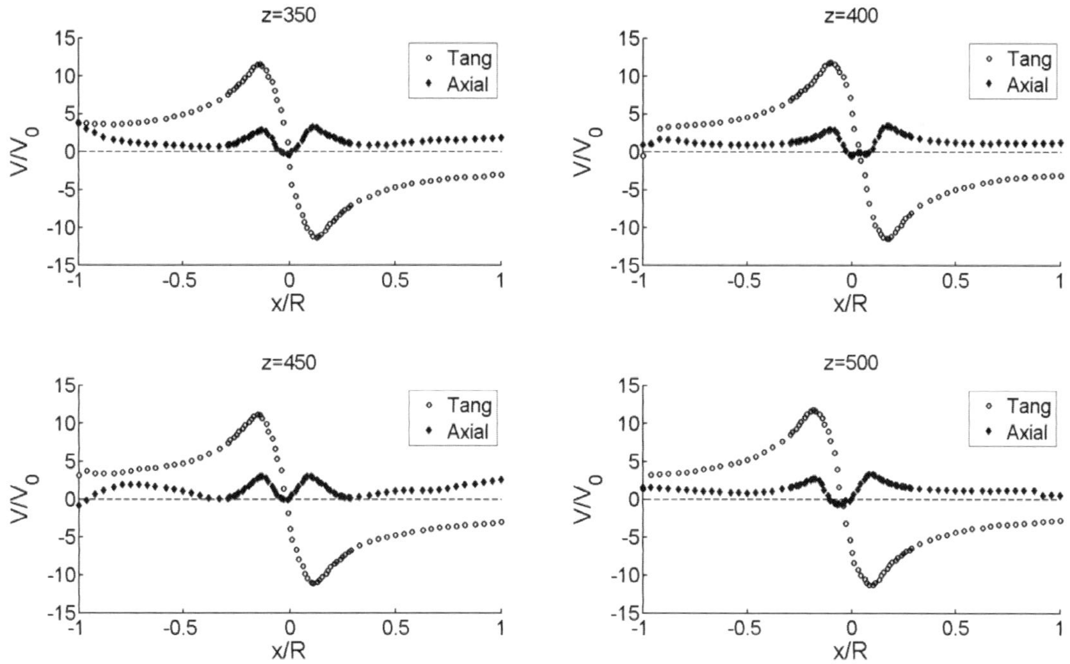

Figure 3-6. Profiles of tangential and axial velocities for the central vortex at different heights of the cross section of the vortex chamber.

values from -0.5 D to 0.5 D, D = 190 mm, and coordinate y = 0 remained constant for both vortex structures. LDA was used to measure two projections of the velocity vector V(y) and V(z) in the range from 0.001 to 30 m/s. The relative error did not exceed 0.1%. The size of the measuring volume was equal to 0.1x0.1x0.5 mm with an accuracy of 0.1 mm. A coordinate-shifting device (CPU) allowed shifting the measuring volume within 250x250x250 mm with an accuracy of 0.1 mm. The number of measurements in each was 10000, 5000 for each projection of the velocity vector. The maximum measurement time in the point was 120 s. The measuring accuracy of the average velocity was guaranteed at 95%. The used tracers for obtaining a Doppler signal were polystyrene particles with a density of 1.069 – 1.225 g / cm^3 and a size of 20-50 microns.

Figures 3-6 show distributions of tangential and axial velocities for the central vortex at different heights of the cross section of the vortex chamber, measured using the LDA method. The design swirl parameter was equal to 6.6, and the flow rate was maintained equal to 15 m^3/ h. For convenience, graphs are presented in a dimensionless form. On the x-axis there is a laid off x-coordinate, related to the chamber radius, and on the y-axis there is the velocity U, related to the average velocity U_0 at the given flow rate. According to figures 3-6, the flow structure remains the same regardless of the height of measurement, since the dimensionless velocity profiles remain unchanged.

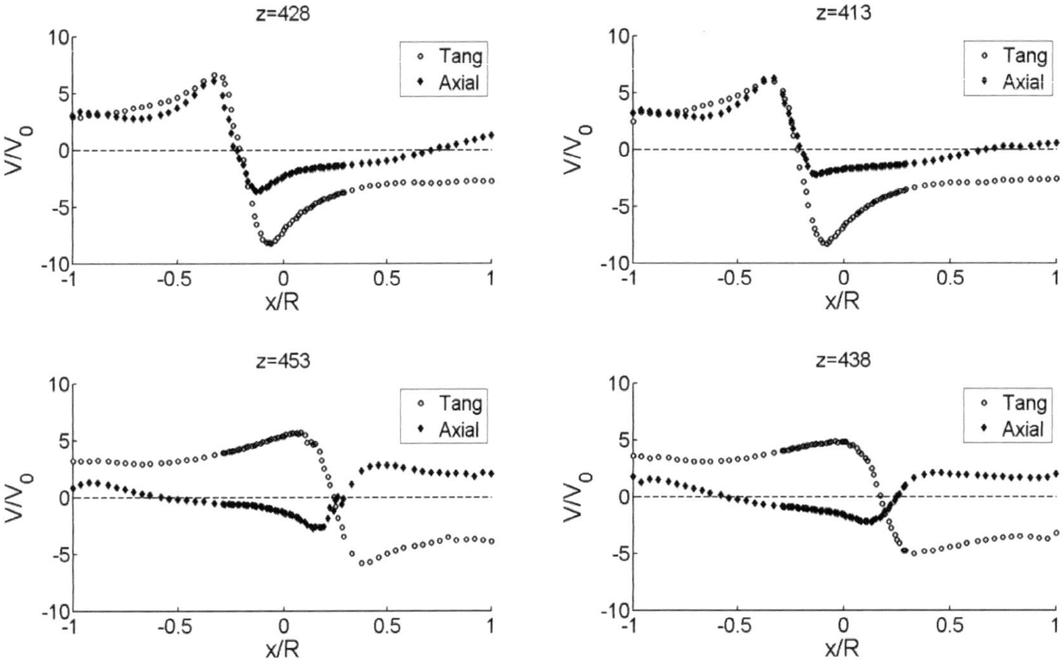

Figure 7-10. Profiles of tangential and axial velocities for the spiral vortex at different heights of the cross section of the vortex chamber.

A similar situation is observed in the case of a spiral vortex. Figures 7-10 illustrate distributions of tangential and axial velocity for a spiral vortex. The liquid flow rate and the swirl parameter do not change. It should be noted that graphs 7, 8 and 9, 10 are built for different positions of the diaphragm at the outlet of the vortex chamber. The difference lies in the rotation of the output orifice of the diaphragm by 90 degrees. The velocity profiles remain unchanged in both measurements. In addition, the figures clearly show the displacement of the vortex core relative to the center of the vortex chamber.

On the basis of the measured velocity profiles it is possible to determine the geometric parameters of the vortex core. The size of the vortex core is defined as the distance between zero and maximum values of the tangential velocity. Thus, from figures 3-6, it follows that the size of the vortex core ε of the central vortex on average is ε = 13.96 mm, and for the spiral vortex ε = 14.89 mm. Knowing the vortex core size and the maximum value of tangential velocity, we can calculate the circulation rate $\Gamma = 2\pi \cdot \varepsilon \cdot U_{max}$. The velocity circulation is 0.111 m^2/ s and 0.075 m^2/ s for the central and spiral vortices, respectively. The velocity circulation in the spiral vortex should have a smaller value than in the central one, as part of it goes for the creation of a complex flow structure. It should also be noted that the calculations do not take into account the inclination angle of the helical line of the vortex. Because of the inclination, the peaks of tangential velocity will be somewhat less; therefore, the circulation for the spiral vortex will also have a smaller value. The central vortex is formed on the central axis of the vortex chamber, so its deviation from the geometric center of the chamber $\boldsymbol{a}$ equals zero. For the spiral vortex, the deviation is determined as the distance between the chamber center and the zero value of tangential velocity, and is 22 mm. The step of the helical line is calculated from the following dependence: $U_z(r) = U_0 r \cdot U_\varphi(r) / L$, where r is the radial coordinate, $U_z(r)$ is the axial velocity at the point r, U_0 is the axial velocity in the center of the chamber, $U_\varphi(r)$ is the tangential velocity, and L is the helix step. For the spiral vortex, the step of the helical line is 193 mm, and for the central vortex, the step of the helix tends to infinity.

Table 1. Vortex structures parameters.

Table 1.	ε / R	$\Gamma / 2\pi R U_0$	a / z	L / R
Central	0.147	1.257	0	∞
Spiral	0.156	0.849	0.234	0.3205
Double	0.211	0.3031	0.296	0.421
	0.212	0.3035	0.338	

Table 1 presents the abovementioned parameters in a dimensionless form. To obtain the parameters of the double vortex the experimental data from [3] were used.

Conclusion

The paper presents an experimental study of large-scale vortex structures in a tangential vortex chamber. At the first stage of the experiment, visualization of a swirling flow was performed. It has revealed the presence of thin concentrated vortex filaments in the flow. The optical technique of laser Doppler anemometry was used for quantitative measurements of vortex flows. The velocity profiles have showed the presence of a vortex core in the flow. The flow pattern remained the same in all sections at a changing height of the measurement. The main parameters of the vortex core have been determined on the basis of the measured profiles.

Acknowledgements

The work was supported by the Russian Science Foundation, project 14-29-00093.

References

[1] Alekseenko S V, Kuibin P A and Okulov V L 2007 *Theory of concentrated vortices. An introduction* (Berlin, Heidelberg, New york: Springer)

[2] Gesheva E S, Litvinov I V, Shtork S I and AlekseenkoS V 2014 *Thermal Engineering* **61**(9) 649–657

[3] Dremov S V, Shtork S I, Skrypkin S G and Kabardin I K 2016 An experimental investigation of the interaction between a pair of precessing vortices in a tangential vortex chamber *Journal of Physics: Conference Series* **754**(022003)

Visualization of vortex structures and analysis of frequency of PVC

E S Gesheva[1], S I Shtork[1,2], S V Alekseenko[1,2]

[1]Kutateladze Institute of Thermophysics, Siberian Branch of Russian Academy of Sciences, 1 Lavrentyev Ave., Novosibirsk, 630090 Russia

[2]Novosibirsk State University, 2 Pirogov Str., Novosibirsk, 630090 Russia

gesheva_es@itp.nsc.ru

Abstract. The paper presents the results of the study of large-scale vortex structures in a model chamber. Methods of forming quasi-stationary vortices of various shapes by changing the geometric parameters of the chamber have been proposed. In the model chamber with a tangential swirl of the flow, a rectilinear vortex, single helical and double helical vortices were obtained. The double helical structure of the vortex is unique due to its immovability around the axis of the chamber. The resulting structures slowly oscillate around their own axes, which is called the vortex core precession; while the oscillation frequency depends linearly on the liquid flow rate. The use of stationary vortex structures in power plants will increase the efficiency of combustion chambers and reduce slagging.

1. Introduction and experimental method

The feature of the swirling flow in the model of a tangential furnace is the formation of spatially complex large-scale vortices that completely determine the global flow structure [1]. Setting up the experiment, the hydrodynamic bench was designed to experimentally study the vortex structures, formed in the working chamber with a tangential swirl of the flow (Fig. 1).

It was earlier shown experimentally in [2, 3] that in the studied vortex tangential chamber with a fully open outlet, a swirling flow has a complex structure. When the boundary conditions are modified, namely, setting the diaphragm at the outlet, the vortex structure becomes stationary, has a rectilinear shape and extends from the bottom of the chamber to the outlet.

Let's consider in more detail the flow in a tangential chamber with the formation of stationary vortex structures. It should be noted that the vortex structures under study are related to concentrated vortex filaments. In such vortices, strong localization (concentration) of vorticity is observed. The term "vortex filament" is used for vortices in which the transverse dimension (core diameter) is much smaller than the chamber dimensions and the vortex length.

The scheme of the hydrodynamic bench is shown in Fig. 1. Water was used as a working fluid, and visualization was carried out with small air bubbles. The swirling flow was generated by means of twelve tangentially directed nozzles, arranged in three layers and combined in four corner blocks. The flow swirl intensity was determined using a design swirl parameter, calculated as S = 12 sin α. The angle α was equal to 45^0 for the flow with rectilinear and spiral vortex; and in the case of double structure, the turning angle of the nozzles α was equal to 30^0, that is in line with the values of S = 8.5 and S = 6, respectively. The height of the chamber from the bottom to the output orifice was 420 and

Content from this work may be used under the terms of the Creative Commons Attribution 3.0 licence. Any further distribution of this work must maintain attribution to the author(s) and the title of the work, journal citation and DOI.
Published under licence by IOP Publishing Ltd

560 mm, depending on the position of the output diaphragm, the chamber width b was 188 mm, and the diameter of the diaphragm orifice d_t equaled 70 mm. For the chamber geometry with a two-ramp bottom insert plates mounted at an angle of 50^0 to the horizon were used.

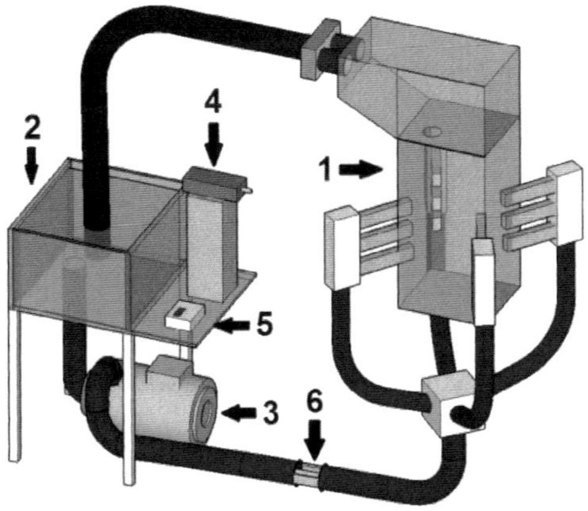

Figure 1. Scheme of the hydrodynamic stand: 1 – work area, 2 – water tank, 3 – pump, 4 – laser illumination, 5 – flowmeter, 6 – measuring section of the flowmeter.

The flow visualization was carried out by small air bubbles, which were fed into the flow in front of the working area. The flow backlighting was performed with a laser sheet of 3 W of a continuous solid-state laser. To register the flow pattern we used digital camera Canon 7d with a number of characteristics typical for professional models.

Although the vortex chamber used in the experiment is not a canonical object due to the square shape of the cross section, it nevertheless proved to be very suitable for studying various vortex structures. Moreover, with the help of cylindrical inserts, it was shown that in the regimes with concentrated vortices the flow patterns in the axial region of the cylindrical and rectangular channels practically do not differ from each other.

2. Visualization of the vortices and frequency characteristics

Suppose that a flat bottom is installed in the lower part of the vortex chamber, and a diaphragm with an outlet d_e is located in the output section at height h_e. The central position of the outlet leads to formation of a stable rectilinear vortex (Fig. 2). The vortex structure is a thin vortex filament running from the bottom of the chamber to the outlet. The vortex filament expands approaching the outlet. The particles in the vortex chamber move spirally. The vortex structure is fixed for all realized liquid flow rates and does not change its shape.

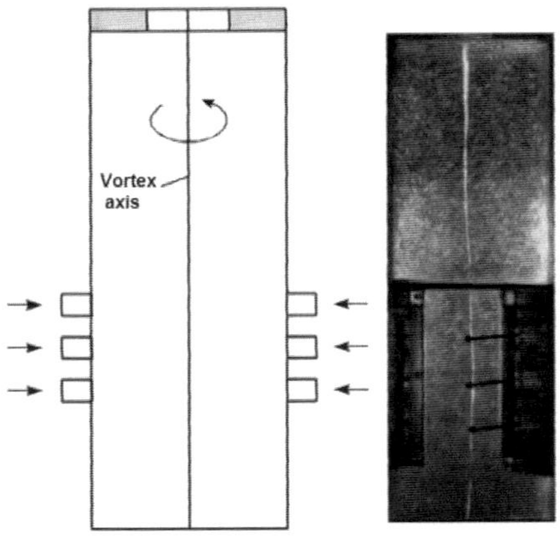

Figure 2. Rectilinear vortex filament: $h_e = 560$ mm, $d_e = 70$ mm, $Re = 20000$, $S = 8,5$.

The displacement of the outlet with a diameter d_e at a distance δ_e relative to the axis of the chamber leads to radical changes in the flow structure (Fig. 3). The air filament, which visualizes the axis of the vortex, is folded into a coil of a spiral. In this case, the helical vortex is stationary and stable, the fluid particles move around the spiral axis, thus making a double spiral motion. Areas with a return flow are also fixed. When the flow rate of the liquid varies, the vortex structure remains stable, the helical shape of the vortex is fixed for all realized liquid flow rates.

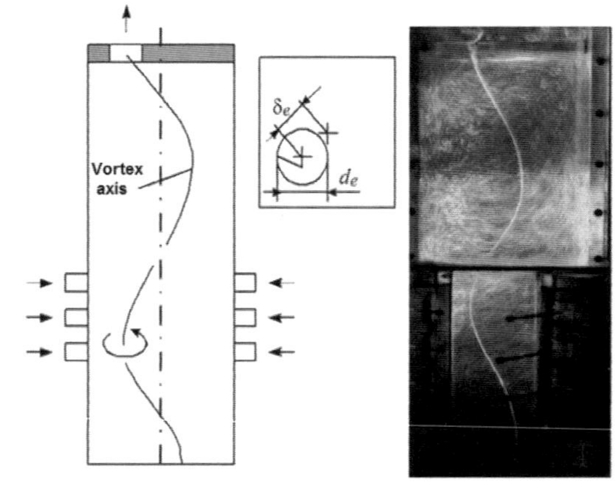

Figure 3. The geometric feature of the chamber and the vortex structure formed in it. h_e = 560 mm, d_e = 70 mm, δ_e = 60 mm, Re = 13 000, S = 8.5.

The experimentally observed structures described above refer to single vortex filaments. Theoretical analysis allows the formation of any number of vortex filaments [4], but in practice the observation of vortex structures consisting of more than one vortex filament is hampered by the instability of the vortex interaction. Indeed, we can assume that the observation of such spirals is very difficult. However, there are a large number of publications on the formation of a nonstationary double helical structure in the decay of a vortex, for example, [5, 6], as well as in a mixing layer with natural convection [7]; however, the existence of a large-scale stationary double-helical vortex is not mentioned in the literature.

Attempts to generate stationary double helical vortex in a tangential chamber were undertaken after observing stable single helical filaments. And these attempts have been successful in a variety of options [2, 3, 8]. Such a double helix appears in the vortex chamber with a centrally located outlet and two flat ramps at the bottom of the chamber (Fig. 4).

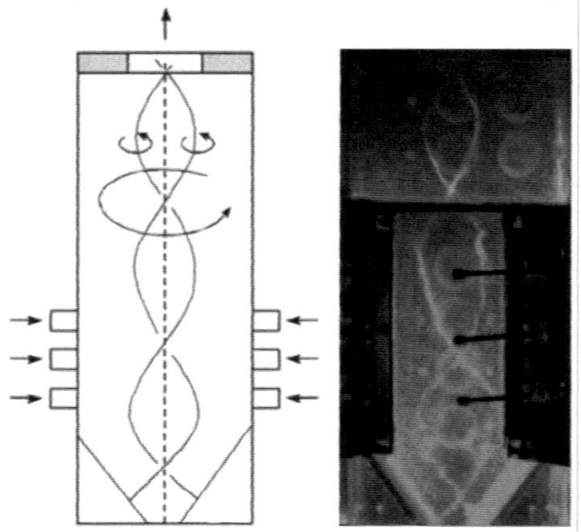

Figure 3. Double helical vortex. h_e = 420 mm, d_e = 70 mm, Re = 40000, S = 6.

The structure is two intertwined helical vortex filaments of the same sign. The presence of two inclined ramps on the bottom is associated with the need to specify the initial deformation of the vortex axes. Unlike a very stable single-helical vortex flow, the double-helix regime should be considered as quasi-stationary for the following reasons. First, the air filaments that visualize the vortex axes are somewhat blurred, chaotically oscillate and sometimes disintegrate. Secondly, one of the threads always tends to dominate. But in general, as can be seen in photo (Fig. 4), the double-helical regime of the vortex motion is fixed distinctly.

The Reynolds number has no noticeable effect on the vortex structure, as in the case of single-helical vortices. However, when the Reynolds number is varied, the decay and formation of the vortex filaments occur at different intervals. So, after analyzing the video recordings, we can say that the best way is a double helix at Re = 40,000 and swirl parameter S = 6. With these parameters, there is no complete decay of the vortex filament, but only the blurring of one or two threads with a periodicity of about 10 seconds.

It should be noted that the investigated vortex filaments are quasistationary and perform small periodic oscillations around the vortex axis; in the case of a rectilinear vortex, the axis of rotation coincides with the axis of the chamber. Let us consider in more details the frequency characteristics for vortices. The frequency of vortex precession was calculated from the analysis of video records by high-speed camera at the rate of 60 frames per second. For single vortices, precession frequencies were also obtained with a hydrophone.

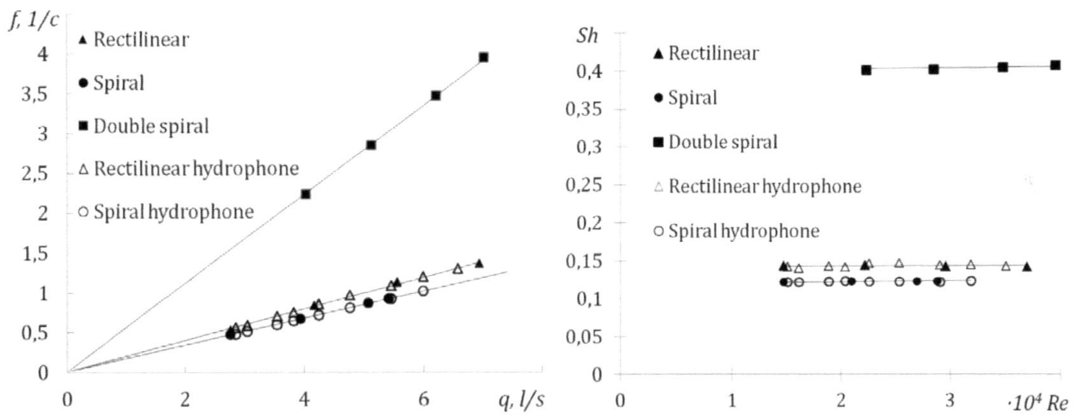

Figure 5. Dependence of precession frequency on liquid flow rate of (a); dependence of Strouhal number on Reynolds number (b).

Figure 5a shows the dependence of the vortex precession frequency on the fluid flow rate. In the case of a double-helical vortex, the precession frequency was defined as the frequency of oscillations of one vortex, which is dominant. The frequency of the precession of the vortices is linearly dependent on the flow rate of the liquid in the chamber, and when the data is extrapolated, it vanishes at zero liquid flow rate.

It is also interesting to look at the dependence of Strouhal number on Reynolds number (Fig. 5b). The Strouhal number is defined as $Sh = fb/W$, where f is the vorticity precession frequency, b is the transverse chamber size, and W is the average flow velocity. In all the vortex flows represented, the Strouhal number does not depend on the Reynolds number.

The Strouhal number characterizes [9] the order of the ratio of local and convective derivatives in the total derivative in the equation of motion. If the Strouhal number is small, $Sh \ll 1$, then the term containing the time derivative can be neglected, and the flow may be approximately treated as stationary or quasistationary [9]. The values of Strouhal number in the case of single vortices Sh = 0.12; 0.14 $\ll 1$, which indicates the stationarity of the investigated regime. For a double-helix regime, the Strouhal number $Sh = 0.41 < 1$, and according to [9], this regime should be considered as quasi-stationary, which, in turn, is confirmed by the data on visualization of the double-helix regime.

Acknowledgments
The study was funded by the grant of the Russian Science Foundation (project No.14-29-00093).

References

[1] Gupta A, Lilley D and Syred N 1984 *Abacus Press* p 488
[2] Alekseenko S V, Kuibin P A, Okulov V L and Shtork S I 1999 *Journal of Fluid Mechanics* **382** 195-243
[3] Alekseenko S V and Shtork S I 1992 *Russ. J. Eng. Thermophys* **2** 231-266
[4] Alekseenko S V, Kuibin P A and Okulov V L 2007 *Springer Science & Business Media* p 477
[5] Chandrsuda C, Mehta R D, Weir A D and Bradshaw P 1978 *J. Fluid. Mech* **85**. 693–704
[6] Faler J H and Leibovich S 1978 *J. Fluid Mech.* **86** 313–335
[7] Boubnov B M and Golitsyn G S 1986 *J. Fluid Mech* **167** 503–531
[8] Alekseenko S V and Shtork S I 1994 *JETP Letters* **59** 746-50
[9] Baranov V B 1987 (Moscow: Izdatelstvo MGU) vol. 1 p 184

IOP Conf. Series: Journal of Physics: Conf. Series **980** (2018) 012015 doi:10.1088/1742-6596/980/1/012015

Heat transfer of coolants in channels with swirling and continuous wall roughness

AV Zlobin, S E Tarasevich, AV Shishkin, A B Yakovlev

Kazan National Research Technical University named after A.N. Tupolev - KAI, Kazan 420111, 10 Karl Marx, Russia

Email: shishkin-andrei@mail.ru

Abstract. The work presents the results of experimental study of heat transfer in rough tubes, including tubes with twisted tape insert, at water and R134A one-phase flow. The roughness of tube inner surface is obtained by thread cutting. The combined influence of turbulizing factors on the heat transfer intensity is analyzed. Generalizing dependences are obtained for heat transfer calculation.

1. Introduction

The research of combined influence of two types of intensifiers is important to enhance processes in the tubes and channels of various heat exchange equipment. The use of swirling in the channels with rough surfaces for the boiling flows allows significantly reducing overall dimensions of the evaporator since the emerging mass forces contribute to the fluid motion to the surface with the developed evaporation centers. The extensive review of heat transfer and hydraulic resistance in channels with rough surfaces is presented in work [1]. It is noted that at present, the number of papers considering hydrodynamics and heat transfer of one and two-phase flows at the combined influence of swirling and artificial roughness of the channel surface is limited.

2. Experimental research of heat transfer

2.1. Heat transfer in straight rough channels at one-phase flow

The research of heat transfer in rough tubes was performed using a water test bench and a test bench with refrigerant R134A. The detailed description of these test benches is presented in [2, 3]. Heat was supplied by electrocontact (ohmic) method, i.e. by passing electric current directly through the tube wall. Distribution of wall temperatures on the tube surface was defined by means of 28 thermocouples arranged on the outer surface of the tube. The roughness of the inner surface of the tube was obtained by thread cutting with pitches of the thread t = 0.3…1.0 mm and average height of projection Δ= 0.07…0.65 mm (Table 1). The inner tube diameter d (about 10 mm) was defined on the roughness cavity.

Studies of heat transfer for distilled water flow were performed at Reynolds number Re=6000…160000 and heat flux q=30…190 kW/m^2, and for refrigerant R134A flow – at Re=26000…100000, q=8…20 kW/m^2.

Content from this work may be used under the terms of the Creative Commons Attribution 3.0 licence. Any further distribution of this work must maintain attribution to the author(s) and the title of the work, journal citation and DOI.
Published under licence by IOP Publishing Ltd

IOP Conf. Series: Journal of Physics: Conf. Series **980** (2018) 012015 doi:10.1088/1742-6596/980/1/012015

Table 1. Parameters of rough surfaces.

Profile 1	Profile 2	Profile 3	Profile 4
t=0.3 mm Δ=0.07 mm Δ/d =0,007	t=0.4 mm Δ =0.07 mm Δ/d =0.007	t=0.5 mm Δ =0.14 mm Δ/d =0.014	t=0.6 mm Δ =0.14 mm Δ/d =0.014
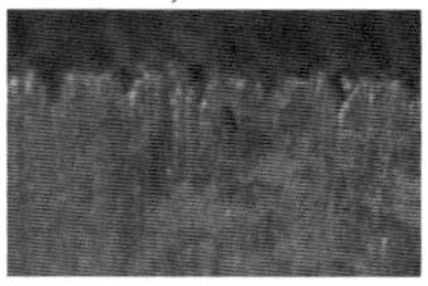	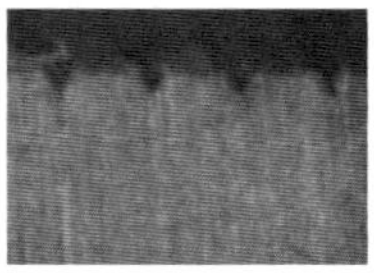	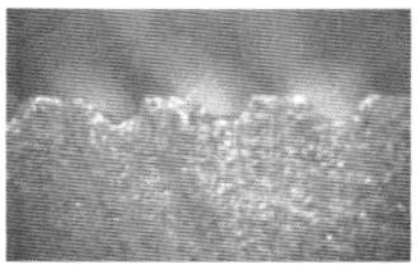	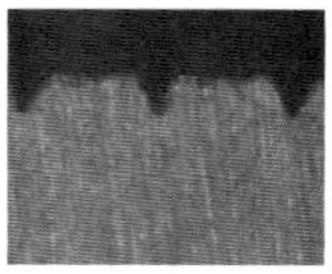
Profile 5	**Profile 6**	**Profile 7**	**Profile 8**
t=0.5 mm Δ=0.25 mm Δ/d =0.025	t=0.75 mm Δ =0.35 mm Δ/d =0.035	t=1.00 mm Δ =0.45 mm Δ/d =0.045	t=1.25 mm Δ =0.65 mm Δ/d =0.065
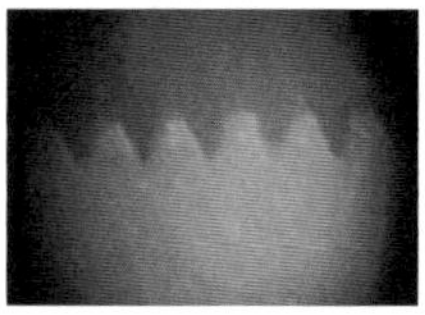	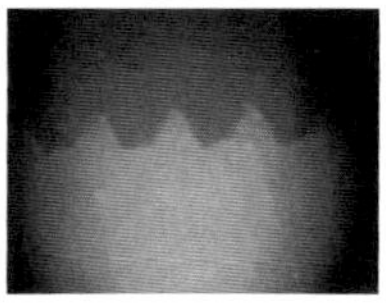	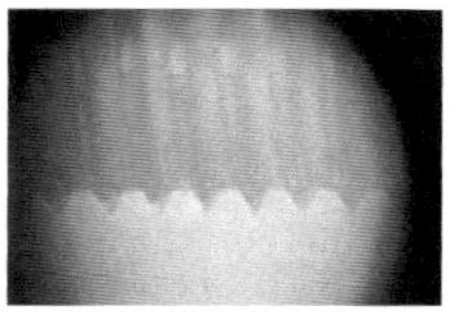	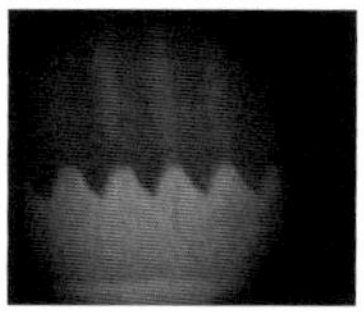

Typical curves of changes of heat transfer coefficient α for one-phase flows of water and refrigerant R134A depending on relative length x/d are shown in Figure 1 a, b.

Figure 2 shows the experimental data on heat transfer in straight rough channels with different depth of cutting for the distilled water flow. As can be seen an increase of roughness height Δ and increment of Reynolds number Re leads to an increase of heat transfer. The values of the Nusselt number Nu for rough tubes with profiles 1 and 2 practically coincide with the Nusselt number for a smooth tube since small roughness peaks (Δ=0.07 mm) disturb the flow only slightly.

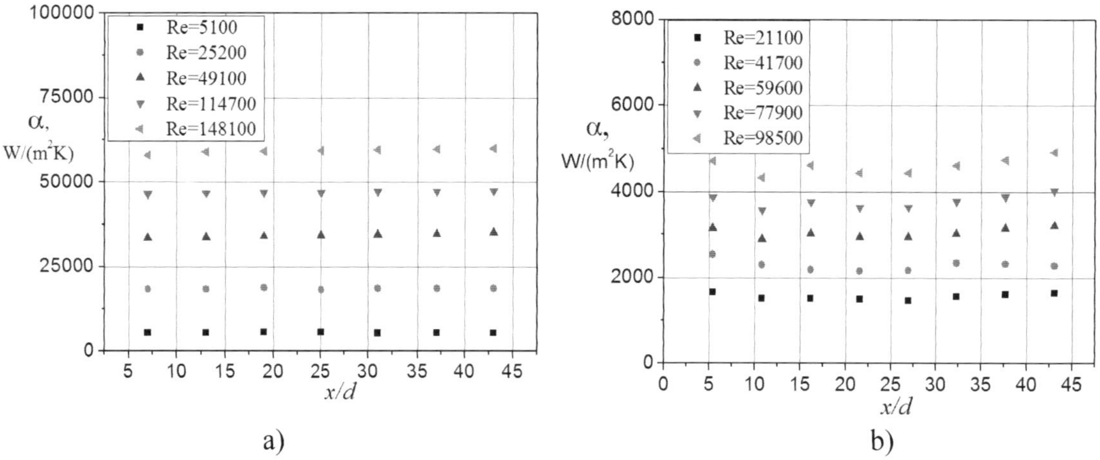

Figure 1. Local values of heat transfer coefficient α in a straight rough tube with Δ=0.45 mm and t=1.0 mm for distilled water (a) and refrigerant R134a (b)

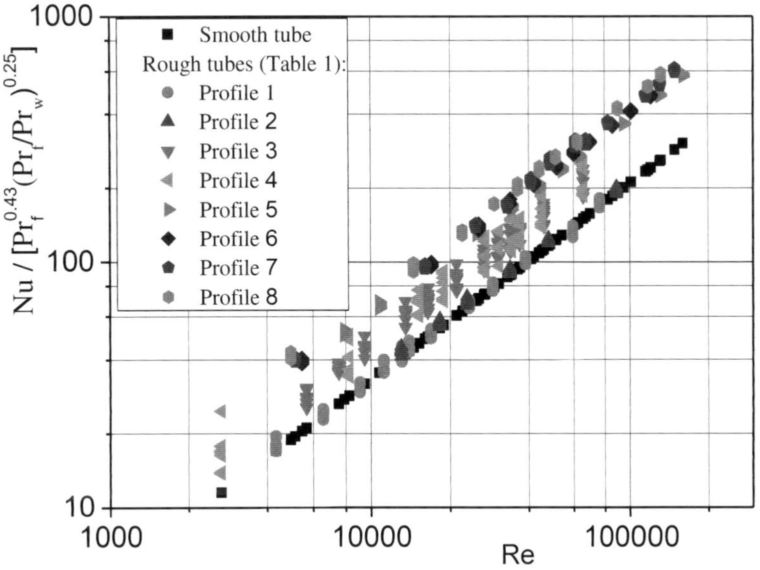

Figure 2. Dependence of heat transfer in straight rough channels with various cutting depth for the distilled water flow

A significant increase of heat transfer (by 38…49%) occurs even in channels with Δ=0.14 mm. The coincidence of the heat transfer level for tubes with profile 3 and profile 4, which have the same cutting depth but different pitch, indicates that the frequency density of the protrusions location has no influence on heat transfer.

Figure 3 presents the comparison of Nusselt numbers for the straight rough tube with Δ=0.35 mm and d=10 mm with Nusselt numbers, calculated by the formulas for heat transfer in rough tubes, proposed by Hausen [5], Yaglom, Kader [6, 7] and Galin [8]. As it can be seen from the graph the heat transfer values calculated by the formula of Hausen [5], Yaglom and Kader [6, 7] as well as the experimental data obtained by Isachenko, Agobabov, Galin [4] are slightly overestimated. However, there is a coincidence of the heat transfer level with the Galin's data [8] at Reynolds numbers Re=10000…40000.

Calculations of heat transfer by other dependences presented in [1] give significant divergences. This once again confirms that there is no universal approach for taking the roughness into account.

To calculate the heat transfer of the tubes with continuous artificial roughness the generalizing dependence of the following form was obtained

$$Nu/Nu_0=1+2.8\cdot(\Delta/d)^{0.3}, \quad (1)$$

where for straight smooth tube

$$Nu_0=0.021\cdot Re^{0.8}Pr_f^{0.43}(Pr_f/Pr_w)^{0.25}, \quad (2)$$

Pr_f and Pr_w are the Prandtl numbers at flow temperature and wall temperature.

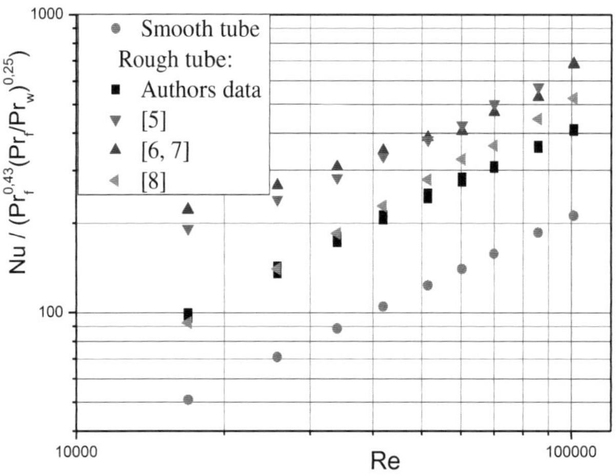

Figure 3. Comparison of heat transfer in a straight rough tube (profile 6) with experimental data of the authors [5-8]

2.2. Heat transfer in rough channels with flow swirling

Flow swirling in rough tubes was organized by twisted tape inserts [9, 10] with relative pitch of twisting s/d=2.5…6 at a turn by 180^0. The thickness of tapes δ was 0.8 mm.

Heat transfer for the distilled water flow in the tubes with twisted tape and continuous artificial roughness in the form of a metric thread with a complete cutting profile was studied at Reynolds numbers calculated on the basis of the roughness cavity diameter from 5000 to 80000 and heat flux q =50…250 kW/m2. The Reynolds numbers for a metric thread with a complete cutting profile for the refrigerant R134A flow were in the range from 10000 to 50000, and the heat flux q =9…21 kW/m^2.

Analysis of the experimental data has shown that the flow swirling leads to heat transfer enhancement, and its intensity increases with an increasing degree of the tape twisting s/d (s – pitch of twist). The heat transfer enhancement in the rough tube with flow swirling for Δ=0.25 mm is 133…173%, for Δ=0.35 mm it is 142…183%, for Δ=0.45 mm it is 152…192%, and for Δ=0.65 mm it is 163…202% relative to the smooth straight tube.

Figure 4 shows experimental dependences of heat transfer in rough channels with flow swirling for a fixed degree of twisting and various cutting depth of the metric thread.

Heat transfer enhancement in the rough tube with flow swirling relative to heat transfer in the smooth tube with flow swirling calculated according to Manglik and Bergles [9] formula for Δ=0.25 mm is about 60%, for Δ=0.35 mm it is 70%, for Δ=0.45 mm – 80%, and for Δ=0.65 mm – 90%.

The experimental study has shown that the heat transfer depends most strongly on the relative height (Δ/d) of roughness peaks and the degree of flow swirling (s/d) for both coolants.

Generalization of experimental data for rough tubes with flow swirling was carried out for roughness with a complete cutting profile in the form of a triangular thread. The dependence has the form

$$Nu/Nu_S=1+2\cdot(\Delta/d)^{0.3}, \quad (3)$$

where Nusselt number for smooth tube with twisted tape insert is according to [9]

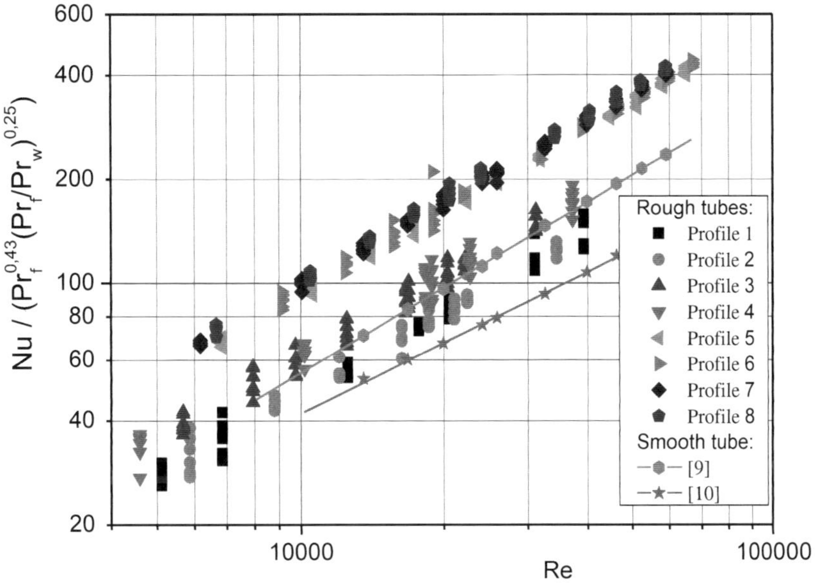

Figure 4. Heat transfer in rough tubes with twisted tape insert (s/d=2.5) for the distilled water flow

$$Nu_s=0.023Re^{0.8}Pr^{0.4}[1+0.769/(s/d)][\pi/(\pi-4\delta/d)]^{0.8}[(\pi+2-2\delta/d)/(\pi-4\delta/d)]^{0.2}(\mu_f/\mu_w)^{0.18}, \quad (4)$$

μ_f and μ_w are the dynamic viscosities at flow temperature and wall temperature, π=3.14...

3. Conclusions

The study of the distilled water and coolant flows in the tubes with rough walls in the form of the metric thread with complete cutting profile have shown a significant influence of the profile height on the heat transfer intensity while the influence of the thread pitch has not been found.

The twisted tape installation in the channel with rough walls gives an additional increment in heat transfer, and this occurs with an increase of the twisting degree. The generalizing dependences for the tubes with and without the tape are obtained. There the roughness influence in the form of complex (Δ/d) was included in equal degree – 0.3, which proves the absence of the wall-adjacent turbulence suppression by the flow swirling.

References

[1]. Tarasevich S E, Zlobin A V and Yakovlev A B 2015 Hydrodynamics and Heat Transfer in Single-Phase Fluid Flows in Artificially Rough Pipes *High Temperature* **53**(6) 908–920
[2]. Giniyatullin A A, Tarasevich S E and Yakovlev A B 2013 Experimental and Numerical Investigation of Heat and Mass Transfer in Tubes with Ribbed Twisted Tape Inserts *Vestnik KSTU named after A.N. Tupolev* **2**(2) 13-18

IOP Conf. Series: Journal of Physics: Conf. Series **980** (2018) 012015 doi:10.1088/1742-6596/980/1/012015

[3]. Tarasevich S E, Shishkin A V, and Yakovlev A B 2013 Stand for Experimental Research of Freon Boiling in Various Channels *Vestnik KSTU named after A.N. Tupolev,* **2**(2) 60-62

[4]. Isachenko V P, Agababov S G and Galin N M 1965 Experimental Study of Heat Transfer and Hydraulic Resistance in Turbulent Water Flow in Tubes with Artificial Roughness *MPEI proc Heat exchange and hydraulic resistance* **53** 27-37

[5]. Hausen H 1976 *Heat transfer in Counterflow, Parallel Flow and Cross Flow* (Moscow, Publishing House Energoizdat) p 384

[6]. Yaglom A M and Kader B A 1974 Heat and Mass Transfer between a Rough Wall and Turbulent Flow at High Reynolds and Peclet Numbers *J. Fluid Mechanics.* **62**(3) 601

[7]. Kader B A and Yaglom A M 1975 Turbulent Heat and Mass Transfer from Wall Coated with Two-Dimensional Roughness *In the book: Wall-adjacent turbulent flow, part 1* (Novosibirsk, Publishing house SB AS USSR) p 203.

[8]. Galin N M 1967 Heat Transfer in Case of Turbulent Gas Flow near Rough Walls *Teploenergetika* **5** 67-71

[9]. Manglik R M and Bergles A E 1993 Heat Transfer and Pressure Drop Correlations for Twisted-Tape Inserts in Isothermal Tubes: part II – Transition and Turbulent Flows *ASME J. of Heat Trans.* **115**(890) 890-896

[10]. Ibragimov M Kh, Nomofilov E V and Subbotin V I 1961 Heat Transfer and Hydraulic Resistance for Fluid Helical Motion in the Tube *Teploenergetika* **7** 57-60

Numerical simulation of a low-swirl impinging jet with a rotating convergent nozzle

K Borynyak[1], M Hrebtov, M Bobrov and N Kozyulin

Institute of Thermophysics of SB RAS, 630090, Ac. Lavrentiev ave. 1, Novosibirsk, Russia

Abstract. The paper presents the results of Large Eddy Simulation of a swirling impinging jet with moderate Reynolds number (10^4), where the swirl is organized via the rotation of a convergent nozzle. The results show that the effect of the swirl in this configuration leads to an increase of axial velocity, compared to the non-swirling case. It is shown that turbulent stress plays an important role in this effect. The vortex structure of the jet consists of multiple pairs of nearly parallel helical vortices with opposite signs of rotation. The interaction of vortices in the near region of the jet leads to radial contraction of the jet's core which in turn, causes an the increase in the axial velocity.

1. Introduction

Jet flows are used in a wide variety of technical systems: jet engines, cooling systems, mixers, chemical reactors, furnaces and combustion chambers. This makes such flows are perspective for further investigations. The application of swirls to the jet flows is often used for flame stabilization in burners and combustion chambers, since it increases in intensity of heat and mass transfer in the near region of the jet. In particular this is due to the formation of large-scale helical vortex structures that have a significant effect on transport and mixing in swirling flows. For jets with a strong swirl the central recirculation zone is formed as a result of breakdown of the vortex core, which also leads to flame stabilization.

As a drawback, flows with a strong swirls usually show an intensive precession of a helical vortex core, which in a number of cases has negative consequences such as thermoaccoustic resonance in combustion chambers and vibrational combustion regimes. The characteristics of large eddies in jets with a weak swirl have not been investigated. The effect of their interaction on the intensity of the heat-mass transfer, mixing and the acoustic noise generated by the jet remains unclear.

Turbulence structure of the flow has significant changes when swirling is applied: instead of ring-structure of the vortices flow in a non-swirling jet case, the variety of helical structures can be seen as the dominant modes. The interaction of helical vortices may be significantly different from the non swirling case, because the mean helicity, being an invariant may affect the spectral energy transfer and dissipation [1]. Swirl is considered as one of the effective ways of passive flow control by many authors[2].

In the recent years due to the development of the optoelectronic technologies and increasing of computational resources for numerical simulation, it has become possible to investigate the effects of the interaction of vortices inside the jet. The identification of parameters that affect the vorticity dynamics enstrophy will allow extending the range of the methods to control the heat-mass transfer in jets.

There are many ways to introduce a swirl to the jet. The most common and well-studied one is by mounting stationary blades inside the nozzle at some angle to the incoming flow. In such setup the drag, induced by the blades in the central part of the flow, leads to the slowing down of the flow in the jet's core. This effect helps in the formation of recirculation bubble. However, it is possible to introduce the swirl without putting obstacles to the flow by using a rotating nozzle. In the present work we studied, by means of numerical simulation, the rotating nozzle configuration of the impinging jet with a low swirl rate. The vortex structure of the flow and the effect of the swirl on the vorticity dynamics were the main objects of the study.

[1] Corresponding author: borisnsk07@gmail.com

Content from this work may be used under the terms of the Creative Commons Attribution 3.0 licence. Any further distribution of this work must maintain attribution to the author(s) and the title of the work, journal citation and DOI.
Published under licence by IOP Publishing Ltd

2. Equations

We use LES approach with dynamic Smagorinsky model, with the following equations for the momentum transport:

$$\frac{\partial \overline{u}_i}{\partial t}+\frac{\partial}{\partial x_j}\left(\overline{u}_i\overline{u}_j\right)=-\frac{1}{\rho}\frac{\partial \overline{p}}{\partial x_i}+\frac{\partial}{\partial x_i}\left(\left(\nu+\nu_{sgs}\right)\overline{S}_{ij}\right). \tag{1}$$

We use classical form of the subgrid viscosity:

$$\nu_{sgs}=\left(C_s\Delta\right)^2\left|S_{ij}\right|, \tag{2}$$

where Δ is a grid spacing and C_S is Smagorinsky coefficient, which should be evaluated dynamically. The classical dynamic model of Smagorinsky [3] model computes C_S after applying the explicit spatial filtering to the velocity field and then minimizing the difference between the subgrid and residual stress tensors. However, this procedure may lead to negative turbulent viscosity (i.e energy backscatter from the subgrid to resolved scales). In turn, this is proved to lead to the blow up of the solution because the amount of energy which may be transferred from the small scales is unlimited. It is common to use clamping of Smagorinsky constant only to positive values. But such procedure is an ad hoc and non-physical in nature. It was shown that the Smagorinsky constant averaging over homogeneous directions is a cure-all solution, but only a few types of flow have homogeneous directions for such averaging. Another approach adopted here was introduced in [4] and consists in averaging of the residual tensor magnitude over the lagrangian trajectories. The model has shown a good agreement with direct numerical simulations and can be adopted for almost every flow.

3. Computational domain and numerical scheme

The computational domain covered a short inlet pipe section continued by the convergent nozzle and a cylindrical domain ending with an impact wall (Fig 1). The ratio of the nozzle inlet to outlet diameter (D_{in}/D) was equal to 4. The inlet pipe section length was 1.6 D, and the convergent nozzle length was 2.5D. The impact wall was placed at the distance of 3D from the nozzle exit. The geometry of the nozzle was built using the Vitoshinsky equation[5]. The domain around the nozzle was cylindrical and had a diameter of 10D The numerical grid consisted of 10^7 hexahedral cells, refined toward the solid boundaries, with the wall-adjacent gridcell width (y+) of about 1 in wall units.

The momentum transport equation Eq.1 was discretized using the finite volume scheme of the second order of accuracy in time and space. The modified LES solver from OpenFoam[6] package was used.

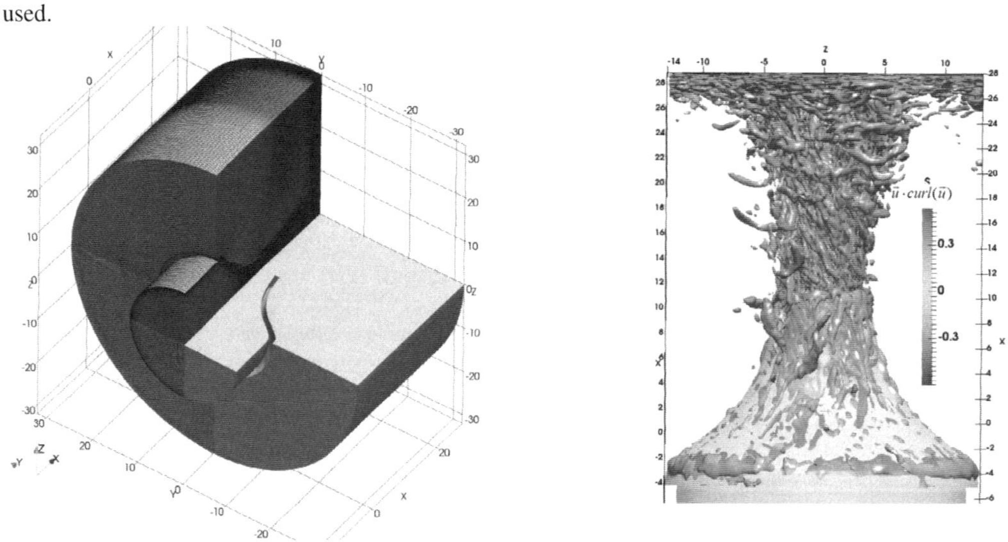

Figure 1. The computational domain for numerical simulation (left); iso-surface of Q-criterion (Q = 5), colored by helicity density.

4. Boundary conditions

The numerical simulation of the complex flow requires correct boundary conditions. Nowadays, there are several ways to generate turbulent initial conditions at the inlet. In our work we used the anisotropic divergence-free synthetic turbulence at inflow boundary, introduced in [7]. This method helps to avoid numerical artifacts, as in random fluctuations generation, and costly preliminary simulations for inlet conditions.

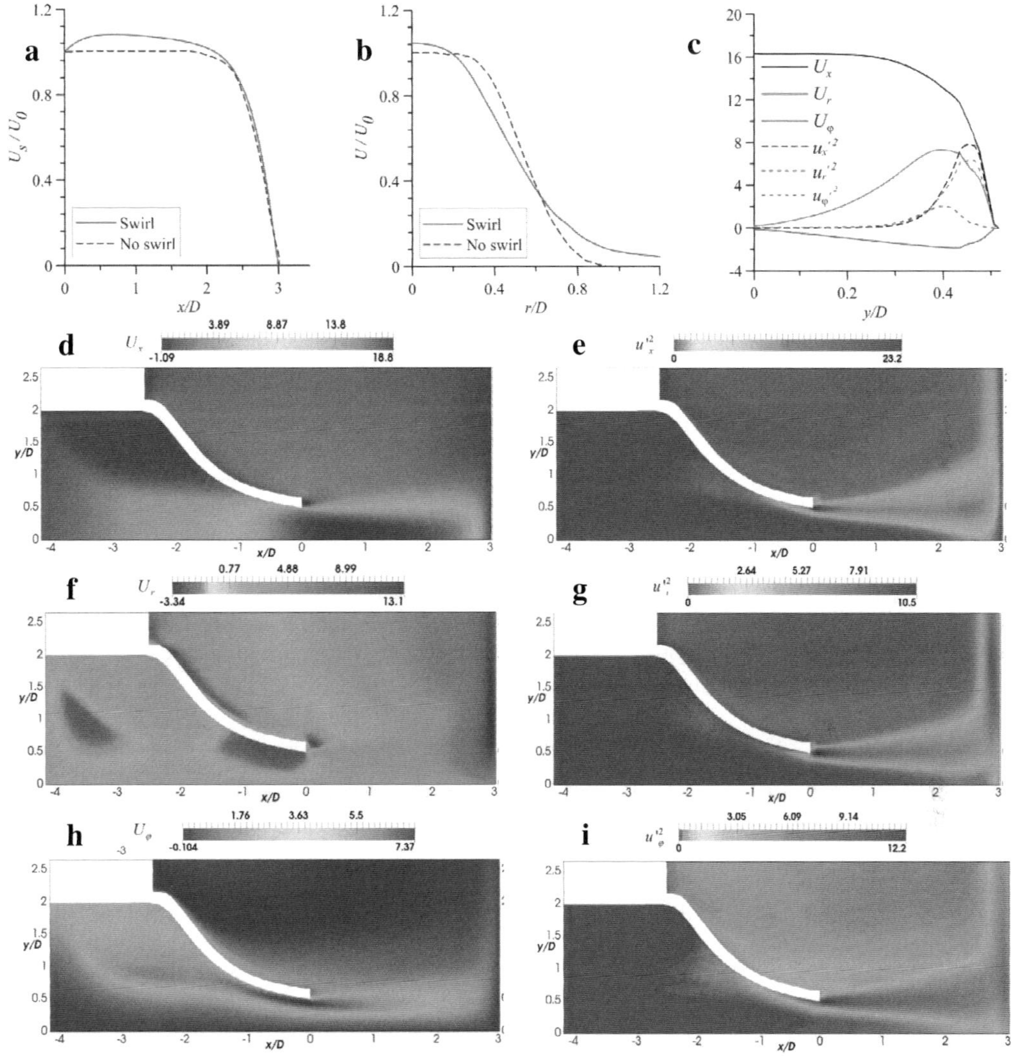

Figure 2 The profile of axial velocity along the jet axis (a), along spanwise cross-sections *x/D*=2 (b) and *x/D*=0 (c); mean-velocity components (d, f, h) and turbulent intensities (e, g, i) in longitudinal cross-section of the jet.

In our case the swirl is introduced by the rotating convergent nozzle with inlet to outlet diameters ratio of 4. It was shown in [8] that for convergent nozzle the azimuthal and axial velocity components will be affected differently by the contraction. From an inviscid analysis one obtains that the axial velocity will increase in proportion to the contraction ratio (CR) whereas the azimuthal one - only as a square root of CR. In our case we expected the 16 times increase of axial velocity and only 4-times increase of azimuthal velocity. We chose the Reynolds number at the nozzle outlet to be 10^4 this means that the nozzle inlet Reynolds number should be 2500. The azimuthal velocity of the nozzle inlet pipe surface was set to be two times larger than the mean axial velocity of the nozzle inlet: that gave the swirl number S~0.3 (based on the ratio of azimuthal and axial fluxes) at the nozzle outlet. We used the rotating pipe

IOP Conf. Series: Journal of Physics: Conf. Series **980** (2018) 012016 doi:10.1088/1742-6596/980/1/012016

velocity and fluctuations amplitude profiles (taken from [9]), for the inlet conditions of the rotating pipe section placed before the nozzle. For the outlet surfaces, the convective boundary conditions were used. The nozzle surface was set to a solid body rotation, consistent with the chosen inlet azimuthal velocity. At the impact wall, simple no-slip conditions were used.

5. Results and discussion

Applying the swirl to impinging jet it is expected to intensify the spanwise spread through the effect of the centrifugal force. This behaviour was observed in experiments with stationary blade swirlers [2]. However the evolution of the jet strongly depends on the radial organization of the swirl. In our case the effect of centrifugal force in the contracting part of rotating nozzle induce the negative axial force component. This leads to the formation of recirculation bubble inside the nozzle, which affects the velocity profiles at the nozzle outlet.

The time-averaged fields of velocity and turbulent fluctuations obtained in the present simulation are presented in Fig.2. The comparison with non-swirling case with the same nozzle (Fig.2 a,b) shows that swirl affects the axial velocity. It can be seen that maximum of axial velocity detaches from the nozzle plane and shifts in streamwise direction to x=0.65*D*.The spanwise profile of axial velocity shows that in the swirling case the mixing layer is wider, which leads to a slower growth of Kelvin-Helmholtz type instabilities, compared to a non-swirling case.

Because of its shape the rotating nozzle accelerates the flow rotation in its widest part, and decelerates in the narrowest part close to the nozzle outlet. This leads to generation of axial vorticity of the opposite sign to the main swirl close to the nozzle surface and the formation of secondary vortices in the boundary layer. This is reflected in the azimuthal velocity profile at the nozzle exit. The maximum of azimuthal velocity is shifted from the wall towards the jet core. The positive part of radial gradient of azimuthal velocity is observed to be lower than the negative part, which is located at the outer boundary of the jet. This behavior is pronounced for the distance of about a jet diameter (D) downstream from the nozzle exit. This means that there exist two concentric vortex systems with opposite signs of axial vorticity, one at the jet core and the other one at the jet periphery. Below it will be discussed in more detail.

The radial velocity profile shows negative values which means that the jet is contracting in the radial direction in the region close to the nozzle exit. This contraction is reflected in the growth of axial velocity after the jet exits the nozzle shown in Fig. 2a.

5.1 Turbulence characteristics

The components of turbulence energy are presented in Fig 2 c, e, g, i. At the nozzle exit the highest are the axial velocity fluctuations, but the azimuthal fluctuations are of almost the same magnitude, while the radial fluctuations are much lower, and their maximum is closer to the jet axis. This reflects the fact that the main modes of fluctuations are helical vortices. While moving downstream the radial fluctuations grow up to the same level as azimuthal ones, which might be the reflection of the development of Kelvin- Helmholtz type instabilities.

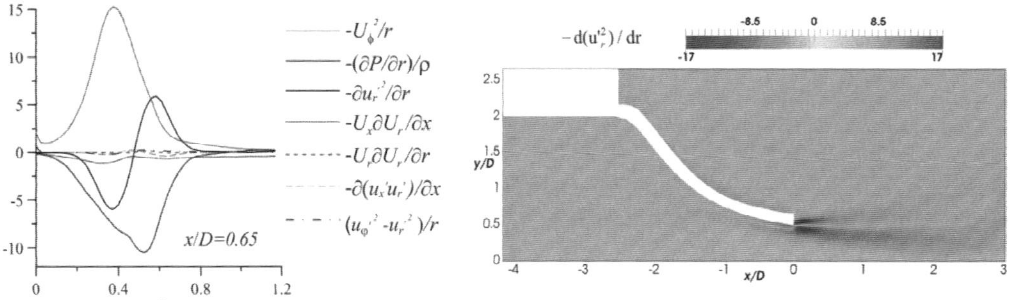

Figure 3 The components of mean radial momentum equation for *x/D*=0.65 (left) and the distribution of $-\partial u_r'^2 / \partial r$ in longitudinal cross-section (right).

One of the most interesting questions of the study is to search for the mechanism of jet radial contraction as it exits the nozzle.

We consider now the equations of mean radial momentum balance to see which term contributes the most to the negative radial force component. Neglecting the viscous terms one can write

$$U_x \frac{\partial U_r}{\partial x} + U_r \frac{\partial U_r}{\partial r} - \frac{U_\varphi^2}{r} = -\frac{1}{\rho_0}\frac{\partial P_m}{\partial r} - \left\langle \frac{\partial (u'_x u'_r)}{\partial x} + \frac{\partial u_r'^2}{\partial r} - \frac{u_\varphi'^2 - u_r'^2}{r} \right\rangle,$$

where the first four terms are the effect of the mean flow characteristics while the last three are the effect of turbulent fluctuations. From Fig. 3a it can be seen that the most positive radial forcing is provided by the centrifugal force, while the significant part of negative forcing is provided by the effect of radial velocity fluctuations. The pressure gradient acts to counterbalance these two parts. This means that the effect of radial contraction at the near region of the jet is caused by the effect of turbulence.

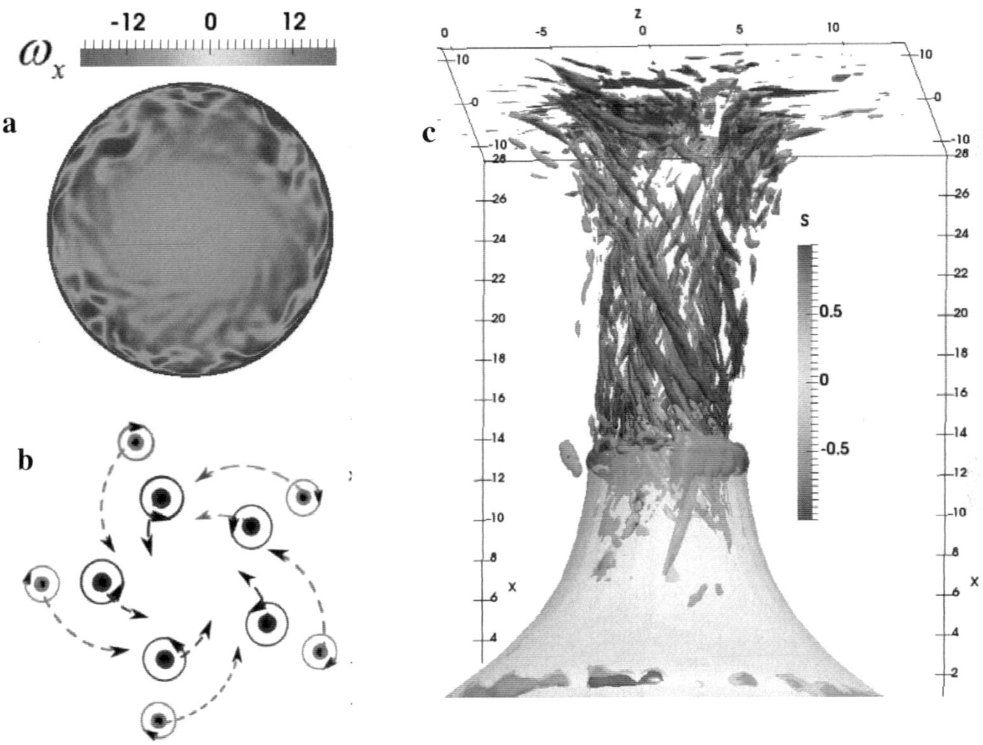

Figure 4 Distribution of axial vorticity component in the nozzle cross-section (a), the schematics of vortex interaction (b), the main vortex modes of the jet shown by iso-surfaces of Q-criterion, colored by helicity density (c)

From the axial distribution of $-\partial u_r'^2 / \partial r$ it could be seen that it causes the positive radial forcing at the jet periphery, and negative forcing closer to the jet's core.

We will now discuss the vortex structure of the jet. As it was mentioned above, the most energetic modes of the mixing layer are helical vortices (Fig 4). The axial vorticity distribution is shown in Fig 4a. It can be seen that the distribution consists of two parts with opposite signs, one with negative (same as the main swirl) sign in the jet core and the other with positive sign close to the nozzle boundary. Due to the growth of instabilities, the substantial part of vorticity volume is organized in separate counter-rotating vortices, represented as red and blue spots in Fig 4a. Spatially these vortices have helical shape. After exiting the nozzle the positive axial vorticity becomes amplified because of the generation of instabilities in the mixing layer. After that the two-part helical structure is established with checkerboard pattern of positive and negative axial vorticity. In three dimensions these leads to generation of parallel helical vortex pairs, in which the vortices with negative axial vorticity are located closer to the jet axis and with positive axial vorticity around them at the jet periphery. The interaction of such vortices may lead to the radial contraction of the jet a shown schematically in Fig 4b.

It is also noticeable that when reaching the impact surface the helical structure of the jet evolves into two separate helical “sleeves”, each having about a half of the jet core vorticity. These sleeves form a “Y” shape and split near the wall.

6. Conclusions

The impinging jet with a rotating convergent nozzle was simulated. It has been found that in the near region of the jet the axial velocity intensifies, compared to the non-swirling case. The point of maximum axial velocity is shifted by 0.65 nozzle diameters downstream from the nozzle edge. The turbulent stress is found to have a sufficient contribution to the effect of radial contraction of the jet that leads to the growth of axial velocity. Possible explanation of such influence of turbulent vortices on the mean flow is proposed based on the dynamics of helical vortex pairs, evident in the simulation.

This work was supported by the Russian Science Foundation (grant № 16-19-10566).

References

[1] Teitelbaum T and Mininni P D 2010 Large-scale effects on the decay of rotating helical and non-helical turbulence *Physica Scripta* **142** 014003
[2] Alekseenko S V et. al. 2007 Experimental study of an impinging jet with different swirl rates *International Journal of Heat and Fluid Flow* **28**(6) 1340-1359
[3] Germano, Massimo et al. 1991 A dynamic subgrid-scale eddy viscosity model *Physics of Fluids A: Fluid Dynamics* **3**(7) 1760-1765
[4] Meneveau C, Thomas S L and William H C 1996 A Lagrangian dynamic subgrid-scale model of turbulence *Journal of Fluid Mechanics* **319** 353-385
[5] Gorlin S M and Slezinger I I 1964 Aeromekhanicheskie *izmereniya: Metody i pribory (Aeromechanical Measurements. Methods and Devices)*
[6] Weller H G et. al. 1998 A tensorial approach to computational continuum mechanics using object-oriented techniques *Computers in physics* **12**(6) 620-631
[7] Kroger H and Kornev N 2015 *Generation of anisotropic divergence free synthetic turbulence at inflow boundaries for LES and DNS using the Turbulent Spot Method* (ICHMT DIGITAL LIBRARY ONLINE, Begel House Inc.)
[8] Komori Satoru and Hiromasa Ueda 1985 Turbulent flow structure in the near field of a swirling round free jet *The Physics of fluids* **28**(7) 2075-2082
[9] Facciolo, Luca 2006 *A study on axially rotating pipe and swirling jet flows* (Diss, KTH.)

IOP Conf. Series: Journal of Physics: Conf. Series **980** (2018) 012017 doi:10.1088/1742-6596/980/1/012017

Investigation of acoustic and gas dynamic characteristics of strongly swirled turbulent jets

S Yu Krasheninnikov[1], VP Maslov[1], AK Mironov[1], PD Toktaliev[1]

[1]Central Institute of Aviation Motors named after P.I. Baranov, Moscow 111116, Aviamotornaya 2, Russia

E-mail: pavel_d_m@mail.ru

Abstract. Generalization of the series of experimental and numerical results for properties and characteristics of swirling jets with high swirling intensity $W_0>1$ is considered. These jets are typically used in gas turbine aviation engines for intensification of mixing process and combustion process stabilization. Flow structures in swirling jets and in the near-field are analyzed. It is shown, that, in the main, the flow structure behind the swirling device can be determined by swirling intensity W_0 and acoustic fluctuations field formed far from the jet boundaries. Experimental measurements and numerical simulation of the noise levels of the highly swirling jet are performed using Ffowcs-Williams-Hawkins analogy. Maximum levels of noise axis are observed at angles of 50°-70° from the jet.

1. Introduction

Most of the modern turbojet aviation engine combustors use swirling flows after frontal devices to intensify mixing, however, a number of mechanisms in such flows are not fully clarified yet, including conditions and mechanisms of the secondary loss of the swirling jet stability. Investigations of swirling turbulent jets show that a swirling jet generates intensive acoustic radiation in the form of discrete tones, when swirling intensity is more than one. The swirling jet generates a tone noise while the noise of the non-swirling turbulent jet appears to be broadband. Thus, swirling jets are a suitable object for acoustic field investigation, because pressure and velocity pulsations spectra of swirling jets have marked frequencies.

The flow behavior in swirling jets is determined by two main parameters: Reynolds number and swirling intensity S, which correlate axial and circumferential angular momentum fluxes [1]:

$$S=\left(2\int_0^\infty \rho U_x U_\tau r^2 dr\right)\left(D\int_0^\infty \rho\left(U_x^2-\frac{U_\tau^2}{2}\right) r dr\right)^{-1}.$$

Here ρ is the density, U_x, U_τ are the axial and rotational velocity components, r is the radial coordinate, and D is the diameter of the swirling device outlet cross-section. The magnitude of swirling intensity defined this way remains constant in axial direction due to the momentum conservation law. Commonly, the swirling magnitude defined as $W_0=U_{\tau,max}/U_{x,avg}$ [2] appears to be more useful for measurements and calculations; here $U_{\tau,max}$ is the maximal circumferential velocity in the current section, and $U_{x,avg}$ is the mass flow averaged velocity in the section. It can be shown that $W_0 \sim S$. In practice, the swirling jet is created in primary combustor zone by means of different types of swirlers

Content from this work may be used under the terms of the Creative Commons Attribution 3.0 licence. Any further distribution of this work must maintain attribution to the author(s) and the title of the work, journal citation and DOI.
Published under licence by IOP Publishing Ltd

with a swirling intensity within the range of W_0=1.1-1.8. These swirling magnitudes correspond to stable reverse flow zone.

2. Flow structure in highly swirling jet behind the swirling device

Detailed investigations of flow structure in the swirling jet are performed in [2-5]. The scheme of highly swirling jet in fig.1 may be used as a generalization of these results; here the instantaneous axial velocity field for the swirling jet with W_0=1.36 in submerged space is presented as well. The flow structure consists of reverse flow zone (RFZ) (1) located near the axis of the device, the zone of mixing layers (2) and ambient ejection zone (3). Other above mentioned investigations prove the presence of linear relation between the average dimensionless length of reverse flow zone and the swirling intensity in the form of [2]:

$$L^0 = L / D \cong 4W_0. \tag{1}$$

Here L is the dimensionless reverse flow zone length, and D is the diameter of the swirling device outlet cross-section. Computational and experimental data for L^0 for different swirling devices, such as atomizers and blade swirlers, are presented in figure 2. Experimental data were obtained by Particle Image Velocimetry (PIV) and Laser Doppler Velocimetry (LDV) techniques, and computational results were obtained by numerical solution of Navier-Stokes equations. Details of measurement technique and experimental facility can be found in [6, 7], details of computational procedure in [8]. Dashed line in figure 2 shows dependence (1). Figure 2 demonstrates that RFZ exists in the flow at $W_0>0.7...0.8$, and it appears to be unstable at $W_0\sim1$ regimes, i.e. small perturbations may lead to its complete disappearance. In the range of $2.5>W_0>1$ the back-coupling mechanism [4] stabilizes RFZ, and it can exist in the average flow even in the presence of large upstream perturbations. At $W_0>2.5$ the swirling jet "breaks up": spiral vortex structures in the flow lose integrity and break into separate large-scale vortexes, which come off from the edge of the swirling device nozzle; and jet loses its periodical motion structure.

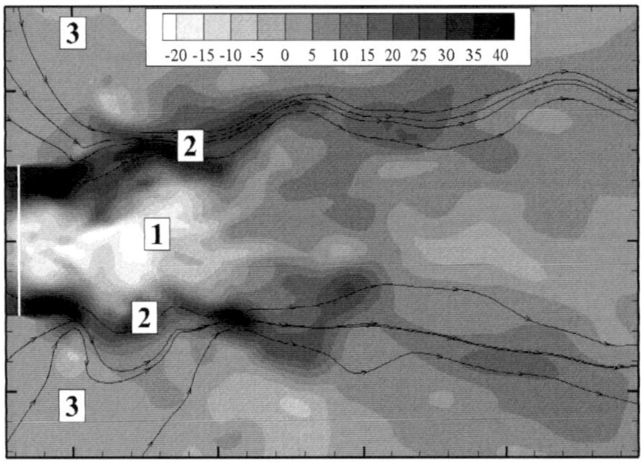

Figure 1. Typical flow scheme in highly swirling jet, contours - instantaneous axial velocity, numbers: 1 - reverse flow zone, 2 - mixing layers, 3 - ambient ejection zone, white vertical line - nozzle exit. Numerical results.

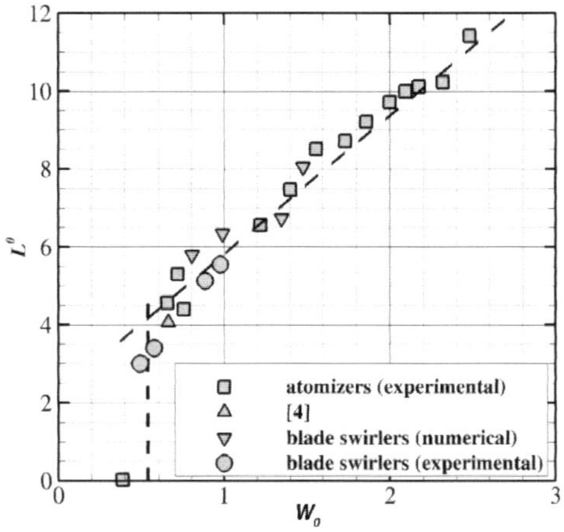

Figure 2. Dimensionless reverse flow zone length for different swirling devices. Measured and calculated values.

Analysis of experimental and computational data of velocity profiles near the swirling device outlet cross-section shows that the fields of averaged axial and rotational velocity components for the submerged jet can be approximated by modified Monkewitz profile [9, 10] with high accuracy:

$$U_x = 4BF_1\left(1 - BF_2\right),$$
$$F_j = \left[1 + \left(\exp(r^2 b_j) - 1\right)^{N_j}\right]^{-1}, j = 1,2,$$
$$B = 0.5\left[1 + \left(1 - U_{x,axis}\right)^{1/2}\right].$$

Here b_j *are the* parameters, and N_1 and N_2 parameters define thicknesses of the inner and outer mixing layers. Such representation of the averaged flow allows applying linear stability analysis of averaged flow field relative to small perturbations [11, 12]. Stability analysis and data obtained in the experiment and from numerical solution of filtered Navier-Stokes equations show that complicated precessional motion of RFZ exists in a jet with $W_0>1$, and this motion appears to be a source of a tone noise. Thus static pressure pulsation spectra in the far-field have a distinct tone, which contains the major part of total acoustic energy radiated by jet (more than 80%).

3. Swirling jet noise

Unsteady flows in highly swirling jets produce acoustic fluctuations outside the jet. In this research, the acoustic field and flow characteristics outside the jet are experimentally measured using microphones and constant temperature anemometry techniques. Computational results are based on numerical solution of unsteady three-dimensional filtered Navier-Stokes equations, using the control volume method and well-known SIMPLEC [13] procedure, reformulated for compressible flows. Precessional motion in the jet with high swirling intensity produces mass flow rate fluctuations in ejected medium (zone 3 in figure 1), however pressure and velocity fluctuations do not correspond to the properties of acoustic disturbances. Numerous studies conducted for different swirling devices, such as atomizers and blade swirlers [2, 3], including current research, show, that for jets with $W_0>1$ the frequency of the first harmonic (f_0) of a distinct tone can be connected with swirling intensity in the following way:

$$Sh = f_0 D / U_{x.avg} \cong 0.7W_0. \tag{2}$$

This frequency coincides with hydrodynamic frequency of RFZ rotation about the symmetry axis of the swirling device. It is also possible to observe the second harmonic of distinct tone in the spectra of pressure and velocity fluctuations; it arises from asymmetry of RFZ form in cross section. Experimental and numerical data for the first harmonic of natural frequency of swirling jets, flowing from different swirling devices, are presented in figure 3. The dashed line in figure 3 indicates the dependence (2) with sufficient accuracy reproducing the calculated and experimental data both for blade swirlers and for centrifugal atomizers [14]. Separate points for blade swirler type devices with predominating second harmonic in the spectrum are given in figure 3 ($2f_0$, marked by triangles). There are two frequencies with prominent amplitude in pressure and velocity fluctuations spectra for such devices. Existence of the distinct tone in highly swirling jet allows studying the acoustic disturbances formation in the far-field by measurements and calculations only for the frequency of the first harmonic. Transition from convective type disturbances in the jet and in the near-field to acoustic disturbances are illustrated in figure 4. Figure 4 shows static pressure fluctuations intensity for the first harmonic frequency for two-stage blade swirler (with vortex tube) at different distances from the swirling jet source along the radius with angle of 60° to symmetry axis. Experimental and numerical points are denoted by square and circle markers, respectively, and the dashed line indicates theoretical relation between the amplitude and the distance for acoustic disturbances. As can be seen in figure 4 pressure fluctuations levels converge to acoustic type line at distances $R/D\approx10$. Numerical results for $R/D\approx20$ also lie near the theoretical dependence.

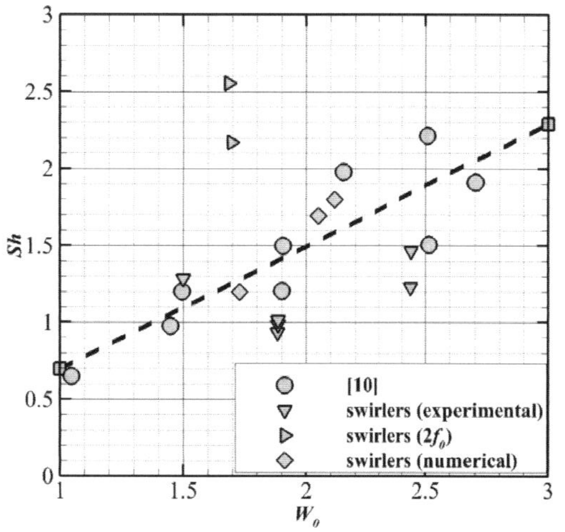

Figure 3. .Dependence of the first main frequency of the swirling jet on swirl intensity. Measured and calculated results.

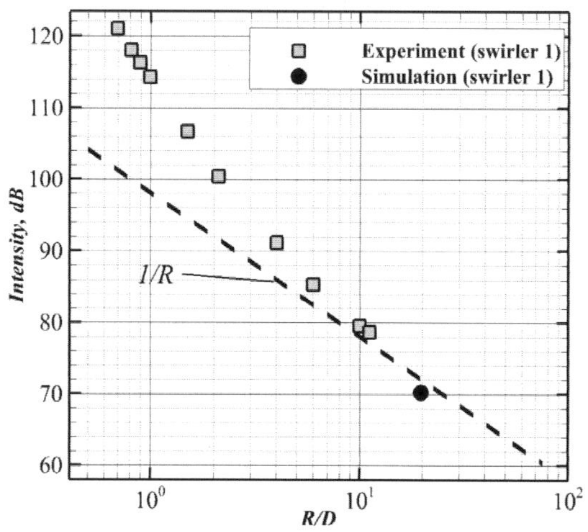

Figure 4. Asymptotic behaviour of pressure fluctuations intensity in far-field of the swirling jet, direction angle from axis - 60°.

To describe swirling jet formation and propagation processes and to investigate acoustic characteristics of highly swirling jets this work applied the numerical procedure based on detached eddy simulation approach (DES)[15]. As can be concluded from previous works, DES approach has a number of principal advantages for highly swirling jet investigations in comparison with RANS approach. One of the main advantages is the potentially more accurate description of turbulent fluctuations spectrum, and consequently, a more appropriate description of unsteady flow parameters. At the same time, simulation accuracy is determined, among others, by the width of directly resolved part of turbulent spectrum that explicitly depends on cell sizes in computational domain. Another factor, which affects the accuracy of unsteady characteristics, is the collected flow statistics; it is especially important for spectral characteristics of the flow. It has been also clarified that the main difficulty for DES approaches is the description of the flow in the transition region from the LES solution to the RANS solution, which implies the complexity of determining the universal type of limiter l_{DES}. Thus, different types of limiters implicitly impose limitations on the characteristics of the computational grid in the transition region and in the boundary layer; often these restrictions explicitly depend on the characteristics of the current flow itself: the thickness of the boundary layer, the pressure drop along the boundary layer, and so on. In current research, a IDDES-type limiter (Improved Delayed DES-IDDES) was used to describe a swirling jet, taking into account the assumed values of the determining characteristics [16]:

$$l_{IDDES} = f_d(1+f_e)l_{RANS} + (1-f_d)l_{LES}.$$

Here f_d, f_e are the damper functions, and l_{RANS}, l_{LES} are the spatial scales of RANS and LES zones respectively. A more detailed description of the form of damping functions and scales can be found in [16]. A comparison of the results of IDDES calculations with experimental data on the mean and pulsation characteristics for swirling jets is presented in [17]. In addition to calculating the flow fields for determining the noise levels in the far field, the integral formulation of the Ffowcs-Williams-Hawkins method is used in current research. The scheme of the experimental model for performing acoustic measurements is shown in figure 5. The model consists of a blade swirler, a vortex tube mounted to the outlet of the swirler and a nozzle. The model creates a swirling jet with W_0> 1. The results of measurements of pressure fluctuations levels for the presented model are shown in figure 4, and the results of calculations are in figure 4 (marked by circle) and in figure 6. Figure 6 shows the directional diagram of the noise of a swirling jet for R = 1 m, indicating that the maximum noise level is observed for the model in question at angles of 50 ° -70 ° and this maximum has the value of 80-85 dB.

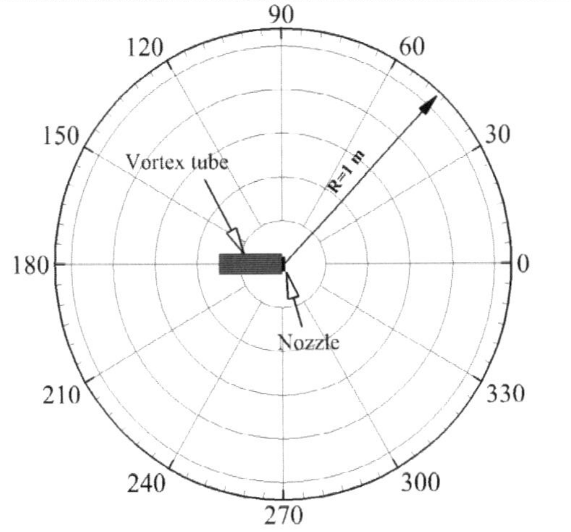

Figure 5. Surrounding hemisphere scheme for pressure fluctuation measurement and calculation.

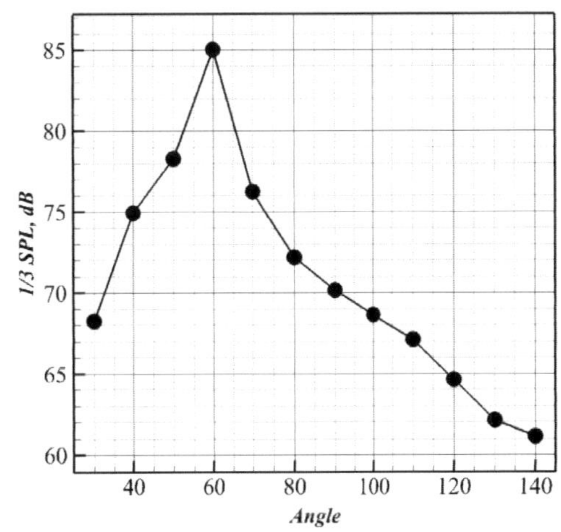

Figure 6. Directional diagram of pressure fluctuation in the far-field of the swirling jet, R=1 m, D=44 mm

4. Conclusions

Numerical and experimental investigation of the properties and characteristics of submerged swirling jets with a large swirling intensity W_0> 1 has been performed. On the basis of available data, the correlations between the size of the RFZ, precession frequencies of the swirling jet and swirling intensity have been generalized and analyzed. It is shown that the criterion dependences proposed in [2] can be applied not only to swirling devices such as atomizers, but also for other types of swirling devices. Based on the features of the spectrum of pressure and velocity fluctuations in highly swirling jets it is shown that acoustic pressure fluctuations are formed for a swirling jet at distances R/D>10. And the maximum intensity for a typical highly swirling jet has a noise in the far-field at the angles of 50° -70° from the device axis.

Acknowledgments

Research has been supported by the RFBR grant No. 17-01-00213.

References

[1] Chigier N and Chervinsky A 1965 *Isr. J. Technol.* vol **4** pp 44–54
[2] Abramovich G N 1984 *Theory of Turbulent Jets* (Moscow: Nauka)
[3] Gupta A K, Lilley D G and Syred N 1984 *Swirl Flows* (Tunbridge Wells: Abacus Press)
[4] Knysh Yu A and Uryvsky A F 1984 *Izv. Vuzov. Aviat. Tekh.* vol **3** pp 41–44
[5] Krasheninnikov S Yu, Mironov A K, Toktaliev P D, Zakharov D L, Maslov V P 2014 *Fluid Dyn.* **49**(1) 51–62
[6] Maslov V P and Mironov A K 2005 *Proc. Int. Conf. «Heat and Mass Transfer and Hydrodynamics in Swirling Flows»* (March Moscow)
[7] Krasheninnikov S Yu, Maslov V P, Mironov A K, Toktaliev P D 2016 *TsAGI Sci. J.* **47**(2) 111-37
[8] Krasheninnikov S Yu, Lyubimov D A, Mironov A K, Pudovikov D E and Toktaliev P D 2014 *TsAGI Sci. J.* **45**(3) 183–213
[9] Monkewitz P A, Huerre P, Chomaz J-M 1993 *J. Fluid Mech.* **251** 1–20
[10] Michalke A 1999 *Eur. J. Mech.* **18** 3–12

[11] Toktaliev P D 2011 *Proc. Int. Conf. «Heat and Mass Transfer and Hydrodynamics in Swirling Flows»* vol 1 (October Moscow)
[12] Oberleithner K, Sieber M, Nayeri C N, Paschereit C O, Petz C, Hege H-C, Noack B R, Wygnanski I 2011 *J. Fluid Mech.* **679** 383-414
[13] Doormaal J P, Raithby G.D 1984 *Numer. Heat Transfer* **7**(2) 147-63
[14] Krasheninnikov S Yu 1971 *Fluid Dyn.* **6**(6) 1039-45
[15] Spalart P R, Jou W H, Strelets M, Allmaras S 1997 *Advances in DNS/LES* pp 137-148
[16] Shur S, Spalart P, Strelets M, Travin A 2008 *Int. J. Heat Fluid Flow* **29** 1638-49
[17] Krasheninnikov S Yu, Mironov A K, Pudovikov D E, Toktaliev P D 2015 *Fluid Dyn.* **50**(3) 371-86

IOP Conf. Series: Journal of Physics: Conf. Series **980** (2018) 012018 doi:10.1088/1742-6596/980/1/012018

Vortex Structure Effects on Impingement, Effusion, and Cross Flow Cooling of a Double Wall Configuration

P M Ligrani[1]

[1] Eminent Scholar in Propulsion, Professor of Mechanical and Aerospace Engineering, Department of Mechanical and Aerospace Engineering, University of Alabama in Huntsville, Huntsville, Alabama 35899 USA. PHONE: 256-824-5173. FAX: 256-824-7205.

Email: pml0006@uah.edu

Abstract. A variety of different types of vortices and vortex structures have important influences on thermal protection, heat transfer augmentation, and cooling performance of impingement cooling, effusion cooling, and cross flow cooling. Of particular interest are horseshoe vortices, which form around the upstream portions of effusion coolant concentrations just after they exit individual holes, hairpin vortices, which develop nearby and adjacent to effusion coolant trajectories, and Kelvin-Helmholtz vortices which form within the shear layers that form around each impingement cooling jet. The influences of these different vortex structures are described as they affect and alter the thermal performance of effusion cooling, impingement cooling, and cross flow cooling, as applied to a double wall configuration.

1. Introduction

To provide insight into the effects of different vortex structures, as they affect and alter the thermal performance of effusion cooling, impingement cooling, and cross flow cooling, as applied to a double all configuration, new cooling data are provided using a unique experimental facility which is designed and assembled at the Propulsion Research Center of the University of Alabama in Huntsville. The facility is designed to provide full coverage film cooling effectiveness data and impingement cooling effectiveness data, wherein the film cooling flow is supplied using either cross flow, impingement flow, or a combination of both together. The facility uses three independent flow channels to provide double wall cooling arrangements which model the same configurations from operating gas turbine engines. Presented are new data which illustrate the performance of a sparse effusion hole array which is supplied by two different configurations: (i) a cross flow arrangement, and (ii) an array of impingement jets.

For the effusion cooled surface, presented are spatially-resolved distributions of surface adiabatic film cooling effectiveness, and surface heat transfer coefficients (measured using transient techniques and infrared thermography). For the impingement or cross flow cooled surface, presented are spatially-resolved distributions of surface Nusselt numbers (measured using steady-state liquid crystal thermography). Of interest are techniques which enhance the cooling capabilities of effusion cooling, especially at cool side locations between entrances to individual effusion cooling holes. To produce

Content from this work may be used under the terms of the Creative Commons Attribution 3.0 licence. Any further distribution of this work must maintain attribution to the author(s) and the title of the work, journal citation and DOI.
Published under licence by IOP Publishing Ltd

this cool side augmentation, impingement jet arrays at different jet Reynolds numbers, from 7930 to 18000, are employed. Experimental data are given for a ratio of jet-to-target plate distance to effusion hole diameter of 14, impingement plate thickness of 3.0 effusion hole diameters, and spanwise and streamwise impingement hole spacing such that coolant jet hole centerlines are located midway between individual effusion hole entrances. For the effusion cooling, streamwise hole spacing and spanwise hole spacing (normalized by effusion hole diameter) are 15 and 4, respectively. Effusion hole angle is 25 degrees, and effusion plate thickness is 3.0 effusion hole diameters. Considered are overall effusion blowing ratios from 3.3 to 7.4, with subsonic, incompressible flow.

Experimental data are presented for a freestream velocity variation, which is adjacent to the effusion cooled boundary layer, which is approximately constant with streamwise distance. With this configuration, the ratio of the inlet area to the exit area of the main stream flow duct is 1. Of particular interest are effects of impingement jet Reynolds number, effusion blowing ratio, streamwise development, and mainstream Reynolds number. Also included are comparisons of impingement jet array cooling results with results associated with cross flow supply cooling. As such, the present experimental results are employed to expand the range of applicability of current correlations, and for development of new design correlations for experimental conditions not previously considered. Also of interest are the effects of a variety of different types of vortices and vortex structures on thermal protection, heat transfer augmentation, and cooling performance of impingement cooling, effusion cooling, and cross flow cooling.

2. Experimental Apparatus and Procedures

Details on the experimental apparatus and procedures which are employed are provided by Rogers et al. [4] and Ren et al. [5,6]. Those discussions provide a complete description of relevant measurement details for the presently presented results. Figure 1 shows a side, cross-sectional view of the test section, including optical instrumentation arrangements, from Rogers et al. [4]. Here, the film cooling plate is double-wall cooled, with effusion cooling on the hot, mainstream side, and either cross flow or impingement cooling on the cold side.

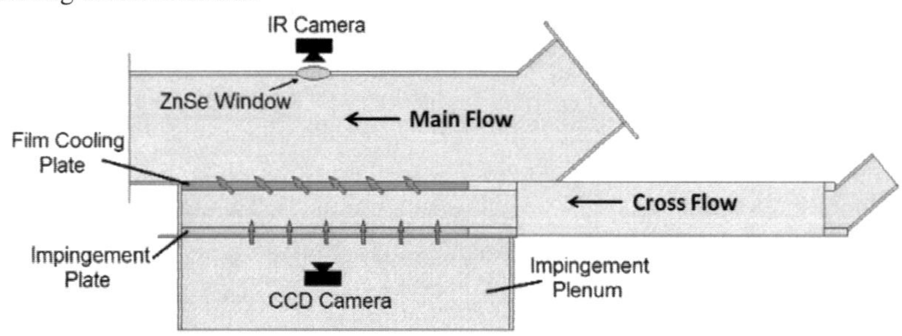

Figure 1. Side, cross-sectional view of the test section, including optical instrumentation arrangements, from Rogers et al. [4].

3. Experimental Results With Cross Flow Cooling – Effusion Cooled / Hot Surface of Effusion Plate

The data within this section are obtained with cross flow cooling. Presented first are measured results for the effusion cooled / hot surface of the effusion plate.

Figure 2 presents the surface distribution of adiabatic film cooling effectiveness for a main flow velocity of 5 m/s, a Reynolds number of about 97,267, and an average blowing ratio of 3.4. The main flow temperature is 306K. These data are obtained for an extent of the test surface which spans more than four rows of effusion holes. The data in Figure 2 show that adiabatic effectiveness values are lower between and away from the holes and higher downstream of the film cooling holes. Figure 2 also shows overall increases in adiabatic effectiveness as x/d_e increases from 0 to 70, due to increased accumulations of film near to the downstream parts of the test surface.

Figure 3 gives a surface distribution of heat transfer coefficient in dimensional form for a main flow velocity of 5 m/s, a Reynolds number of about 93,642, and an average blowing ratio of 5.5. Here, heat transfer coefficients range from approximately 20 W/m^{2}K to 140 W/m^{2}K along the test surface. Higher heat transfer coefficient values are generally observed just upstream and around the holes. Lower heat transfer coefficient values are observed away from the holes. These variations are tied local magnitudes of three-dimensional turbulent transport, which is associated with jet mixing and shear near jet edges.

Within Figures 4 and 5, the effects of varying blowing ratio on adiabatic film cooling effectiveness distributions and heat transfer coefficient distributions are presented and described for a sparse effusion cooling hole array supplied by a cross flow arrangement. The data in Figure 4 show that line-averaged adiabatic cooling effectiveness generally increases as blowing ratio increases, when compared at a particular x/d_e location, indicating that surface protection is improved as film concentration increases near the surface. Also evident are larger variations with blowing ratio BR, and overall increases of adiabatic effectiveness, as the flow advects downstream, and x/d_e increases. The average blowing ratios for the present data are 3.4, 4.4, 5.5, 6.4, and 7.3.

The data in Figure 5 show that line-averaged heat transfer coefficient values also generally increase as the blowing ratio increases from 3.4 to 7.3, when compared at a particular x/d_e location. Lower values of heat transfer coefficient indicate less mixing between the film cooling gas and the hot main flow gas, which leads to better thermal protection of the surface from the hot main flow gas.

4. Experimental Results With Cross Flow Cooling – Cross Flow Cooled / Cold Surface of Effusion Plate

The data within this section are obtained with cross flow cooling. Presented here are measured results for the cross flow cooled / cold surface of the effusion plate.

Figure 6 gives the local, spatially resolved Nusselt number distribution with main flow velocity of 5 m/s, main flow Reynolds number of approximately 95,000, and a blowing ratio of 3.4. Figure 7 then presents the streamwise variation of line-averaged Nusselt numbers. Here, line-averaged Nusselt numbers increase with the blowing ratio from 3.4 to 6.4, and then remain approximately constant as the blowing ratio increases from 6.4 to 7.3, when compared at a particular x/de location. Note the increases of local and line-averaged Nusselt numbers in the vicinity of effusion hole entrances. Such behavior is due to increases of local flow advection speeds as cross flow fluid approaches and then enters into effusion hole entrances. Here, line-averaged Nusselt numbers are determined as the average of local, spatially resolved Nusselt numbers at a constant streamwise (x/d_e) location, for a range of spanwise locations from $y/d_e = 0$ to $y/d_e = 8.2$.

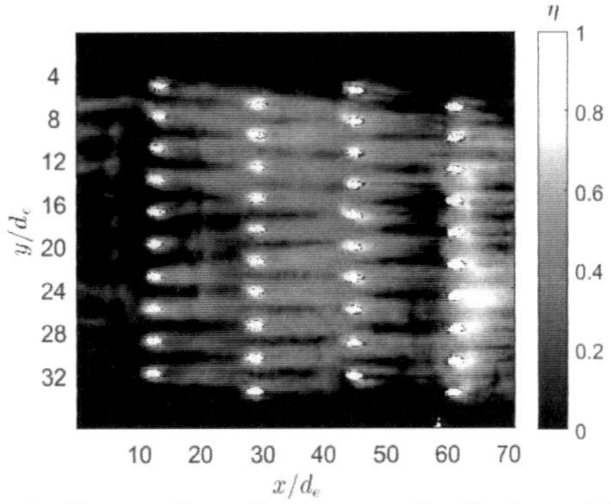

Figure 2. Local, spatially-resolved surface adiabatic film cooling effectiveness distribution with main flow velocity of 5 m/s and cross flow coolant supply, with a Reynolds number of 97,267, a main flow temperature of 306 K, and a blowing ratio of 3.4, for hot / effusion cooled side of effusion plate, with coolant supplied by cross flow [5].

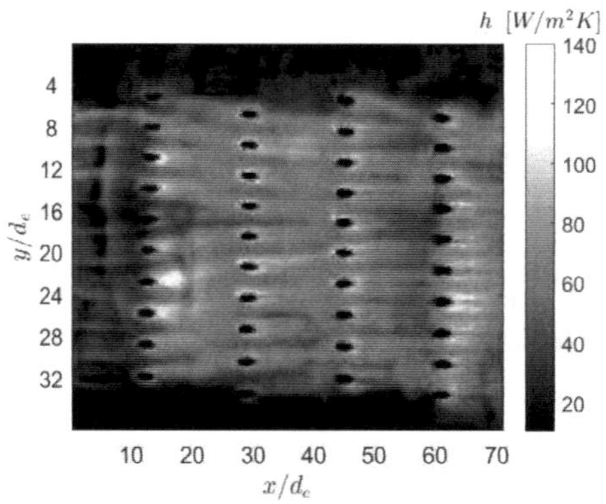

Figure 3. Local, spatially-resolved surface heat transfer coefficient distribution with main flow velocity of 5 m/s, with a Reynolds number of around 93,642, a main flow temperature of 307 K, and a blowing ratio of 5.5, for hot / effusion cooled side of effusion plate, with coolant supplied by cross flow [5].

5. Vortex Structure Associated With a Single Film Cooling Jet in Cross Flow

Figures 8 and 9 show schematic diagrams of the vortices, which are associated with a single film cooling jet in cross flow. Included are jet wake vortices, jet shear-layer vortices, horseshoe vortices, and counter-rotating vortices (or kidney vortices). Note that an upwash region is located between the two vortices within the counter-rotating vortex pair. This upwash region often results in advection of film coolant concentrations away from the test surface, a phenomenon referred to as film coolant lift-off. One consequence of lift-off is decreased film coolant concentrations, and decreased thermal protection near surfaces, at locations downstream of film cooling holes. According to Walters and Leylek [7], this counter-rotating secondary flow vortex structure downstream of the jet exit is the most significant mechanism affecting filrn cooling performance. The dominant vorticity within this structure emanates from the film hole. Its origin is a result of streamwise oriented vorticity contained within the boundary layers which develop within individual film holes.

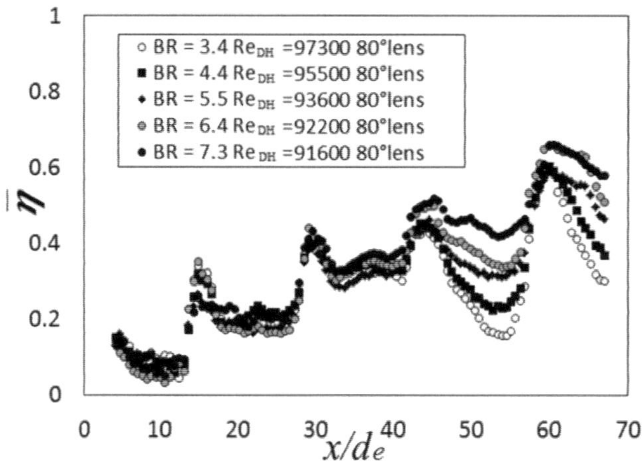

Figure 4. Streamwise variation of line-averaged adiabatic film cooling effectiveness with main flow velocity of 5 m/s, with a Reynolds number of around 95000, main flow temperature of 307 K, and the blowing ratios of 3.4, 4.4, 5.5, 6.4, and 7.3, for hot / effusion cooled side of effusion plate, with coolant supplied by cross flow [5].

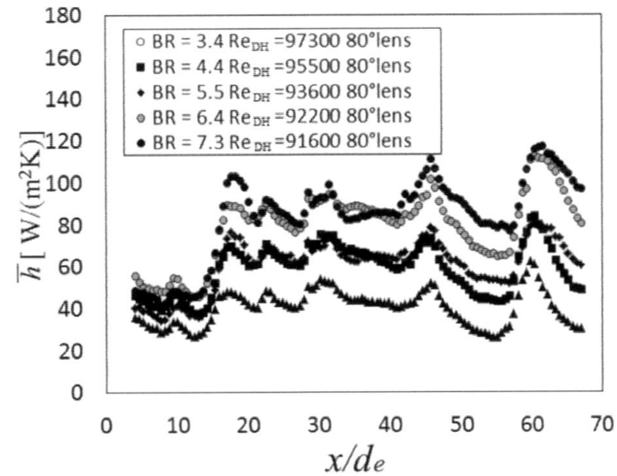

Figure 5. Streamwise variation of line-averaged heat transfer coefficient with main flow velocity of 5 m/s, with a Reynolds number of around 95000, main flow temperature of 307 K, and the blowing ratios of 3.4, 4.4, 5.5, 6.4, and 7.3, for hot / effusion cooled side of effusion plate, with coolant supplied by cross flow [5].

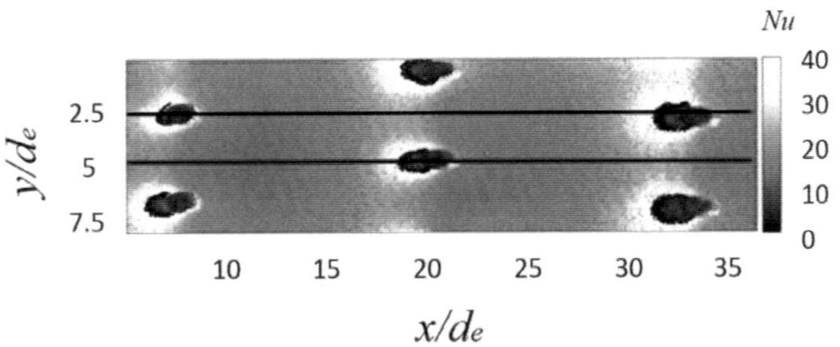

Figure 6. Local, spatially resolved Nusselt number distribution with main flow velocity of 5 m/s, main flow Reynolds number of approximately 95,000, crossflow temperature of 294 K, and blowing ratio of 3.4, for coolant / cross flow side of effusion plate, with coolant supplied by cross flow [5].

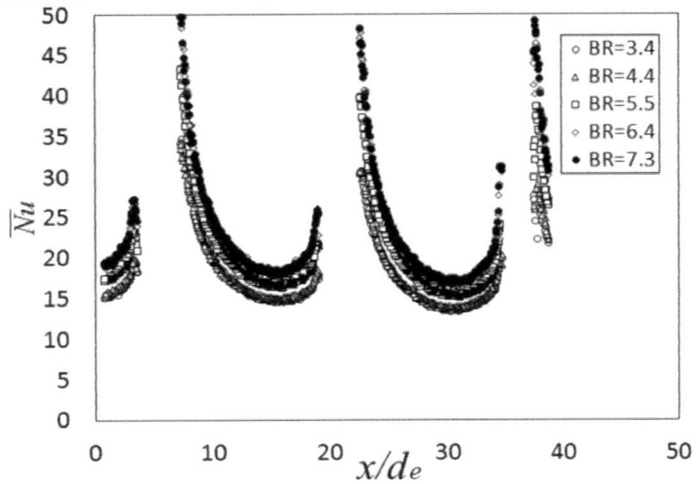

Figure 7. Streamwise variation of line-averaged, spatially resolved Nusselt number for different blowing ratios with main flow velocity of 5.1 m/s, main flow Reynolds number of approximately 95,000, and crossflow temperature of 294 K, for coolant / cross flow side of effusion plate, with coolant supplied by cross flow [5].

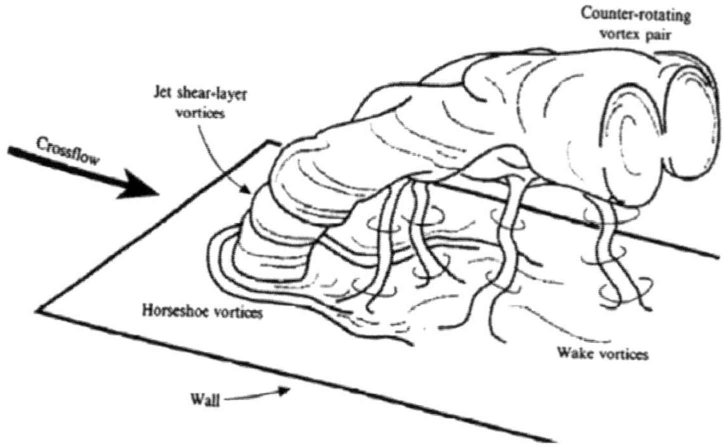

Figure 8. Different vortex structures associated with a single film cooling jet.

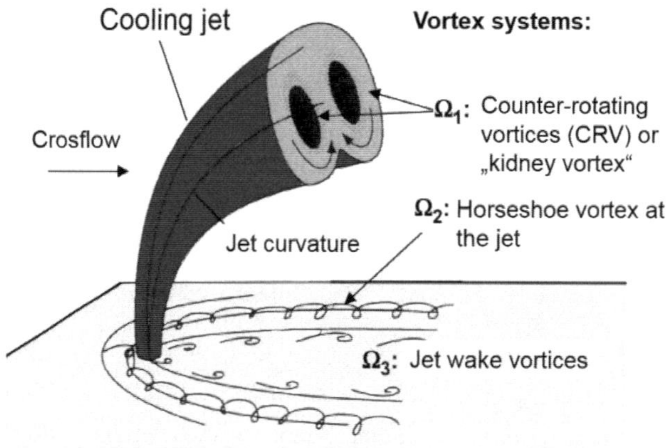

Figure 9. Film cooling jet vortex systems from Margason [3].

6. Experimental Results With Impingement Jet Array Cooling – Impingement Cooled / Cold Surface of Effusion Plate

The data within this section are obtained with impingement jet array cooling. Presented first are measured results for the impingement cooled / cold surface of the effusion plate.

Figure 10 provides a comparison of local, spatially-resolved surface Nusselt number variations for the coolant / cross flow side of the effusion plate, for different blowing ratios, for Re_{MS}=142,000-155,000. Here, data are given for BR values of 3.3, 4.3, 5.5, 6.3, and 7.4. Note that x/de = 0 is defined at the upstream edge of the area of present spatially-resolved measurements. Also included within Figure 10a are locations of the impingement holes as well as the effusion hole entrances.

According to the results within Figure 10, regardless of the value of blowing ratio BR, the highest local Nusselt numbers are present at smaller x/de locations, relative to the locations of the impingement hole centerlines. Such Nu augmentation regions are associated with impingement jet stagnation locations, which are positioned at different x/de locations, relative to the impingement holes. This is because of turning and re-direction of the impingement jets as they cross the impingement passage. The re-direction is a result of the static pressure variations within the passage, with lowest values near the entrances of the effusion holes. With this physical situation, coolant within impingement jets exits the impingement holes, turns to smaller x/de locations, and then impacts upon the coolant side of the effusion plate. Afterwards, the coolant is then re-directed along the plate surface. Afterwards, it enters into individual effusion holes, which are also located at smaller x/de locations, relative to adjacent impingement hole locations.

The resulting coolant advection trajectory is likely aided by secondary flows within the impingement cooling passage. According to Al Dabagh, et al. [1], internal flows within the impingement jet passage are complex, and are dominated by the counter-rotating vortices between individual impingement jets, with flows directed opposite to the jets, at locations between individual jets. Cho and Rhee [2] describe similar jet interactions, with secondary vortices which form between these strong primary vortices. According to these investigators, the strength and complexity of such interactions change dramatically, as the locations of impingement and effusion holes are altered, relative to each other.

Figure 11 shows line-averaged surface Nusselt numbers (determined from the results given in Figure 10), as they vary with streamwise development for different blowing ratios for Re_{MS}=142,000-155,000. Here, line-averaged magnitudes are determined over y/de values from 0.0 to 8.0. Within this figure, solid rectangles denote effusion hole entrance streamwise locations, and dashed rectangles denote impingement hole streamwise locations. Here, line-averaged Nusselt numbers often increase as the blowing ratio increases from 3.3 to 7.4, when compared at a particular x/de location. As for the results which are presented within Figure 10, Figure 11 shows that highest local Nusselt numbers are present at smaller x/de locations, relative to the locations of the impingement hole centerlines. The lowest local Nusselt numbers are then present at and near effusion hole entrance locations. Responsible is the turning of the impingement jets through the impingement passage, the collection of coolant along the surface (at x/de locations just larger relative to effusion hole entrances), followed by the entrance of the coolant into individual adjacent effusion holes.

Figure 12 then compares spatially-averaged surface Nusselt numbers, as they vary with streamwise development. These data are also presented for the coolant / cross flow side of the effusion plate, for different blowing ratios and Re_{MS}=142,000-155,000. Each spatial-average is determined over one complete period of impingement array hole pattern, which is equivalent to a streamwise extent of 15 effusion hole diameters. Here, spatially-averaged Nussult numbers associated with x/de=42.5 increase continually as the blowing ratio increases from 3.3 to 7.4. Values associated with x/de=12.5 and x/de=27.5 increase as the blowing ratio increases from 3.3 to 5.5. However, afterwards, spatially-averaged Nussult numbers are nearly invariant as the blowing ratio increases further from 5.5 to 7.4.

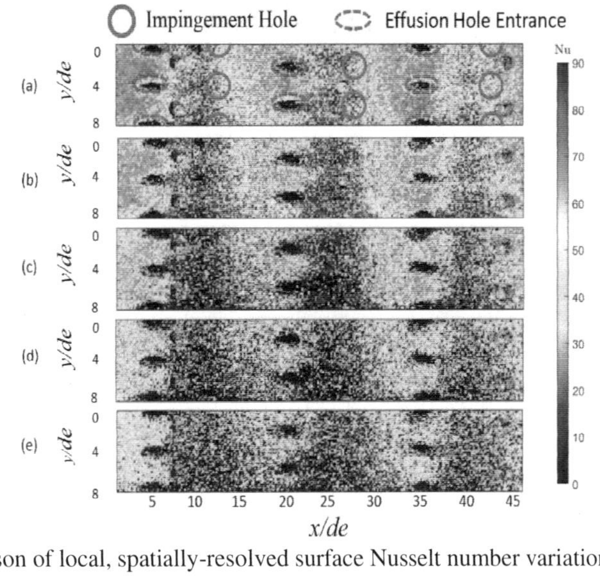

Figure 10. Comparison of local, spatially-resolved surface Nusselt number variations for different blowing ratios for Re_{MS}=142,000-155,000 for coolant side of effusion plate, with coolant supplied by an impingement jet array [6]. (a) BR=3.3. (b) BR=4.3. (c) BR=5.5. (d) BR=6.3. (e) BR=7.4.

\

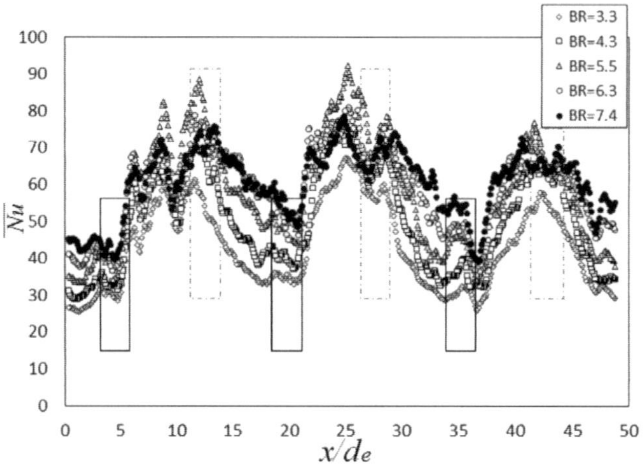

Figure 11. Comparison of line-averaged surface Nusselt numbers with streamwise development for different blowing ratios for Re_{MS}=142,000-155,000 for coolant side of effusion plate, with coolant supplied by an impingement jet array [6]. Solid rectangles denote effusion hole entrance streamwise locations. Dashed rectangles denote impingement hole streamwise locations.

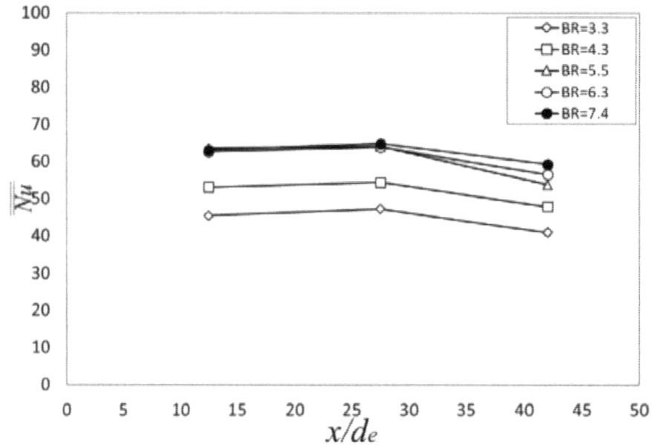

Figure 12. Comparison of spatially-averaged surface Nusselt numbers with streamwise development for different blowing ratios for Re_{MS}=142,000-155,000 for coolant side of effusion plate, with coolant supplied by an impingement jet array [6].

7. Experimental Results With Impingement Jet Array Cooling – Effusion Cooled / Hot Surface of Effusion Plate

The data within this section are obtained with impingement jet array cooling. Presented are measured results for the effusion cooled / hot surface of the effusion plate.

Figure 13 gives surface, local heat transfer coefficient variations in dimensional form for BR=7.4 and Re_{MS}=142,000. Here, heat transfer coefficients range from approximately 20 W/m^2K to 140 W/m^2K along the test surface. Higher heat transfer coefficient values are generally observed just upstream, around, and along a trajectory downstream of each hole. Lower heat transfer coefficient values are generally present away from the holes. These variations are tied to jet mixing, vortex development around jet coolant concentrations, increased shear near jet edges, and the resulting augmentations of local three-dimensional turbulent transport.

Figure 14 presents distributions of surface, local adiabatic film cooling effectiveness variation for BR=6.3 and Re_{MS}=146,000. These data show that adiabatic effectiveness values are increased along the entire test plate, including around and downstream of the film cooling holes. The overall high values of adiabatic effectiveness are due to abundant effusion coolant, resulting from efficient insertion of coolant at the entrance of each effusion hole from an adjacent impingement jet.

Figure 15 shows line-averaged, dimensional heat transfer coefficient values, as they vary with streamwise development, for different blowing ratios for Re_{MS}=142,000-155,000. Here, line-averaged values are determined over y/d_e from 13 to 25. These data are provided for the effusion cooled / hot surface of the effusion plate for a sparse effusion cooling hole array supplied by an impingement jet array. Here, line-averaged heat transfer coefficient values increase as the blowing ratio increases from 3.3 to 7.4, when compared at a particular x/d_e location. Lower values of heat transfer coefficient indicate less mixing between the film cooling gas and the hot main flow gas, which leads to better thermal protection of the surface from the hot main flow gas.

Figure 16 shows the variation of line-averaged adiabatic film cooling effectiveness values with streamwise development for different blowing ratios for Re_{MS}=142,000-155,000. Within this figure, values given in Figure 16a represent coolant stagnation temperatures determined using cross flow stagnation conditions, and values given in Figure 16b represent coolant stagnation temperatures determined using impingement plenum stagnation conditions. For both sets of results, line-averaged adiabatic cooling effectiveness values generally increase as blowing ratio increases from 3.3 to 5.5, when compared at a particular x/d_e location, indicating that surface protection is improved as film concentration increases near the surface. However, as the blowing ratio increases from 5.5 to 6.3, and then to 7.4, less variation with blowing ratio is evident. This is believed to be due to abundant coolant along and near to the test surface, which is provided by efficient insertion of coolant into effusion hole entrances by individual impingement jets.

8. Vortex Structure Associated With a Single Impingement Jet

Figures 17 and 18 show numerically-predicted, and experimentally measured development of Kelvin-Helmholtz vortices, which are associated with a single jet. The results in Figure 17 are provided for a turbulent jet, whereas the results in Figure 18 are provided for a laminar jet.

Figure 17 shows vortex formation occurs along the edge of a free axisymmetric jet. This is caused by Kelvin-Helmholtz instability from the shear force between the high velocity air jet and the surrounding stagnant air. Small vortices are created close to the nozzle and grow in size as they are carried along the flow until they are big enough to fully interrupt the jet. The vortices then collapse creating a fully turbulent flow. This visualization is produced by injecting smoke into the nozzle air flow as the fluid is forced through a 100 mm diameter nozzle. A laser sheet provides a two-dimensional view of the flow. The associated turbulent Reynolds number is approximately 10,000.

Figure 18 presents numerical prediction results for laminar jet flow with Kelvin-Helmholtz instability and vortices. Provided are contour variations of specific entropy through the center of the computational domain, at times of 0.0, 3.0, 6.0 and 10.0 seconds, with Kelvin-Helmholtz dynamic time scale of 3.4 seconds.

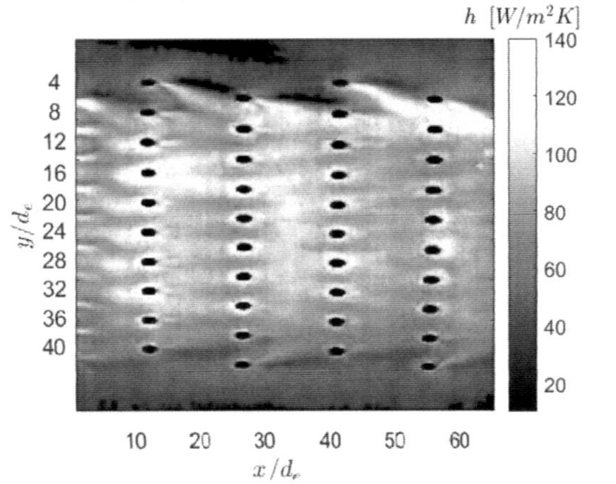

Figure 13. Surface, local heat transfer coefficient variation for BR=7.4 and Re_{MS}=142,000 for hot / effusion cooled side of effusion plate, with coolant supplied by an impingement jet array [6].

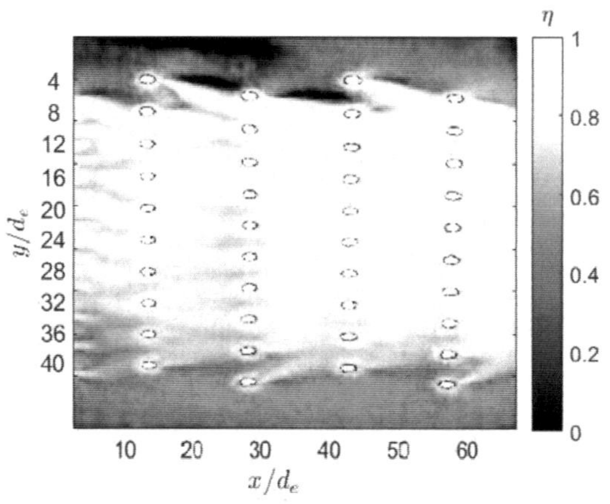

Figure 14. Surface, local adiabatic film cooling effectiveness variation for BR=6.3 and Re_{MS}=146,000 for hot / effusion cooled side of effusion plate, with coolant supplied by an impingement jet array [6].

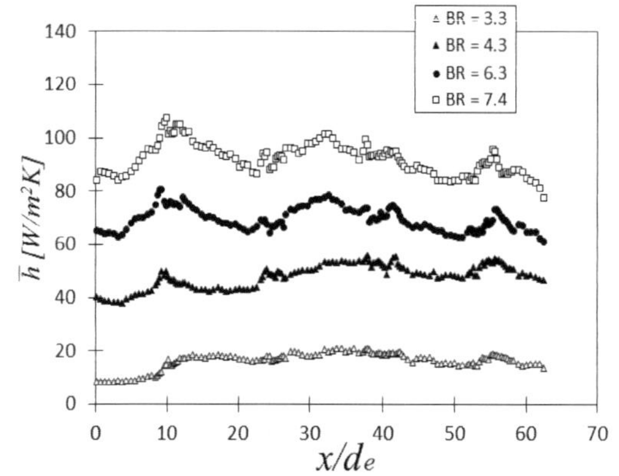

Figure 15. Comparisons of line-averaged heat transfer coefficient values with streamwise development for different blowing ratios for Re_{MS}=142,000-155,000 for hot / effusion cooled side of effusion plate, with coolant supplied by an impingement jet array [6].

According to Lee et al. [8], local and spatially-averaged Nusselt numbers beneath impingement jets with different jet-to-target plate distances are generally the result of the competing influences of two different phenomena: (i) the coherence of individual jets and the strength of adjacent shear layers, and (ii) the development and advection of vortices generated by Kelvin-Helmholtz instabilities. The coherence of individual jets and the intensity of the adjacent shear layers, are strongest just after each jet emerges from an impingement plate hole. With this situation, maximum local jet velocities are distributed over the central part of each jet, with the largest gradients of velocity present in the thinnest shear layers which surround each jet. As each jet advects downstream and surrounding fluid is entrained to become part of each jet, the local velocity profile becomes more rounded with adjacent hear layer velocity gradients which are less intense as jet fluid diffuses and advects in lateral directions. If such jets impinge upon a surface, the highest local and spatially-averaged surface Nusselt number augmentations are expected to be present just after jets emerge from originating holes, when they are most coherent with the most intense adjacent shear layers [8].

Another mechanism for local heat transfer augmentation is formation, development, and surface impact of Kelvin-Helmholtz generated vortices. In contrast to the diminishing jet coherence with streamwise development, these vortices continue to develop as they advect in the streamwise direction. Their ability to augment thermal transport is a result of the mixing they induce between the jet and surrounding non-jet fluid. The resulting gradients of density (if the flows are compressible) and stagnation temperature within the vortices are key elements in their capability to augmention local surface heat transfer coefficients and Nusselt numbers [8].

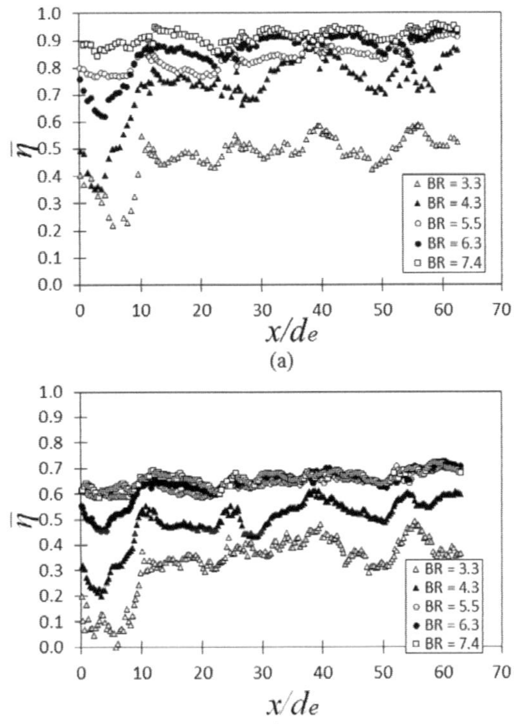

Figure 16. Comparisons of line-averaged adiabatic film cooling effectiveness with streamwise development for different blowing ratios for Re_{MS}=142,000-155,000 for hot / effusion cooled side of effusion plate, with coolant supplied by an impingement jet array [6]. (a) Effectiveness based upon stagnation cross flow conditions. (b) Effectiveness based upon stagnation impingement plenum conditions.

The relative influences of the jet/shear layer coherence and Kelvin-Helmholtz generated vortices are reflected in the *Z/D* values associated with the highest Nusselt numbers. For higher jet Reynolds numbers Re_j, jet/shear layer coherence seems to be more important, since *Z/D*=1.5 gives optimal surface Nusselt numbers for an impingement array. For lower Reynolds numbers, the Kelvin-Helmholtz vortices have greater influences since *Z/D* from 3.0 to 5.0 give optimal impingement Nusselt numbers [8].

9. Comparisons of Experimental Results With Impingement Jet Array Cooling and Cross Flow Cooling – Effusion Cooled / Hot Surface of Effusion Plate

Figure 19 shows comparisons of line-averaged heat transfer coefficient values with streamwise development for impingement flow coolant supply and cross flow coolant supply [4]. Figure 20 then shows comparisons of line-averaged adiabatic film cooling effectiveness with streamwise development for impingement flow coolant supply and cross flow coolant supply. Here, data in Figures 19a and 20a are provided for BR=4.3, and data in Figures 19b and 20b are provided for BR=6.3. These data comparisons are provided for main flow Reynolds numbers of 142000 to 155000.

When compared at particular values of streamwise location x/de and blowing ratio BR, results in Figures 19a, 19b, 20a, and 20b generally show lower heat transfer coefficients, and significantly higher adiabatic effectiveness for the impingement supply arrangement, relative to the cross flow supply arrangement. Thus, for the same Reynolds number, blowing ratio, and streamwise location, significantly increased thermal protection is provided when the effusion coolant is provided by an array of impingement cooling jets. Within the present investigation, this is due to increased abundance of effusion coolant along the test surface, which is provided by efficient insertion of coolant into effusion hole entrances by individual impingement jets.

10. Summary and Conclusions

A variety of different types of vortices and vortex structures form near to and around individual effusion coolant concentrations, including jet wake vortices, jet shear-layer vortices, horseshoe vortices, and counter-rotating vortices (or kidney vortices). Of these different types of vortices, the counter-rotating secondary flow vortex structure downstream of the jet exit is the most important in regard to influences on local and overall film cooling performance. This is a result of the secondary advection and augmented turbulence transport influences of individual vortices within the pair, as well as the upwash region, which is located between the two vortices within the counter-rotating vortex pair.

Local and spatially-averaged Nusselt number magnitudes beneath impingement jets are generally the result of the competing influences of two different phenomena: (i) the coherence of individual jets and the strength of adjacent shear layers, and (ii) the development and advection of vortices generated by Kelvin-Helmholtz instabilities. The coherence of individual jets and the intensity of the adjacent shear layers, are strongest just after each jet emerges from an impingement plate hole. With this situation, maximum local jet velocities are distributed over the central part of each jet, with the largest gradients of velocity present in the thinnest shear layers which surround each jet. In contrast to the diminishing jet coherence with streamwise development, Kelvin-Helmholtz generated vortices continue to develop as they advect in the streamwise jet direction. Their ability to augment thermal transport is a result of the mixing they induce between the jet and surrounding non-jet fluid. The resulting gradients of density (if the flows are compressible) and stagnation temperature within the vortices are thus key elements in their capability to augment local surface heat transfer coefficients and Nusselt numbers, as impingement jets impact upon a target surface.

Within the present paper, the influences of these different vortex structures are described as they affect and alter the thermal performance of effusion cooling, impingement cooling, and cross flow cooling, as they are employed within a double wall cooling configuration. Presented are data for a sparse effusion cooling hole array. Non-dimensional streamwise and spanwise film cooling hole spacing, relative to effusion hole diameter, are 15 and 4, respectively. The facility is designed to provide full coverage film cooling data and impingement cooling effectiveness data, wherein the film cooling flow is supplied using either cross flow, impingement flow, or a combination of both together. Here, data are given with a cross flow supply only (with no impingement cooling), as well as for an impingement jet array supply only (with no cross flow). For both arrangements, experimental results are provided for 5 different crossflow Reynolds numbers, which are associated with 5 different values of blowing ratio BR. Presented are distributions of surface adiabatic cooling effectiveness and surface heat transfer coefficients for the hot side of the effusion test plate, and spatially-resolved distributions of surface Nusselt numbers for the coolant side of the effusion test plate.

With a cross flow only coolant supply, for the hot mainstream / effusion cooled side of the test plate, line-averaged adiabatic film cooling effectiveness generally increases, and line-averaged heat transfer coefficient also generally increases, as the blowing ratio becomes larger, when compared at a particular streamwise location and mainstream Reynolds number. On the coolant / cross flow side of the effusion test plate, increases of local and line-averaged Nusselt numbers occur in the vicinity of effusion hole entrances, due to increases of local flow advection speeds as cross flow fluid approaches and then enters into effusion hole entrances.

With an impingement jet array coolant supply, for the hot mainstream / effusion cooled side of the test plate, when compared at particular values of streamwise location and blowing ratio BR, lower heat transfer coefficients, and significantly higher adiabatic effectiveness are often present for the impingement supply arrangement, relative to the cross flow supply arrangement. Thus, for the same impingement jet Reynolds number, mainstream Reynolds number, blowing ratio, and streamwise location, significantly increased thermal protection is provided when the effusion coolant is provided by an array of impingement cooling jets (compared to a cross flow channel supply arrangement). On the coolant / cross flow side of the effusion test plate, regardless of the value of blowing ratio BR, the highest local Nusselt numbers are present at smaller x/de locations, relative to the locations of the impingement hole centerlines. Such Nusselt number augmentation regions are associated with impingement jet stagnation locations, and re-positioning of individual impingement jets, relative to impingement hole exits, from turning and re-direction of coolant concentrations as they advect across the impingement passage.

11. Acknowledgements
The research presented in this paper was sponsored by Solar Turbines Inc., located in San Diego, California, USA.

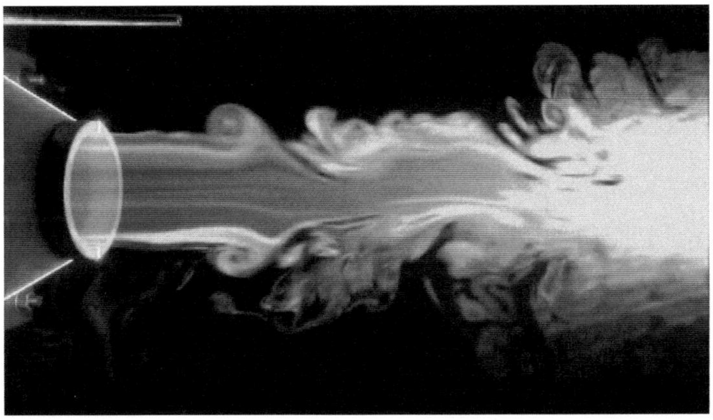

Figure 17. Visualization of turbulent jet induced Kelvin-Helmholtz vortices, including initial formation and subsequent vortex distortion and break-up.

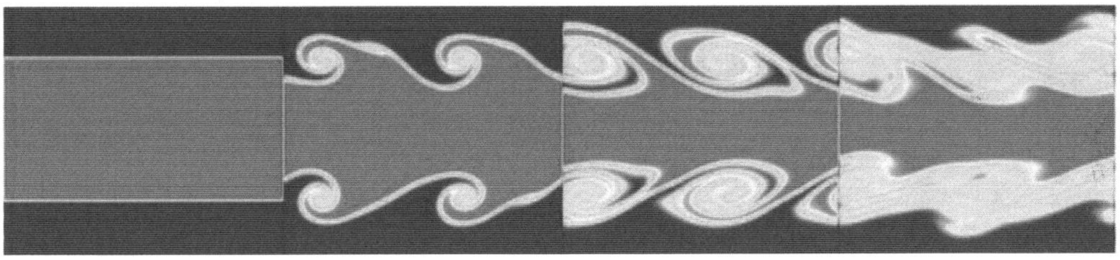

Figure 18. Numerical prediction of laminar jet flow with Kelvin-Helmholtz instability and vortices. Contour variations of specific entropy through the centre of the computational domain at times of 0.0, 3.0, 6.0 and 10.0 seconds, with Kelvin-Helmholtz dynamic time scale of 3.4 seconds.

12. Nomenclature

BR	film cooling blowing ratio, $\frac{\rho_{ef}}{\rho_{ms}}\frac{v_{ef}}{v_{ms}}$
DH	hydraulic diameter
d_e	film cooling effusion hole diameter
h	heat transfer coefficient
$\bar{h}$	line-averaged heat transfer coefficient
Nu	local Nusselt number
$\overline{Nu}$	line-averaged Nusselt number
$\overline{\overline{Nu}}$	spatially-averaged Nusselt number
Re	Reynolds number
v	time-averaged velocity
x	streamwise coordinate

IOP Conf. Series: Journal of Physics: Conf. Series **980** (2018) 012018 doi:10.1088/1742-6596/980/1/012018

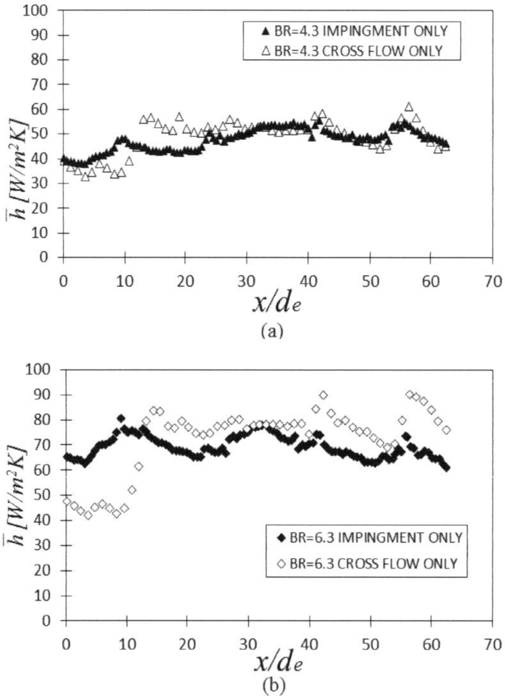

Figure 19. Comparisons of line-averaged heat transfer coefficient values with streamwise development for hot / effusion cooled side of effusion plate for impingement flow coolant supply and cross flow coolant supply [6]. (a) BR=4.3. (b) BR=6.3.

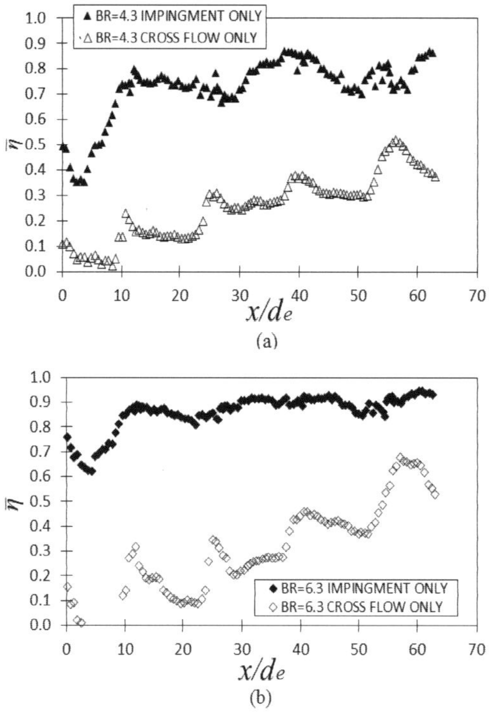

Figure 20. Comparisons of line-averaged adiabatic film cooling effectiveness with streamwise development for hot / effusion cooled side of effusion plate for impingement flow coolant supply and cross flow coolant supply [6]. (a) BR=4.3. (b) BR=6.3.

y	spanwise coordinate
Greek symbols	
η	adiabatic film cooling effectiveness
$\bar{\eta}$	line-averaged adiabatic film cooling effectiveness
ρ	density
Subscripts	
DH	main flow value, determined using hydraulic diameter of main flow passage
ef	effusion jet value
ms	main flow value
MS	main flow value, determined using hydraulic diameter of main flow passage

References

[1] Al Dabagh A M, Andrews G E, Abdul Husain R A A, Husain C I, Nazari A and Wu J 1990 Impingement/effusion cooling: the influence of the number of impingement holes and pressure loss on the heat transfer coefficient *ASME Transactions - Journal of Turbomachinery* **112** (3) 467-476

[2] Cho H H and Rhee D H 2001 Local heat/mass transfer measurement on the effusion plate in impingement/effusion cooling systems *ASME Transactions-Journal of Turbomachinery* **123** (3) 601-608

[3] Margason R J 1993 Fifty years of jet in cross flow research *AGARD CP-534-1*

[4] Rogers N, Ren Z, Buzzard W, Sweeney B, Tinker N, Ligrani P M, Hollingsworth K D, Liberatore F, Patel R, Ho S and Moon H.-K 2016 Effects of double wall cooling configuration and conditions on performance of full coverage effusion cooling *ASME TURBO EXPO 2016: Turbomachinery Technical Conference and Exposition*, Paper Number GT2016-56515, Seoul, South Korea

[5] Ren Z, Vanga S R, Rogers N, Ligrani P M, Hollingsworth K D, Liberatore F, Patel R, Srinivasan R and Ho Y 2017 Internal and external cooling of a full coverage effusion cooling plate: effects of double wall cooling configuration and conditions *ASME TURBO EXPO 2017: Turbomachinery Technical Conference and Exposition*, Paper Number GT2017-64921, Charlotte, North Carolina, USA

[6] Ren Z, Ligrani P M, Liberatore F, Patel R, Srinivasan R and Ho Y 2017 Double wall cooling of a full coverage effusion plate, including internal impingement array cooling *IMECE 2017, ASME International Mechanical Engineering Congress and Exposition*, Paper Number IMECE2017-72066, Tampa, Florida, USA

[7] Walters D K and Leylek J H 1997 A detailed analysis of film-cooling physics part I: streamwise injection with cylindrical holes *International Gas Turbine and Aeroengine Congress and Exhibition*, Paper Number 97-GT-269, Orlando, Florida, USA

[8] Lee J, Ren Z, Haegele J, Potts G, Jin J S, Ligrani P M, Fox M and Moon H.-K 2014 Effects of jet-to-target plate distance and Reynolds number on jet array impingement heat transfer *ASME Transactions-Journal of Turbomachinery* **136** (5) 051013-1 to 051013-13

IOP Conf. Series: Journal of Physics: Conf. Series **980** (2018) 012019 doi:10.1088/1742-6596/980/1/012019

Modeling air and fire non-stationary whirls in laboratory conditions

A Y Varaksin [1,2]

[1] Joint Institute of High Temperatures, RAS, Moscow, 125412, Russia
[2] Bauman Moscow State Technical University, Moscow, 105005, Russia

E-mail: varaksin_a@mail.ru

Abstract. This study demonstrates generation of wall-free non-stationary air and fire whirls under laboratory conditions without using mechanical swirling devices and estimates their integral parameters. A simple experimental facility for the generation of concentrated air and fire vortex structures by creating unstable stratification and solid fuel combustion (urotropine) is described. With the use of photography, some data on the generation conditions and integral parameters (lifetime, height, and diameter) of air and fire whirls have been determined.

1. Introduction

Swirling (vortex) flows are widely spread in nature (atmospheric cyclones, tornadoes, dusty devils) and find numerous engineering applications [1, 2]. Simulating dust devils and tornadoes in laboratory environments is not a new concept. Many laboratory simulator designs were built for meteorological purposes to understand the parameters influencing dust devil and tornado formation [3-9].

In the simulation of a standing tornado near ground, there are two essentials factors: circulation and updraft. In the earliest works by Ying and Chang [3] and Wan and Chang [4] at the Catholic University of America, the circulation was produced by bringing air radially inward through a rotating cylindrical screen, and the updraft was generated by an exhaust fan at the opening of the top hood along its axis. In this way, both the circulation and the updraft can be controlled separately. The strength of circulation is controlled by the rotational speed of the cylindrical screen and the rate of the updraft can be controlled by the exhaust flow rate of the fan.

Ward [5] was the first to simulate main characteristic features of tornado in the laboratory such as a characteristic surface pressure profile, a bulging deformation on the whirl core and multiple whirls in a single convergence system. The apparatus consists of a cylindrical container of 8 ft in diameter in combination with a rotating mesh wire on the side and a fan affixed on the top of the cylinder. The fan power can regulate the volumetric flow rate, simulating the strength of the updraft. The rotation of the mesh wire is intended to simulate the ambient rotation in the atmosphere.

In the later works at the Purdue University [6, 7] the similar tornado generator was used. In the Purdue simulator, the four critical quantities, i.e., depth of inflow, radius of updraft hole, volume flow rate and far-field tangential velocity, are variable over a wide range of values. The primary goal of these studies (like in above mentioned works) is to produce the maximum stable whirl to produce quantitative measurements.

The simulator of wall-free whirls was created in Iowa State University [8, 9]. The fan motor and turning vanes were used to create the vertical flow with dry ice supplied for its visualization. The simulator design allowed moving the generated whirl structure along the ground. So, this enabled testing the interaction of tornado-like vortices with the built structures.

Fire whirls, as compared to general fires, are characterized by significant enhancement in burning rates, flame temperature and flame heights, in addition to a strong swirling motion of the flame itself. These peculiarities allow intensifying the heat and mass transfer in different engineering devices. In a

Content from this work may be used under the terms of the Creative Commons Attribution 3.0 licence. Any further distribution of this work must maintain attribution to the author(s) and the title of the work, journal citation and DOI.
Published under licence by IOP Publishing Ltd

fire whirl, the hot gases generated by the fire itself serve as a fluid sink which entrains the ambient air with angular momentum from the eddy to the vortex core (flame). Fire whirls simulation in a laboratory environment is not a new concept either. Various types of facilities may be used to produce single steady fire whirls at laboratory conditions. They are based on the concepts of generating eddy and a fluid sink. These facilities may be generally classified into two types, depending on whether the generating eddy is imposed mechanically by a rotating screen (Emmons-type or rotating screen type [10-13]) or induced by the entrained air flowing through the well-arranged spiral path (fixed-frame type [14-17]). Being similar in many aspects, since both produce a fire swirl, they differ in details.

The above mentioned experimental works dealing with the stationary whirl flows are convenient for detailed experimental description; however, their characteristics and, especially, behavior may significantly differ from the parameters of real non-stationary whirl structures, observed in Earth atmosphere.

The study of wall-free non-stationary concentrated (the vorticity is localized in space) air and fire vortices is complicated by a number of reasons, such as spontaneity of formation, space-time instability, practically impossible control over the characteristics, and so on.

This paper continues the previous investigations [18-22]. The aim of this work is an experimental study of generation of non-stationary air and fire without using mechanical swirling devices and estimation of their integral parameters.

2. The experimental setups.

2.1. The experimental setup for generation of non-stationary air whirls.

In this paper we use the simple experimental setup for controlled heating of aluminum plate top surface (here referred to as underlying surface) and for generating the wall-free non-stationary swirl structures over the plate due to unstable air stratification (the warm air near the top surface and the cold air over it). It is strictly emphasized that contrary to previous studies no mechanical swirling devices (ventilators, guide swirl vanes, screws, internal spiral ribbing, etc.) or underlying surface rotation were used for vortices generation.

The experimental setup is schematically shown in Figure 1 (left). It was located in a room with floor *1* of 6x6 m^2 in area and 3.3 m high ceiling *2* at a distance of 0.5 m from one of the walls *3*. The experimental setup included a 0.35 m high deck *4* with three legs *5*. The horizontal surface of the deck *4* was a sheet of aluminum (Grade D16AM) with a diameter of 1100 mm and thickness of 1.5 mm. The top (underlying) surface of the aluminum sheet was blackened with heat-resistant paint. The placed under the deck was an electrically ignited gas burner *6* with maximal thermal power of 3.5 kW. The diameter of flame *7* of the burner varied (for different modes of thermal power) from 200 to 300 mm. A liquefied propane-butane mixture required for the operation of the gas burner was placed in a 27-liter vessel *8*.

The experiments were performed in six different thermal modes, characterized by different heating times (1, 2 and 3 minutes), cooling times (10, 15 and 20 minutes) and values of maximal temperature at the underlying surface center (from 420 up 610 K). The "soft" modes (1-3) with relatively low temperatures of aluminum sheet surface were realized at low rates of heating. The "harder" modes (nos. 4-6) are characterized by higher temperatures of underlying, which are attained at high values of heat flux.

This experimental setup enables the controlled heating of the underlying surface of aluminum sheet, which leads to the generation of unsteady vortex structures 9 as a result of development of unstable air stratification. The formed vortex structures were visualized using tracer particles (micrometer-sized particles of magnesia, chemical formula $4MgCO_2Mg(OH)_24H_2O$) or vapor of special easily-boiling fluid (VDLSL5, Velleman company, Belgium) which were applied in a thin layer onto the underlying surface prior to experiments. A digital video camera (Sanyo VCC- 6572P) was used for video filming of the generated vortices.

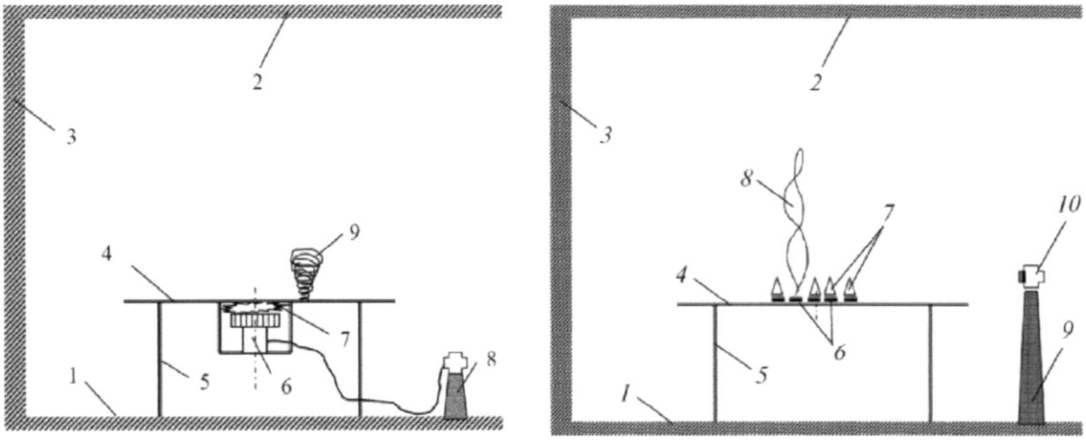

Figure 1. Schematic of the experimental facility. Experimental setup for air whirls generation (left). Experimental setup for fire whirls generation (right).

2.2. The experimental setup for generation of non-stationary fire whirls.

To generate concentrated vortices, we used a simple setup, the layout of which is shown in Figure 1 (right). It was arranged in a room having a floor *1* with an area of 6 x 6 m^2 and ceiling *2* of 3.3 m in height at a distance of 0.5 m from one of the walls *3*. The experimental setup was a 0.35 m high table *4*, which had three legs *5*. The horizontal surface of the table *4* was an aluminum sheet (D16AM brand) with a diameter of 1100 mm and thickness of 1.5 mm. The upper (underlying) surface of the aluminum sheet was blackened with heat resistant paint. Pellets of urotropine *6* (hexamethylenetetramine, chemical formula C6H12N4) were arranged in the central part of the underlying surface before the experiment. The weight of each pellet was 21 g (40 mm in diameter). The combustion heat of urotropine is 30 MJ/kg.

Experiments were carried out with different amounts (1, 7 and 19) of the combustible pellets and different arrangements (distance between the pellets 65 and 130 mm) on the underlying surface.

In the course of the experiment, the pellets were lit forming the flames 7. During combustion of pellets, generation of fire vortices *8* was observed. Their height considerably exceeded the flame height over the fuel arrangement region. Video filming of the combustion process and generating vortex structures was performed using a digital photocamera *10* arranged on a tripod *9*.

3. Results.

3.1. The conditions of generation and integral parameters of non-stationary air whirls.

The frame-by-frame analysis (see figure 2) of video records in different thermal modes provides information about the following parameters of the generation of whirls and their characteristics: (1) the values of temperature, at which the vortexes are generated; (2) the region of underlying surface, where the whirls are generated; (3) the direction of rotation of vortex structure; (4) the number of whirls observed per experiment; (5) the trajectory of travel of the vortex structure base; (6) the length of the vortex base trajectory; (7) the velocity of the vortex base; (8) the lifetime of vortex structure; (9) the height of whirls; (10) the diameter of whirls, etc.

Repeated experiments in different modes gave rise to the following inferences. A stable generation of whirls was observed in all modes except for mode no.1 [20]. Vortex structures began to form in the mode of the underlying surface heating after reaching the temperature of 470 K at its center. The largest whirls were generated at temperatures at the surface center over 570 K. Vortex structures were largely generated in a circular region of abrupt rise of temperature gradient. The direction of the observed rotation of the majority of whirls was counterclockwise (if the underlying surface was viewed from above). Up to ten vortex structures were observed per experiment. Two types of trajectories of vortex base motion were identified. The majority of vortex structures moved in spirals (trajectories of the first type). Some vortexes moved in fact along the shortest, almost rectilinear trajectories (trajectories of the second type) from the region of their generation to the edge of underlying surface where they disintegrated.

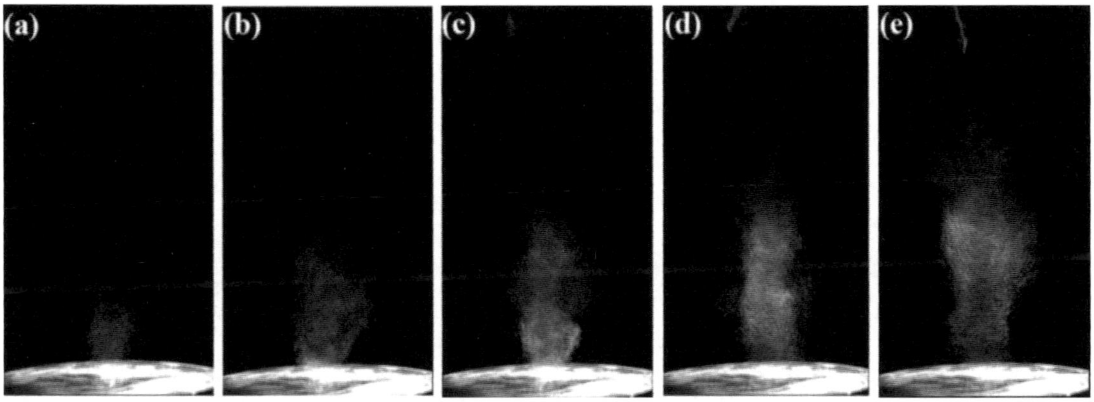

Figure 2. The typical air whirl: (a) whirl appearance, (b), (c) and (d) whirl growth. The time between frames is 0.36 s. The visualization was made by magnesia tracer particles (mode no.6).

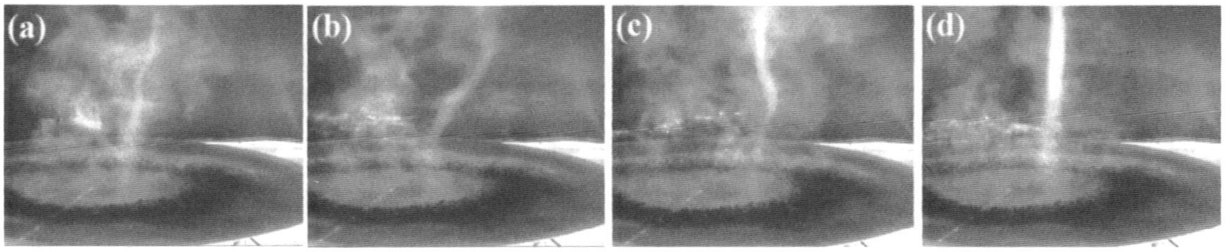

Figure 3. The typical air whirl: (a) whirl appearance, (b) and (c) whirl development, (d) maximal whirl development. The time between frames is 0.48 s. The visualization was made by vapor of special easily-boiling fluid (mode no.3).

The maximal length of trajectory of the vortex structures base was 50-100 cm, and the velocity was 5-10 cm/s. Therefore, the limiting lifetime of observed vortexes was about 20 s. The maximal height of generated vortex structures was 1.5 m, and their maximal diameter was 0.1 m. In the case of operation in "softer" modes (nos. 2 and 3), the number of whirls observed per experiment and their geometric dimensions were smaller (see figure 3), and the lifetime was longer than those in the case of operation in "hard" modes (nos. 4, 5, and 6).

3.2. The conditions of generation and integral parameters of non-stationary fire whirls.

The frame-by-frame analysis (see figure 4) provided information on the following parameters of the generation of flaming whirls and their integral characteristics: (1) the temporal range of the formation of flaming vortex structures, (2) the region of emergence of flaming whirls, (3) the number of observed framing whirls for one experiment, (4) the lifetime of flaming vortex structures, (5) the height of flaming whirls, and (6) the diameter of the flaming whirl.

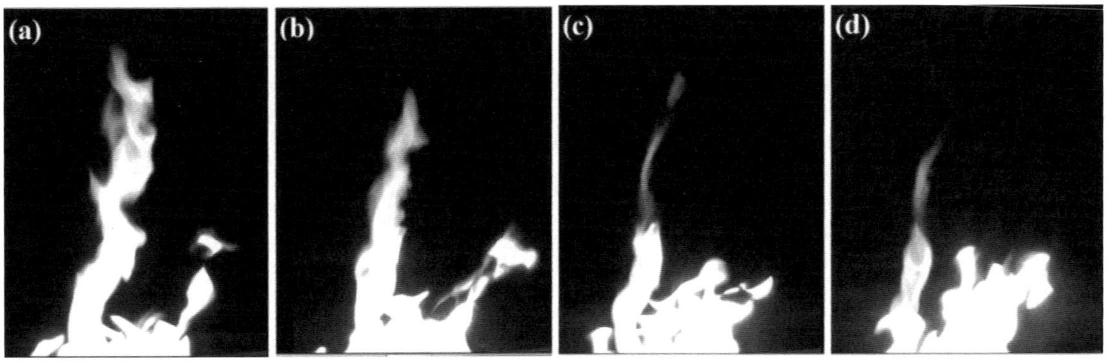

Figure 4. The typical fire whirl: (a) maximal development, (b), (c) and (d) whirl decaying. The time between frames is 0.09 s.

Multiple repetition of the experiments allowed drawing the following conclusions. The first flame vortex structures started to form in 3 min, and the last ones, 12 min after lighting the pellets. The pellets combusted completely in 15-17 min. The flame whirls formed both in the center of the fuel arrangement and on its periphery. The number of observed whirl structures was up to 15 for one experiment. The lifetime of the overwhelming majority of generated whirls was from 1 to 5 s. The largest height of flaming vortex structures reached 0.7 m, while their maximal diameter was 0.05 m.

4. Conclusion

The generation of wall-free non-stationary air and fire whirls under laboratory conditions has been demonstrated. It is strictly emphasized that contrary to previous studies no mechanical rotation devices or underlying surface rotation were used for the whirls generation. A simple experimental facility for the generation of concentrated air and fire vortex structures by creating the unstable air stratification and solid fuel combustion (urotropine) has been described. Some data on the generation conditions and integral parameters (lifetime, height, and diameter) of air and fire whirls have been obtained.

Acknowledgements

The work was supported by the Russian Foundation for Basic Research, grant no. 18-08-01382.

References

[1] Alekseenko S V, Kuibin P A and Okulov V L 2007 Theory of concentrated vortices. An introduction (Berlin, Heidelberg, New York: Springer)
[2] Varaksin A Yu, Romash M E and Kopeitsev V N 2015 Tornado (New York: Begell House)
[3] Ying S J and Chang C C 1970 *J. Atmos. Sci.* **27** 3-14
[4] Wan C A and Chang C C 1972 *J. Atmos. Sci.* **29** 116-127
[5] Ward N B 1972 *J. Atmos. Sci.* **29** 1194-1204
[6] Church C R, Snow J T and Agee E M 1977 *Bull. Amer. Meteorol. Soc.* **58** 900-908
[7] Baker G and Church C R 1979 *J. Atmos. Sci.* **36** 2413-2424
[8] Kuai L, Haan F L, Gallus W A and Sarkar P P 2008 *Wind Struct.* **11** 75-96
[9] Haan F L, Sarkar P P and Gallus W A 2008 *Eng. Struct.* **30** 1146-1159
[10] Emmons H W and Ying S J 1967 *Proc. Combust. Inst.* **11** 475-488
[11] Chuah K H and Kushida K 2007 *Proc. Combust. Inst.* **31** 2599-2606
[12] Lei J, Liu N, Zhang L and Satoh K 2015 *Combust. Flame* **162** 745-758
[13] Lei J, Liu N and Satoh K 2015 *Proc. Combust. Inst.* **35** 2503-2510
[14] Byram G M and Martin R E 1970 *Forest Sci.* **16** 386-389
[15] Martin R E, Pendleton D W and Burgess W 1976 *Fire Technol.* **16** 33-40
[16] Kuwana K, Morishita S, Dobashi R, Chuah K H and Saito K 2011 *Proc. Combust. Inst.* **33** 2425-2432
[17] Chuah K H, Kuwana K, Saito K and Williams F A 2011 *Proc. Combust. Inst.* **33** 2417-2424
[18] Varaksin A Yu, Romash M E, Kopeitsev V N and Gorbachev M A 2010 *High Temp.* **48** (6) 918-925
[19] Varaksin A Yu., Romash M E and Kopeitsev V N 2010 *AIP Conf. Proc.* vol **1207** pp 342-347
[20] Varaksin A Y, Romash M E, Kopeitsev V N and Gorbachev M A 2012 *Int. J. Heat Mass Transfer* **55** 6567-6572
[21] Varaksin A Y, Romash M E and Kopeitsev V N 2013 *Int. J. Heat Mass Transfer* **64** 817-828
[22] Varaksin A Yu, Protasov M V and Teplitskii Yu S 2014 *High Temp.* **52** (4) 554-559

IOP Conf. Series: Journal of Physics: Conf. Series **980** (2018) 012020 doi:10.1088/1742-6596/980/1/012020

Large-eddy simulations of the near field of a swirling turbulent annular jet

V. Ryzhenkov[1,2], **R. Mullyadzhanov**[1,2]

[1] Institute of Thermophysics SB RAS, Novosibirsk, Russia

[2] Novosibirsk State University, Novosibirsk, Russia

E-mail: v.o.ryzhenkov@gmail.com

Abstract. We perform Large-eddy simulations of an annular swirling turbulent jet for $S = 0$, $S = 0.3$ and $S = 1$, where S is swirl number. The Reynolds number based on the bulk velocity in the feeding channel and the outer diameter was 8900. The analysis showed that the region around the stagnation point in the end of the recirculation zone has a high level of turbulent kinetic energy. The length of the recirculation zone decreases in case for low swirl ($S = 0.3$) compared to the non-rotating case, while for the higher rotation rate ($S = 1$) the recirculation bubble becomes enormously long. Spectral analysis showed two peaks corresponding to the Kelvin-Helmholtz instability and some low-frequency oscillations of the recirculating zone.

1. Introduction

The flow in a channel between two coaxial cylinders issuing into a larger volume forms an annular jet. This class of flows combine features of jets and wakes due a recirculation zone past the central bluff body and a couple of mixing layers close to the nozzle exit. The first mixing layer is between the recirculation zone and the jet, while the second one is between the jet and the co-flow (ambient fluid). This combination is traditionally employed for burner devices design where the recirculation zone helps to stabilize the flame [1].

Chigier and Beer [2] performed one of the first experiments on annular jets comparing their self-similarity properties with the round counterpart. Ko and Chan [3, 4, 5] considered the near field for various configurations of the geometry of the central bluff body. While the dynamics of the outer shear layer is similar to the one of round jets, the recirculation zone usually introduces low-frequency oscillations of its length and amplitude of turbulent fluctuations [6, 7]. Similar observations were also made for non-swirling annular jets [8].

While for round swirling jets there is a detailed description of the effect of rotation on the flow topology and dynamics including the typical instabilities and the vortex breakdown phenomenon [9, 10, 11] and large-scale coherent structures [12, 13, 14, 15, 16, 17, 19], for annular jets it is much more limited. Sheen et al. [20] performed first experiments and identified several typical patterns depending on the flow parameters including the formation of the vortex breakdown bubble at an intermediate swirl number, which moves upstream as the swirl number increases. The bluff-body recirculation zone and the breakdown bubble merge to form a single recirculation zone at a sufficiently high swirl instensity as was also recently observed using Particle image velocimetry (PIV) [21]. In this paper we use well-resolved Large-eddy simulations (LES) of

Content from this work may be used under the terms of the Creative Commons Attribution 3.0 licence. Any further distribution of this work must maintain attribution to the author(s) and the title of the work, journal citation and DOI.
Published under licence by IOP Publishing Ltd

the annular turbulent jet with different several swirl numbers to study the coherent structures populating the near-nozzle area.

2. Computational details and problem

We consider a flow of an incompressible fluid governed by the spatially filtered Navier–Stokes equations:

$$\frac{\partial \boldsymbol{u}}{\partial t} + (\boldsymbol{u}\nabla)\boldsymbol{u} = -\nabla p + \frac{1}{Re}\Delta \boldsymbol{u} - \nabla \cdot \tau, \quad (1)$$

$$\nabla \cdot \boldsymbol{u} = 0. \quad (2)$$

where $\boldsymbol{u}$ and p are the non-dimensionl filtered velocity and pressure field, respectively. An extra term τ represents the subgrid-scale stresses appearing in the Large-eddy simulation framework. The above described governing equations are implemented in the Nek5000 code [22] with a spatial discretisation based on the spectral-element method using Lagrange polynomials. The Navier–Stokes equations are cast in a weak form and discretised in space by means of Galerkin approximation using Nth-order Lagrange polynomial interpolants on the Gauss-Lobatto-Legendre points for the velocity field and pressure ($\mathbb{P}_N - \mathbb{P}_N$ formulation). In the present work the polynomial order N is set to 7. The semi-implicit time-stepping scheme is of the third-order accuracy. In the present LES simulations the two highest modes are filtered with parabolic transfer function, and the filter amplitude was 5% in the last mode [23]. The accuracy of simulations has been previously assessed in a number of configurations including the channel flow and the non-swirling annular jet [24, 25].

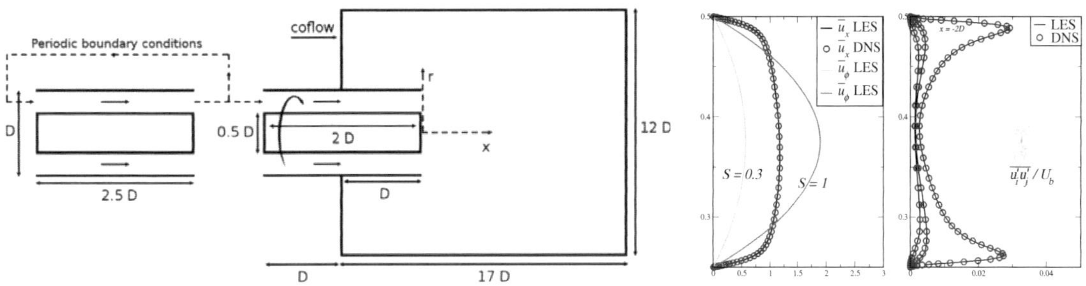

Figure 1. Left: computational domains and boundary conditions. Right: the verification of the inflow boundary conditions produced in the simulation of the flow in a periodic annular channel where the time-averaged axial velocity profile and normal Reynolds stress components are compared with the DNS data [28]. Red and green profiles show the profile of the tangential velocity with is added to the turbulent inflow velocity field in swirling cases.

Numerical simulations of turbulent annular swirling jets of incompressible fluid are performed for the Reynolds number $Re = 8900$, based on the bulk velocity U_b and the outer diameter D, while the ratio of the outer to inner diameter is equal to $D/d = 2$. Figure 1 (left) shows two computational domains used for simulation. A periodic annular channel with the length $L = 2.5D$ and $8 \times 32 \times 20 = 5120$ spectral elements in r, θ, and x directions is used to generate a fully-developed turbulent inflow velocity fields for the main computation of the annular jet. To validate this simulation radial profiles of the time-averaged axial velocity and different components of Reynolds stresses are compared with DNS by Chung et al. [28] showing excellent agreement, see Fig. 1 (right). The main domain represents a cylinder with a diameter $D_{main} = 12D$ and length $L_{main} = 17D$, which includes a feeding annular channel with length equals to $2D$, while the thickness of the outer wall of the channel is $0.03D$. The instantaneous velocity field in some fixed $r-\theta$ plane from a periodic channel simulation was copied to the feeding channel of the main domain every time step. At this cross-section $x/D=-2$ we also add a parabolic tangential velocity profile according to the formula $u_\phi=\alpha(1-r^2)$ where α is a parameter defined below. Previous studies showed that the fully developed rotating annular flow provided by the rotating wall is not efficient to arrange high swirl rates [26, 27], thus, leading to the present way of setting up the rotation. The swirl number is defined as

$$S=\frac{\int_0^R \overline{u}_x\overline{u}_\phi r^2 dr}{R\int_0^R \overline{u}_x^2 r dr}, \tag{3}$$

where $R=D/2$. In the present study we consider $S=0$, 0.3 and 1.0 calculated at the inflow section $x/D=-2$ corresponding to $\alpha=0$, 0.45 and 1.91, respectively, see Fig. 1 (right). The co-flow in the main domain was set to $0.04U_b$ while on other boundaries Neumann boundary conditions were imposed. The main domain was meshed with about 80 thousand spectral elements ($35\times32\times70$ elements in r, θ and x). Since we used polynomials of the seventh order, there were 8^3 computational nodes inside each spectral element, so that the total number of computational points was about 41.3×10^6 for the main simulation and about 2.6×10^6 for the periodic annular channel. The distance between computational nodes at the outer wall inside the periodic channel was $\Delta r^+=0.76$, $(r\Delta\theta)^+=6.4$ and $\Delta x^+=64$ in radial, azimuthal and axial direction, thus our computational grid satisfied the near-wall well-resolved LES criteria [29]. The superscript '+' denotes the non-dimensional units using the friction velocity and viscosity.

3. Results

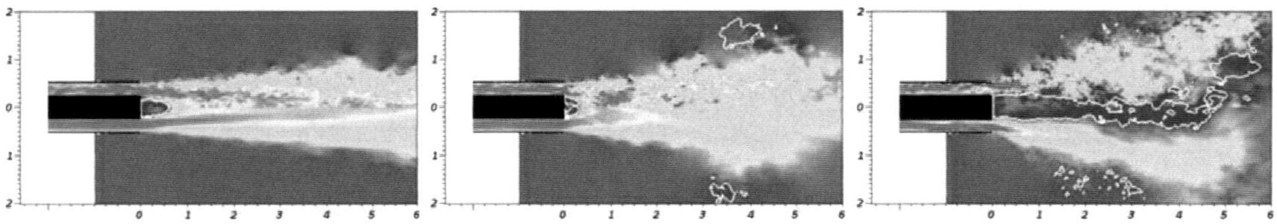

Figure 2. The instantaneous (top half) and time-averaged axial velocity field is shown for $S=0$ (left), $S=0.3$ (middle) and $S=1$ (right)in $x-r$ plane.

Figure 2 shows the instantaneous and time-averaged axial velocity fields in $x-r$ plane for all swirl numbers. For $S=0$ inner and outer mixing layers merge at $x\approx2D$, while with an increase of swirl to $S=0.3$ this process goes faster and occurs at $x\approx0.8D$. However, close to zero and slightly negative values of axial velocity appears at $x\approx3D$ which is the sign of the vortex breakdown as described in the literature. The new bubble grows in size with the further increase of S and merges with the recirculation zone close to the nozzle. This result is well represented by the instantaneous and time-averaged velocity field for $S=1$ where one can see a very long bubble with the length around $4D$.

Figure 3 shows the comparison of radial profiles of the time-averaged axial velocity and its fluctuations at several axial stations along x from $x/D=0$ to $x/D=2.25$ with step equals to 0.25. The recirculation zone is characterized by the high level of fluctuations in inner mixing layer, especially close to the stagnation point. The peaks of axial velocity and fluctuations move away from the centerline with an increase of swirl. For the case $S=0.3$ the axial velocity decrease after $x/D=1.25$. This fact is the evidence of emergence of the second recirculation zone.

Figure 4 shows the energy spectra for individual velocity components in the inner and outer shear layer at $x/D=0.25$ computed for the time interval around 40 non-dimensional time units.

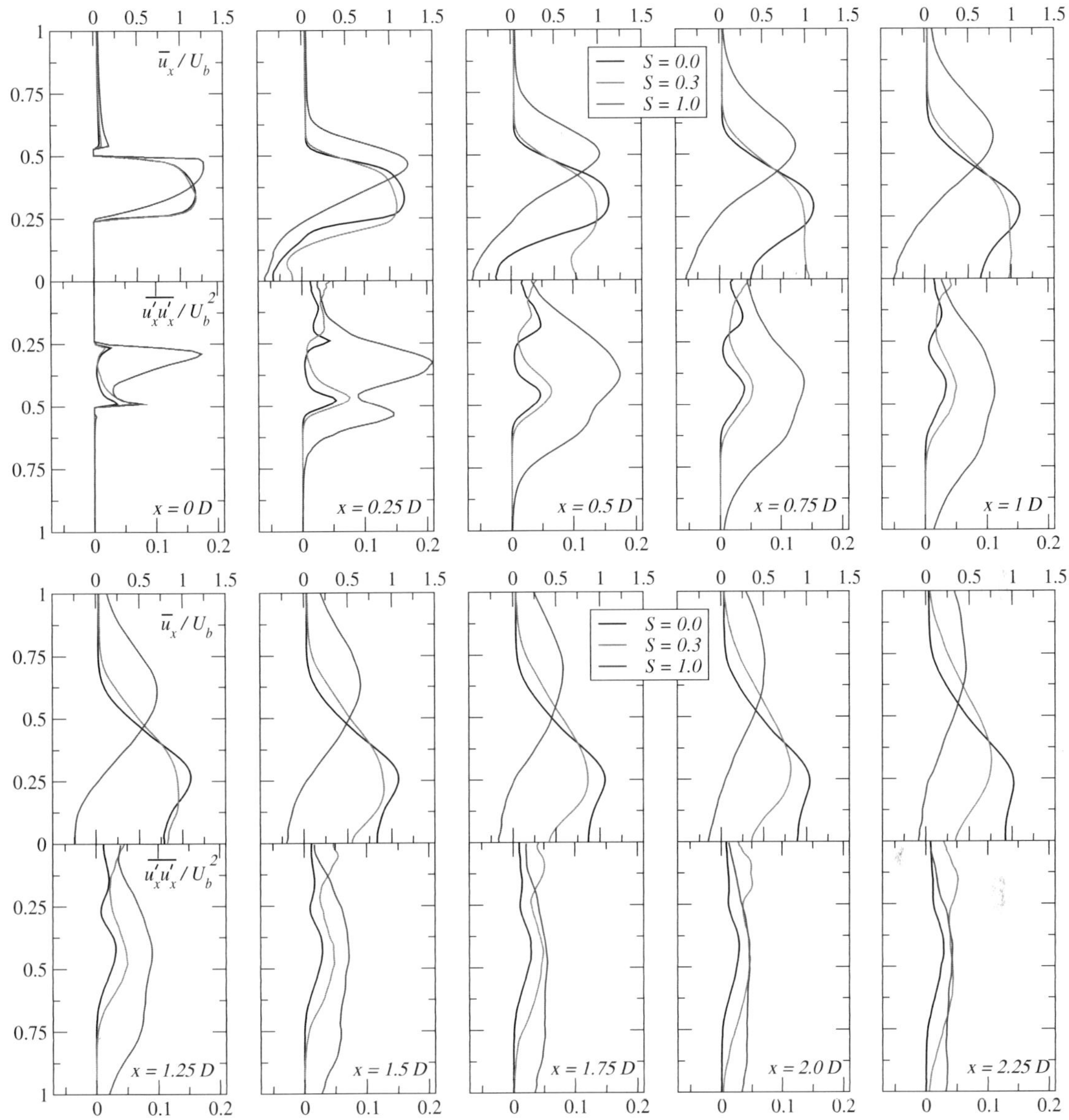

Figure 3. The radial profiles of the time-averaged axial velocity (top) and axial velocity fluctuations (bottom) for different axial stations.

Two peaks are clearly seen. The first peak corresponds to the Kelvin-Helmholtz instability, which is better pronounced for the radial and azimuthal velocity components with a non-dimensional frequency $f_0 = fD/U_b \approx 1$. A similar effect was observed in the mixing layer for a fully developed round jet [30]. The second peak is connected to the low-frequency oscillations of the recirculating zone and is better detected for axial and radial components of velocity. Further study will concentrate on the spatial form of the velocity fluctuations with low frequency.

4. Conclusion

We investigated the influence of the swirl intensity on statistical characteristics of the near field of the turbulent annular jet. The Reynolds number based on the bulk velocity in the feeding

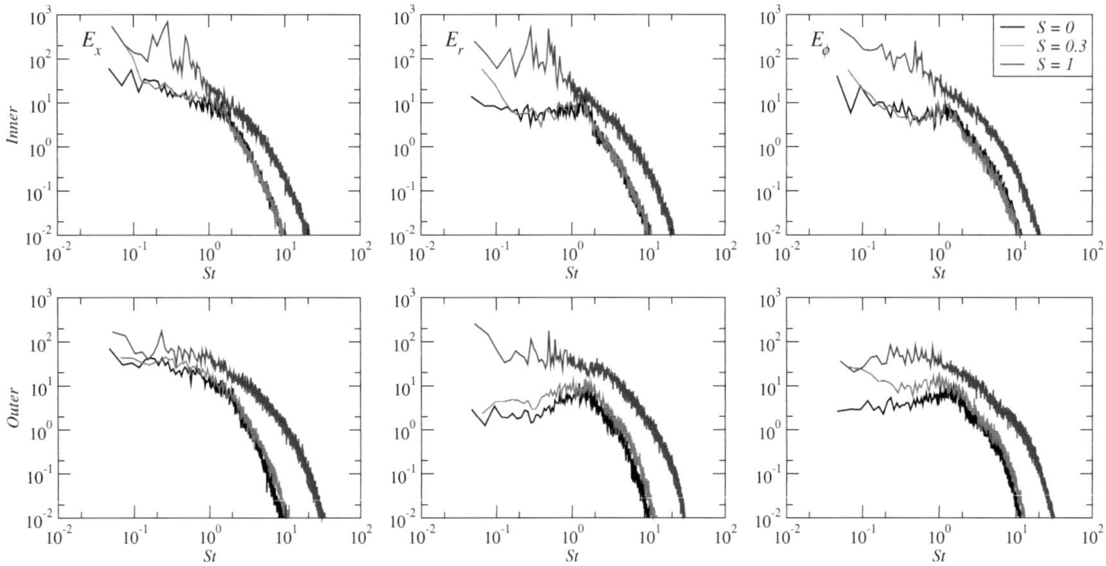

Figure 4. Energy spectra for separate velocity components (left: axial, middle: radial, right: tangential). Upper row corresponds to the inner mixing layer, while bottom row is for the outer mixing layer.

channel and the outer diameter was 8900. The analysis showed that the region around the stagnation point in the end of the recirculation zone has a high level of turbulent kinetic energy. The length of the recirculation zone decreases in case for low swirl ($S = 0.3$) compared to the non-rotating case, while for the higher rotation rate ($S = 1$) the recirculation bubble becomes enormously long. Spectral analysis showed two peaks corresponding to the Kelvin-Helmholtz instability and some low-frequency oscillations of the recirculating zone.

Acknowledgments

This work is funded by the Russian Science Foundation grant No. 14-19-01685. The computational resources are provided by Siberian Supercomputer Center SB RAS and Supercomputer Center of Novosibirsk State University.

References

[1] Gupta A, Lilley D, Syred N 1984 *Swirl flows* (Tunbridge Wells: Abacus Press)
[2] Chigier N A and Beer J M 1964 The flow region near the nozzle in double concentric jet *J. Basic Engnrg.* **86** 797–804
[3] Ko N W M and Chan W T 1978 Similarity in the initial region of annular jets: three configurations *J. Fluid Mech.* **84** 641–656
[4] Chan W T and Ko N W M 1978 Coherent structures in the outer mixing region of annular jets *J. Fluid Mech.* **89** 515–563
[5] Ko N W M and Chan W T 1979 The inner region of annular jets *J. Fluid Mech.* **93** 549–584
[6] Berger E, Scholz D and Schumm M 1990 Coherent vortex structures in the wake of a sphere and a circular disk at rest and under forced vibrations *J. Fluid Struct.* **4** 231–257
[7] Palkin E, Mullyadzhanov R, Hadziabdic M and Hanjalic K 2016 Scrutinizing URANS in shedding flows: The case of cylinder in cross-flow in the subcritical regime *Flow Turbul. Combust.* **97** 1017–1046
[8] Vanierschot M and Van den Bulck E 2011 Experimental study of low precessing frequencies in the wake of a turbulent annular jets *Exp. Fluids* **50** 189–200
[9] Panda J and McLaughlin D K 1994 Experiments on the instabilities of a swirling jet *Phys. Fluids* **6**(1) 263–276

[10] Billant P, Chomaz J M and Huerre P 1998 Experimental study of vortex breakdown in swirling jets *J. Fluid Mech.* **376** 183–219

[11] Lucca-Negro O and O'Doherty T 2001 Vortex breakdown: a review *Prog. Energ. Combust. Scie.* **27**(4) 431–481

[12] Cala C E, Fernandes E, Heitor M V and Shtork S I 2006 Coherent structures in unsteady swirling jet flow *Exp. Fluids* **40**(2) 267–276

[13] Oberleithner K, Sieber M, Nayeri C N, Paschereit C O, Petz C, Hege H C, Noack B R and Wygnanski I 2011 Three-dimensional coherent structures in a swirling jet undergoing vortex breakdown: stability analysis and empirical mode construction *J. Fluid Mech.* **679** 383–414

[14] Litvinov I V, Shtork S I, Kuibin P A, Alekseenko S V and Hanjalic K 2013 Experimental study and analytical reconstruction of precessing vortex in a tangential swirler *Int. J. Heat Fluid Flow* **42** 251–264

[15] Markovich D M, Abdurakipov S S, Chikishev L M, Dulin V M and Hanjalic K 2014 Comparative analysis of low-and high-swirl confined flames and jets by proper orthogonal and dynamic mode decompositions *Phys. Fluids* **26**(6) 065109

[16] Oberleithner K, Paschereit C O, Seele R and Wygnanski I 2012 Formation of turbulent vortex breakdown: Intermittency, criticality, and global instability *AIAA Journal* **50** 1437–1452

[17] Markovich D M, Dulin V M, Abdurakipov S S, Kozinkin L A, Tokarev M P and Hanjalic K 2016 Helical modes in low-and high-swirl jets measured by tomographic PIV *J. Turbul.* **17**(7) 678–698

[18] Mullyadzhanov R, Abdurakipov S and Hanjalic K 2017 Helical structures in the near field of a turbulent pipe jet *Flow Turbul. Combust.* **98** 367–388

[19] Mullyadzhanov R, Abdurakipov S and Hanjalic K 2017 On coherent structures and mixing characteristics in the near field of a rotating-pipe jet *Int. J. Heat Fluid Flow* **63** 139–148

[20] Sheen H J, Chen W J and Jeng S Y 1996 Recirculation zones of unconfined and confined annular swirling jets *AIAA Journal* **34**(3) 572–579

[21] Vanierschot M and Van den Bulck E 2008 Influence of swirl on the initial merging zone of a turbulent annular jet *Phys. Fluids* **20** 105104

[22] S. G. Kerkemeier, P. F. Fischer, and J. W. Lottes. Nek5000: Open source spectral element cfd solver. available at http://nek5000.mcs.anl.gov, 2008

[23] Marin O, Vinuesa R, Obabko A V and Schlatter P 2016 Characterization of the secondary flow in hexagonal ducts *Phys. Fluids* **28** 1–26

[24] Ryzhenkov V, Ivashchenko V, Vinuesa R and Mullyadzhanov R 2016 Simulation of heat and mass transfer in turbulent channel flow using the spectral-element method: effect of spatial resolution *J. Phys.: Conf. Series* **754**(6) 062009

[25] Ryzhenkov V and Mullyadzhanov R 2017 Large-eddy simulations of the near field of a turbulent annular jet *J. Phys.: Conf. Series* **899**(2) 022014

[26] Hadziabdic M, Hanjalic K and Mullyadzhanov R 2013 LES of turbulent flow in a concentric annulus with rotating outer wall *Int. J. Heat Fluid Flow* **43** 74–84

[27] Mullyadzhanov R, Hadziabdic M and Hanjalic K 2015 LES investigation of the hysteresis regime in the cold model of a rotating-pipe swirl burner *Flow Turbul. Combust.* **94**(1) 175–198

[28] Chung S Y, Rhee G H and Sung H J 2002 Direct numerical simulation of turbulent concentric annular pipe flow. Part 1: Flow field *Int. J. Heat Fliud Flow* **23** 426–440

[29] Piomelli U and Chasnov J R 1995 *Turbulence and transition modeling. Large-Eddy Simulations: Theory and Applications* Springer-Science

[30] Mullyadzhanov R, Abdurakipov S and Hanjalic K 2017 Helical structures in the near field of a turbulent pipe jet *Flow Turbul. Combust.* **98** 367–388

Thermo-aerodynamic efficiency of non-circular ducts with vortex enhancement of heat exchange in different types of compact heat exchangers

V Ya Vasilev and S A Nikiforova

Astrakhan State Technical University, Astrakhan 414056, Tatischeva 16, Russia

E-mail: vasiliev-vy@mail.ru

Abstract. Experimental studies of thermo-aerodynamic characteristics of non-circular ducts with discrete turbulators on walls and interrupted channels have confirmed the rational enhancement of convective heat transfer, in which the growth of heat transfer outstrips or equals the growth of aerodynamic losses. Determining the regularities of rational (energy-saving) enhancement of heat transfer and the proposed method for comparing the characteristics of smooth-channel (without enhancement) heat exchangers with effective analogs provide new results, confirming the high efficiency of vortex enhancement of convective heat transfer in non-circular ducts of plate-finned heat exchange surfaces. This allows creating heat exchangers with much smaller mass and volume for operation in energy-saving modes.

1. Introduction

A demonstrative example of the comparison of heat exchangers (HE) with cores based on tube-and plate smooth-duct (TPsm), with ridges and grooves (TPrg) and plate-and fin interrupted (PFint) heat-exchange surfaces (HES), various possible variants of efficiency comparisons: with smooth ducts (without enhancement of heat transfer) and HE with ducts of identical geometry with artificial turbulence of the coolant flow; heterogeneous HE with smooth ducts and channels of other sizes, forms and methods of enhancement of heat exchange; heterogeneous HE with heat exchange enhancement by different methods in channels of various types; groups of heterogeneous HEs with determining the best and worst HE based on the ranking by the volume of the HE cores. The efficiency comparison, in the most general case, distinguishes the following: a reference HE_{st}, presumably less effective (in a particular case, with a smooth-duct heat exchange surface), all characteristics of which are known from the preliminary calibrated heat calculation or the results of thermal engineering tests; compared to HE_{cnf}, with an artificial turbulence of the coolant flow (its characteristics are determined from the design heat calculation). The comparison is carried out at the thermal and aerodynamic boundary conditions of the equality of the thermal loads $Q_{HE\text{-}st}/Q_{HE\text{-}cnf} = 1$, the mass flow rates of the heat carrier $M_{HE\text{-}st}/M_{HE\text{-}cnf} = 1$, the temperature headings $\theta_{HE\text{-}st}/\theta_{HE\text{-}cnf} = 1$ and the ratio $\Delta p_{HE\text{-}st}/\Delta p_{HE\text{-}cnf} \geq 1$ total losses of coolant pressure on friction in HES ducts, input to and exit from HE. In determining the best heat exchanger from a group of all possible analogues, HE can serve as a reference both without and with enhancement of heat exchange in HES ducts. If, in considering the task of comparing the two HE, the deviation from the observance of the thermal and aerodynamic boundary conditions of the realization of the rational enhancement of convective heat transfer (RECH) process at the rationality exponent $\{Ꝁ = [(Nu_{cnf}/Nu_{st})/(\zeta_{cnf}/\zeta_{st})]_{Re=idem}\} \geq 1$, in the part $\Delta p_{HE\text{-}cnf}/\Delta p_{HE\text{-}st} \leq 1$ [1], it is possible to obtain the comparison result in the form of a decrease in the ΔL of the length of the HE in the course of air under the condition $Ꝁ < 1$ with the determination of the calculated value $\Delta p_{HE\text{-}cnf}/\Delta p_{HE\text{-}st} > 1$ or observance of the specified condition for its limitation $(\Delta p_{HE\text{-}cnf}/\Delta p_{HE\text{-}st})_{max} \leq [\Delta p_{HE\text{-}cnf}/\Delta p_{HE\text{-}st}]_{lim}$.

2. Calculations and comparisons of the cores of compact heat exchangers

Content from this work may be used under the terms of the Creative Commons Attribution 3.0 licence. Any further distribution of this work must maintain attribution to the author(s) and the title of the work, journal citation and DOI.
Published under licence by IOP Publishing Ltd

2.1. Plate- and fin heat exchangers with smooth and interrupted ducts of identical geometry in cross sections

An important case of the general problem of comparing the volumes of the same type of heat exchangers is considered by the example of plate- and fin heat exchangers. This ensures reasonable decisions on the expediency of upgrading the smooth-channel plate-and fin (figure 1, a, below) and tube-plate heat exchangers (figure 1, b) in various engineering facilities, replacing them with efficient designs with artificial turbulence of the coolant in the PFint HES channels (figure 1, a, top) and TPrg HES (figure 1, b, right) during the RECH process: $\{K = [(Nu_{cnf}/Nu_{st})/(\zeta_{cnf}/\zeta_{st})]_{Re=idem}\} \geq 1$.

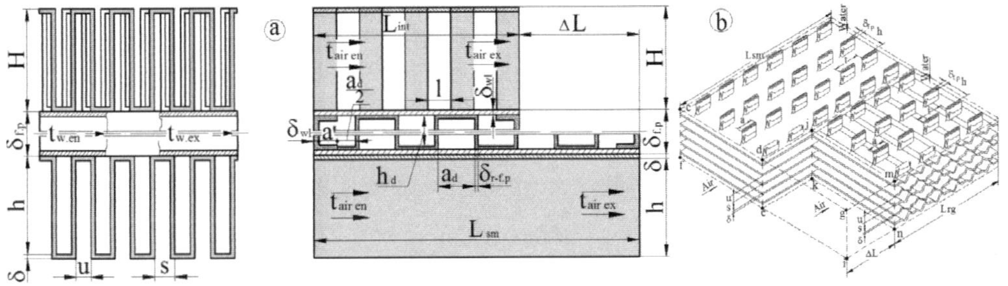

Figure 1. Illustrations of the reduced volume of the HE cores to be compared due to the use of enhancement of convective heat transfer: a – dissection of long smooth ducts into many short ones; B – stamping of discrete turbulators (ridges and grooves) on the walls of ducts.

The results of calculations of the cut-off parameter $(l/d)_x$ and the coefficient ζ_x of the frictional pressure loss in the HES HE ducts on the calculated values of the Nusselt Nu criterion are selectively presented in Table 1 and Figure 2 (graphical solutions).

Table 1. Results of comparison of the reference HE with the PFsm HES and the matched HE with the PFint HES at $s = 5 \cdot 10^{-3}$ m, $d = 7.66 \cdot 10^{-3}$ m, $h/u = 6.750$, $\delta/d = 0.0783$, $\Omega = 428$ m^2/m^3

$W_{a-\infty}$	1.90	2.64	4.70	5.20	7.50	11.22
Q_a	23450	27120	34700	36685	44535	59255
Re_a	1069	1500	2686	2973	4300	6442
α_a	22.65/*49.72*	24.94/*54.94*	30.26/*72.40*	31.90/*66.27*	38.46/*75.47*	52.25/*97.78*
Nu_a	6.33/*13.90*	7.01/*16.62*	8.57/*20.50*	9.04/*18.78*	10.93/*21.45*	14.87/*27.83*
$(l/d)_x$	–/*11.413*	–/*11.333*	–/*11.816*	–/*13.018*	–/*13.947*	–/*14.127*
ζ_{a-x}	0.108/*0.226*	0.082/*0.191*	0.054/*0.120*	0.051/*0.096*	0.042/*0.076*	0.038/*0.066*
Δp_a	6.42/*6.34*	9.40/*19.71*	19.39/*19.13*	22.31/*21.25*	38.52/*37.09*	77.09/*75.77*
V_{st}/V_{cnf}	2.153	2.209	2.314	2.099	1.955	1.812
M_{mt}, kg	22.56/*10.67*	22.56/*9.97*	22.56/*9.97*	22.56/*11.37*	22.56/*12.06*	22.56/*12.76*

To observe the dynamics of changes in the values of the main results of the HE comparison and the flow of the coolant in Table 1, etc. are also given (in the numerator for the reference HE, in the denominator for the matched HE): $w_{a-\infty}$, is the speed of the unperturbed airflow before entering the live cross section of the HE core, m/s; Q_a is the thermal load, W; Re_a is the Reynolds criterion; α_a is the heat transfer coefficient, W/(m^2·K); Δp_a is the resistance of the HE core to the cooled cooling air, Pa; V_{st}/V_{cnf} is the ratio of the core volumes of the compared reference V_{st} (or smooth V_{sm}) and the correlated V_{cnf} HE; M_{mt} is the mass of the metal of the HE core, kg. From Figure 2, it is seen that for

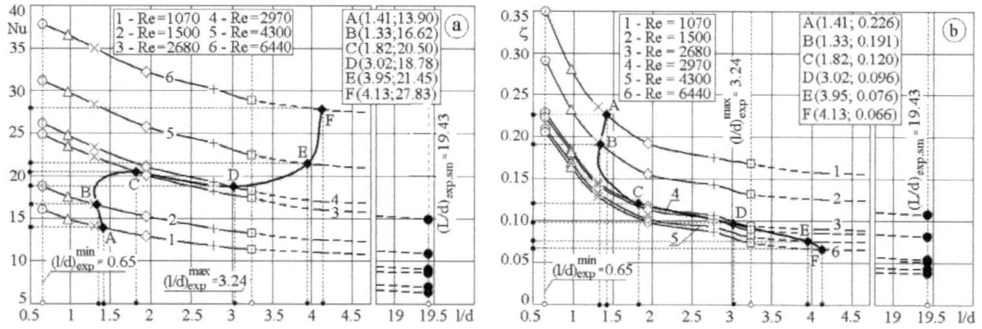

Figure 2. To the definition (graphical solution) of the values $(l/d)_x$ and $\zeta_{a\text{-}x}$ of the heat exchanger with PFint HES at $s = 5\cdot10^{-3}$ m and $w_{a\text{-}\infty} = 1.90$; 2.64; 4.70; 5.20;7.50; 11.22 m/s (the corresponding values of the Reynolds test are shown in the figure)

the extended to the side of the lower values of the regime criterion Re of the transition region of the coolant flow (due to artificial turbulence), the minimum value of the argument of the dependences Nu = f(l/d) and ζ = f(l/d) at Re = variable, indicating the need to use turbulators with a more enhancing ef

Table 2. Results of comparison of the reference HE with the PFsm HES – $s = 5\cdot10^{-3}$ m, $d = 7.66\cdot10^{-3}$ m, $\Omega = 428$ m²/m³ and compared HE with the PFint HES – $s = 4\cdot10^{-3}$ m, $d = 6.12\cdot10^{-3}$ m, $\Omega = 524$ m²/m³ at $(h/u)_m = 6.860$ and $(\delta/d)_m = 0.0799$

$W_{a\text{-}\infty}$	1.90	2.64	4.70	5.20	7.50	11.22
Q_a	23450	27120	34700	36685	44535	59255
Re_a	1069/*869*	1500/*1219*	2686/*2183*	2973/*2416*	4300/*3495*	6442/*5237*
α_a	22.65/*50.14*	24.94/*60.61*	30.26/*74.06*	31.90/*78.23*	38.46/*86.60*	52.25/*109.3*
Nu_a	6.33/*11.20*	7.01/*13.61*	8.57/*16.75*	9.04/*17.71*	10.93/*19.66*	14.87/*24.85*
$(l/d)_x$	–/*2.348*	–/*2.052*	–/*2.655*	–/*2.589*	–/*3.240*	–/*3.240*
$\zeta_{a\text{-}x}$	0.108/*0.210*	0.082/*0.173*	0.054/*0.116*	0.051/*0.112*	0.042/*0.085*	0.038/*0.076*
Δp_a	6.42/*6.15*	9.40/*8.96*	19.39/*18.78*	22.31/*22.25*	38.52/*38.01*	77.09/*82.32*
V_{st}/V_{cnf}	2.770	3.040	2.993	2.907	2.697	2.305
M_{mt}, kg	22.56/*8.92*	22.56/*8.19*	22.56/*8.19*	22.56/*8.19*	22.56/*8.92*	22.56/*8.65*

Table 3. Results of comparison of the reference HE with the PFsm HES – $s = 5\cdot10^{-3}$ m, $d = 7.66\cdot10^{-3}$ m, $\Omega = 428$ m²/m³ and compared HE with the PFint HES – $s = 3.5\cdot10^{-3}$ m, $d = 5.39\cdot10^{-3}$ m, $\Omega = 598$ m²/m³ at $(h/u)_m = 6.714$ and $(\delta/d)_m = 0.0762$

$W_{a\text{-}\infty}$	1.90	2.64	4.70	5.20	7.50	11.22
Q_a	23450	27120	34700	36685	44535	59255
Re_a	1069/*766*	1500/*1074*	2686/*1923*	2973/*2129*	4300/*3079*	6442/*4614*
α_a	22.65/*52.48*	24.94/*64.60*	30.26/*89.34*	31.90/*83.33*	38.46/*101.3*	52.25/*112.2*
Nu_a	6.33/*10.32*	7.01/*12.78*	8.57/*17.80*	9.04/*16.61*	10.93/*20.24*	14.87/*22.47*
$(l/d)_x$	–/*2.546*	–/*2.073*	–/*1.682*	–/*2.622*	–/*2.506*	–/*3.500*
$\zeta_{a\text{-}x}$	0.108/*0.225*	0.083/*0.188*	0.054/*0.145*	0.051/*0.117*	0.042/*0.104*	0.038/*0.075*
Δp_a	6.42/*6.25*	9.40/*9.06*	19.39/*19.50*	22.31/*21.62*	38.52/*39.84*	77.09/*78.10*
V_{st}/V_{cnf}	3.264	3.663	3.947	3.647	3.420	2.865
M_{mt}, kg	22.56/*7.08*	22.56/*6.39*	22.56/*5.70*	22.56/*6.39*	22.56/*6.39*	22.56/*7.77*

fect on the flow of the coolant (PFint HES, that is, with smaller values of the cut-off parameter l/d or l) than in the developed turbulent regime [2]. The reduction in the volume and mass of the cores compared to HE with PFint HES is estimated in $V_{HE\text{-}st}/V_{HE\text{-}cnf} = 1.81\ldots\ 2.31$ times $= L_{HE\text{-}st}/L_{HE\text{-}cnf} = M_{HE\text{-}st}/M_{HE\text{-}cnf}$.

The effectiveness of reducing the volume of the surfaces to be compared with the interrupted ducts with decreasing the equivalent duct diameters of their surfaces is very significant (see Tables 2 and 3). The reduction in the volume, mass of the cores and the surface length along the air of the matched HE with the PFint HES is estimated with respect to HE with a smooth-channel PFsm HES – $s = 5\cdot10^{-3}$ m, $d = 7.66\cdot10^{-3}$ m, $\Omega = 428$ m^2/m^3: $V_{HE\text{-}st}/V_{HE\text{-}cnf} = 2.30\ldots2.99$ раза – $s = 4\cdot10^{-3}$ m, $d = 6.12\cdot10^{-3}$ m, $\Omega = 524$ m^2/m^3 (see table 2); $V_{HE\text{-}st}/V_{HE\text{-}cnf} = 2.86\ldots3.95$ times – $s = 3.5\cdot10^{-3}$ m, $d = 5.39\cdot10^{-3}$ m, $\Omega = 598$ m^2/m^3 (see table 3).

2.2. Plate-and fin heat exchangers with interrupted ducts of similar cross-sectional geometry

Reducing the equivalent diameter of the channel heat exchange surface increases its compactness. However, when using plate-and fin heat exchange surfaces in unclosed systems, the operational limitations imposed by the danger of clogging and expressed by the smallest step-edges – $s_{min} \approx (4\ldots3.5)\cdot10^{-3}$ m are imposed on the reduction in the dimensions of their ducts. The results of calculations in table 4 and 5 for comparing parameters of the same type heat exchangers with

Table 4. Results of comparison of the reference HE with the PFint HES – $s = 5\cdot10^{-3}$ m, $d = 7.66\cdot10^{-3}$ m, $\Omega = 428$ m^2/m^3 and compared HE with the PFint HES – $s = 4\cdot10^{-3}$ m, $d = 6.12\cdot10^{-3}$ m, $\Omega = 524$ m^2/m^3 for $(h/u)_m = 6.860$ and $(\delta/d)_m = 0.0799$

$W_{a\text{-}\infty}$	1.90	2.64	4.70	5.20	7.50	11.22
Q_a	23450	27120	34700	36685	44535	59255
Re_a	1069/*869*	1500/*1219*	2686/*2183*	2973/*2416*	4300/*3495*	6442/*5237*
α_a	49.72/*46.15*	54.94/*60.61*	72.40/*74.06*	66.27/*78.23*	75.47/*79.48*	97.78/*100.8*
Nu_a	13.90/*10.31*	16.62/*13.61*	20.50/*16.75*	18.78/*17.71*	21.45/*18.04*	27.83/*22.47*
$(l/d)_x$	1.413/*3.240*	1.333/*2.052*	1.816/*2.655*	3.018/*2.586*	3.947/*4.050*	4.127/*4.350*
$\zeta_{a\text{-}x}$	0.226/*0.196*	0.191/*0.173*	0.120/*0.116*	0.096/*0.112*	0.076/*0.076*	0.066/*0.065*
Δp_a	6.34/*6.20*	9.71/*8.96*	19.13/*18.78*	21.25/*22.25*	37.09/*36.87*	75.77/*75.66*
V_{st}/V_{cnf}	1.180	1.376	1.293	1.372	1.312	1.286
M_{mt}, kg	10.67/*9.66*	9.97/*8.19*	9.97/*8.19*	11.37/*8.19*	12.06/*9.65*	12.76/*10.38*

interrupted rectangular ducts of differing sizes ($s_{st} = 5\cdot10^{-3}$ and $s_{cnf} = 4.0\cdot10^{-3}$ m – table 4 and $s_{st} = 5\cdot10^{-3}$ and $s_{cnf} = 3.5\cdot10^{-3}$ m – table 5) of the similar geometry of the cross-sections ($h/u \cong$ idem, $\delta/d \cong$ idem)

Table 5. Results of comparison of the reference HE with the PFint HES – $s = 5\cdot10^{-3}$ m, $d = 7.66\cdot10^{-3}$ m, $\Omega = 428$ m^2/m^3 and compared HE with the PFint HES – $s = 3.5\cdot10^{-3}$ m, $d = 5.39\cdot10^{-3}$ m, $\Omega = 598$ m^2/m^3 at $(h/u)_m = 6.860$ and $(\delta/d)_m = 0.0799$

$W_{a\text{-}\infty}$	1.90	2.64	4.70	5.20	7.50	11.22
Q_a	23450	27120	34700	36685	44535	59255
Re_a	1069/*766*	1500/*1074*	2686/*1923*	2973/*2129*	4300/*3079*	6442/*4614*
α_a	49.72/*52.48*	54.94/*64.60*	72.40/*78.90*	66.27/*83.33*	75.47/*90.47*	97.78/*112.2*
Nu_a	13.90/*10.32*	16.62/*12.78*	20.50/*15.72*	18.78/*16.61*	21.45/*18.08*	27.83/*22.47*
$(l/d)_x$	1.413/*2.550*	1.333/*2.080*	1.816/*2.680*	3.018/*2.620*	3.947/*3.490*	4.127/*3.700*
$\zeta_{a\text{-}x}$	0.226/*0.226*	0.191/*0.188*	0.120/*0.123*	0.096/*0.117*	0.076/*0.085*	0.066/*0.072*
Δp_a	6.34/*6.27*	9.71/*9.09*	19.13/*18.54*	21.25/*21.62*	37.09/*36.09*	75.77/*74.97*
V_{st}/V_{cnf}	1.512	1.653	1.597	1.722	1.739	1.648
M_{mt}, kg	10.67/*7.08*	9.97/*6.39*	9.97/*6.39*	11.37/*6.39*	12.06/*7.08*	12.76/*7.77*

uniquely determine the advantages of the heat exchangers compared to the RECH process for $K \cong 1$: the reduction in the volumes and masses of the cores in $V_{HE\text{-}st}/V_{HE\text{-}cnf} = L_{HE\text{-}st}/L_{HE\text{-}cnf} = M_{HE\text{-}st}/M_{HE\text{-}cnf} = 1.18\ldots1.37$ times – see table 4 and 1.51...1.74 – see table 5.

2.3. Group of different types of heat exchangers

A variant of calculations and comparisons of a group of heat exchangers of different designs is considered: with a reference tube-plate smooth-duct surface – PFsm HES (Figure 1, b on the left); with comparable effective tube-and plate surface with ridges and grooves – TPrg HES (Figure 1, b on the right) and plate-and fin with interrupted duct surface – PFint HES (Figure 1, a top). The initial data package contains all values of the values with index "st" (or "sm"), the parameters of the elements of the all-welded multi-channel flat tube, the dimensions of the constructive front of the HE and the results of studies of the thermal Nu = f(Re) and aerodynamic characteristics $\zeta = f(Re)$ HES cores of the reference HE with TPsm HES and matched HE with TPrg and PFint HES [3]. Thermal load HE Q = $52.82 \cdot 10^3$ W ($Q_{HE\text{-}HESst} = Q_{HE\text{-}HESrg} = Q_{HE\text{-}HESint}$). Values at entry to HE are: air velocity $w_{a\text{-}\infty} = 4.70$ m/s; air temperature $t_{a\text{-}\infty} = +30$ °C; water flow rate $V_{w.en} = 0.004167$ m^3/s; water temperature $t_{w.en} = +90$ °C. Core heat exchanger – reference with TPsm HES and correlated with TPrg and PFint HTS (see figure 1): $B = 645 \cdot 10^{-3}$ m; $H = 635 \cdot 10^{-3}$ m; $L_{TPsm} = 138 \cdot 10^{-3}$ m (6-digit HE), $L_{TPrg} = 115 \cdot 10^{-3}$ m (5-row HE) and $92 \cdot 10^{-3}$ m (4-row HE). The air paths of the HE cores: with PFint HES (see figure 1, a) – $H = 30.3 \cdot 10^{-3}$ m, $h = 29.7 \cdot 10^{-3}$ m, $s = 5 \cdot 10^{-3}$ m, $\delta = 0.6 \cdot 10^{-3}$ m, $d = 7.66 \cdot 10^{-3}$ m; c TPsm and TPrg HES (see figure 1, b) – $h = 12 \cdot 10^{-3}$ m, $s = 5 \cdot 10^{-3}$ m, $t_1 = 23 \cdot 10^{-3}$ m, $t_2 = 15 \cdot 10^{-3}$ m, $u = 4.9 \cdot 10^{-3}$ m, $\delta = 0.1 \cdot 10^{-3}$ m, $d = 6.99 \cdot 10^{-3}$ m.

The calculated core parameters of the reference HE with TPsm HES are: $Nu_{TPsm} = 12.47$; $Re_{TPsm} = 2500$; $V_{HE\text{-}TPsm} = 0.056521$ m^3; $\Delta p_{HE\text{-}TPsm} = 58.47$ Pa; $\eta_{TPsm} = 0.978$; $\xi_{HE\text{-}TPsm} = 0.1355$; $\sigma_{ж.HE\text{-}TPsm} = 0.784$; $\Omega_{HE\text{-}TPsm} = 451.62$ m^2/m^3; and $M_{HE\text{-}TPsm} = 17.139$ kg.

Comparison of the results of calculations carried out by the approved method has determined the following indices of reduction in the volume of cores compared to HE: $V_{HE\text{-}TPsm}/V_{HE\text{-}TPrg} = 1.167$; $V_{HE\text{-}TPsm}/V_{HE\text{-}PFint} = 1.640$; $V_{HE\text{-}TPrg}/V_{HE\text{-}PFint} = 1.405$.

3. Conclusion

The implemented method of design calculation of heat exchangers is based on the results of systematic experimental studies of rational enhancement of convective heat transfer in non-circular ducts of plate-and fin and tube-and plate HE in two effective ways.

Comparison of the results of calculations of the same type of HE with artificial turbulence of heat carrier flow in and without ducts (for example, PFint and PFsm HES, TPrg and TPsm HES), performed in compliance with the conditioned thermo-aerodynamic boundary conditions, allows reliably determining the value of the rational convective heat transfer $(Nu_{cnf}/Nu_{st})^{max}$ with the value of the defining complex Ꝁ ≅ 1.

Comparison of the calculated data shows that for the RECH mode, the core of the 5-row HE with TPrg HES occupies 16.7% less volume than the 6-digit HE with the TPsm HES at practically equal air resistance ($\Delta p_{HE\text{-}TPrg} = 60.26$ Pa) ≅ ($\Delta p_{HE\text{-}TPsm} = 58.47$ Pa).

An even greater reduction in the volume or length of the HE core in the air (by 40.4%) ensures the use of PFint HES at ($\Delta p_{HE\text{-}PFint} = 58.81$ Pa) ≅ ($\Delta p_{HE\text{-}TPsm} = 58.47$ Pa).

The obtained results of calculations indicate a sufficiently high thermo-aerodynamic efficiency of tube-and plate heat exchange surfaces with discrete turbulators (ridges and grooves on the fins), and, especially, of plate-and fin HES with interrupted ducts.

Comparison of HE core volumes with enhanced heat exchange and smooth-channel HE based on calculations by the proposed algorithm allows determining the decrease in the length of HES channels with artificial coolant turbulence.

All compared heat exchangers with TPsm, TPrg and PFint HES had the same dimensions of the construction front. Consequently, the changes in the values of the volumes of cores compared to HE were provided by changing the depth of HE in the course of the air with respect to $L_{TPsm} = 138 \cdot 10^{-3}$ m of the reference HES accepted as 100 %: $L_{TPrg} = 115 \cdot 10^{-3}$ m or 83.3 %; $L_{PFint} = 82.3 \cdot 10^{-3}$ m or 59.6 %. It should be specially noted that compared to HE with TPsm HES, the depth of HE with PFint HES decreased noticeably – by $\Delta L_2 = 55.7 \cdot 10^{-3}$ m (40.4 %) against HE with TPrg HES – by $\Delta L_1 = 23 \cdot 10^{-3}$ m (16.7 %).

The available calculation results show that the volume and depth of HE with PFint HES decreased with respect to HE with intensified TPrg HES by $\Delta L_3 = 32.7 \cdot 10^{-3}$ m or by 28.4 %, which is noticeably better than the decrease in volume or depth of HE with TPrg HES with respect to the reference heat exchanger with TPsm HES.

If you depart from the RECH condition and in the $[(Nu_{TPrg}/Nu_{TPsm})/(\xi_{TPrg}/\xi_{TPsm})]_{Re=idem} < 1$ mode, compare the 6-bit smooth-duct HE 4-row HE with TPrg HES, then the calculations show $L_{TPrg} = 92 \cdot 10^{-3}$ м or 66.7 % of L_{TPsm}, and $\Delta L_1 = L_{TPsm} - L_{TPrg} = 138 \cdot 10^{-3} - 92 \cdot 10^{-3} = 46 \cdot 10^{-3}$ m or 33.3 % of L_{TPsm}. In this case, at the same time, the resistance of the core of the HE to the cooled blowing air is noticeably increased (by 37.8%): ($\Delta p_{HE\text{-}TPrg} = 80.54$ Па) > ($\Delta p_{HE\text{-}TPsm} = 58.47$ Па).

Efficiency and high accuracy of calculation results allow this method to become a reliable tool when comparing the heat and aerodynamic efficiency of heat exchangers of various types of structures, as well as evaluating the process of enhancement of convective heat transfer in the ducts of heat exchange surfaces.

4. References

[1] Vasilev V and Zhatkin A 2015 Promising rational enhancement of heat exchange by section of long smooth ducts of plate-fin surfaces with purposes of creation of highly effective compact heat exchangers. *The 24th IIR International Congress of Refrigeration* (Yokohama: Japan) p 8

[2] Vasilyev V and Vinokurova S and Nikiforova S 2014 Control in scale of generated vortexes at an input in not round interrupted ducts with the purpose of realization of a rational enhancement of convective heat exchanger *Proceedings of the International Symposium on Convective Heat and Mass Transfer* (Kuşadasi: Turkey) Begell House, Inc. Redding, CT 06896 (USA) p 14

[3] Vasiliev V Ya and Zhatkin A M and Nikiforova S A 2016 Comparison of heat and aerodynamic efficiency of processes of energy-saving heat transfer enhancement in non-circular ducts of plate-and fin surfaces. *Collection of reports of the All-Russian scientific and practical conference "Modern technologies in shipbuilding and aviation education, science, and production"* (Nizhny Novgorod) p 10

IOP Conf. Series: Journal of Physics: Conf. Series **980** (2018) 012022 doi:10.1088/1742-6596/980/1/012022

Experimental investigation of hydrodynamics and heat exchange in the ring channel with heat exchangers in the modes of single-phase convection and bubble boiling

B Y Agishev[1], E A Boltenko[2], A N Varava[1], A V Dedov[1], A V Zakharenkov[1], A T Komov[1], Y V Smorchova[1]

[1]National Research University "Moscow Power Engineering Institute", Moscow 111250, 14 Krasnokazarmennaya St., Russia
[2]Physics Laboratory, Elektrogorsk Research Center on the Safety of Nuclear Power Plants, Moscow Area, 142530, Russia, Elektrogorsk, 6 Constantine St., Russia

Email: ZakharenkovAV@mpei.ru

Abstract. The effectiveness of the heat exchange intensifier "rib-twisted wire" is considered in this paper. The main goal is to study the influence of the wire coiling step t on heat transfer and hydraulic resistance for different values $\dot{H}$ of the dimensionless height of the edge $\dot{H}$, as well as some results on heat exchange during bubbly boiling in an annular channel.
Show:

- a brief description and an image of the heat exchange intensifier "rib-twisted wire";
- generalized results of studies of heat exchange and hydraulic resistance in the annular channel in the single-phase convection with different geometric characteristics of the intensifier;
- empirical correlations of the generalized experimental results that allow to calculating the coefficient of hydraulic resistance and heat transfer in the range of regime parameters in the single-phase convection that is being studied.
- some results of experiments in bubbly boiling regimes and near-critical thermal loads.

1.Introduction

Today various methods for heat exchange intensification have been proposed and are used effectively [1, 2]. However, the application of these methods to a convex heated surface does not lead to a noticeable increase in heat transfer, and sometimes a negative result is observed. The work carried out earlier by the team of authors [3, 4] showed the effectiveness of using the method of heat exchange intensification on a convex heated surface founded the interaction of swirling and axial (transit) flows [5]. The point of the method is the organization of two interacting streams with the help of the "rib-twisted wire" intensifier in the annular channel. Analysis of extensive experimental data [3, 4] showed that the use of intensifiers of this type provides a significant increase in the heat transfer coefficient in comparison with a smooth annular channel with an acceptable increase in hydraulic resistance. The maximum value of the heat transfer coefficient was obtained with the value of the dimensionless height of the edge $\dot{H}$ = 0.35, and the maximum of the hydraulic resistance coefficient was obtained with $\dot{H}$ = 0.23. Since all the experiments were carried out at the same step of the intensifier wire coiling t = 50 mm, it seemed appropriate to expand the studies in which the value of the coiling step would change for different values $\dot{H}$ of the dimensionless height of the edge $\dot{H}$.

Content from this work may be used under the terms of the Creative Commons Attribution 3.0 licence. Any further distribution of this work must maintain attribution to the author(s) and the title of the work, journal citation and DOI.
Published under licence by IOP Publishing Ltd

IOP Conf. Series: Journal of Physics: Conf. Series **980** (2018) 012022 doi:10.1088/1742-6596/980/1/012022

2. Description of the intensifier "rib-twisted wire"

The research cycle was carried out at the "TVS MEI" test-bed, operating in the range of technological parameters corresponding to the VVER-1000 reactor. Its description and main technical characteristics are presented in [6] and were discussed at conferences [7].

Fig. 1 shows the image of the used intensifier. Longitudinal ribs (2) were attached to the outer surface of the heating element (1), on which a twisting device (wire coiling) was mounted (3). All the elements of the intensifier were made of steel grade 08X18H10T, and their fastening was carried out with laser welding. The intensifiers were installed to provide the clearance of 0.2 mm between the unheated concave surface and the wire coiling, conditioned by the structural features of the installation and the convenience of mounting.

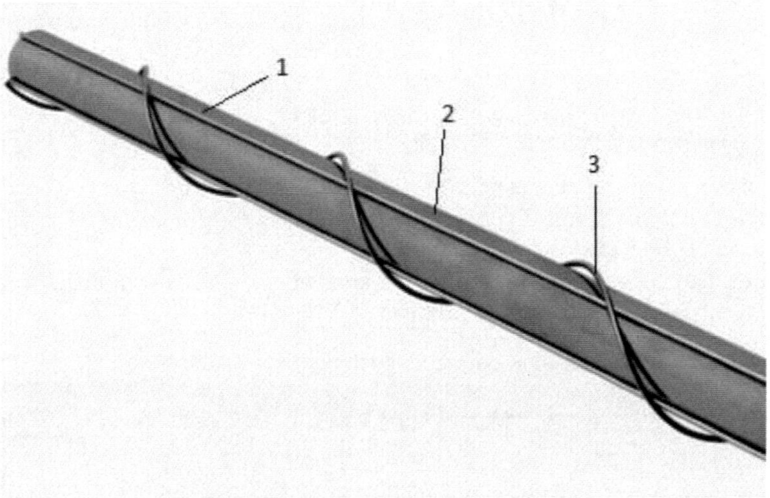

Fig. 1. Elements of the intensifier and their location on the working area: 1 – heating element, 2 – longitudinal rib, 3 – wire coiling.

The detailed description of the working area design and the intensifier is presented in [4].

The main geometric characteristics of the intensifier are the height of the longitudinal edge *h*, the diameter of the twisted wire *d* and the coiling step *t*. For the convenience of generalization of the experimental data, the dimensionless height of the edge $\dot{H}$ and the twist coefficient *k* related to the step *t* are introduced:

$$H^{\&} = \frac{h}{(d_2 - d_1)/2}, k = \pi \bar{d} / t,$$

where d_1 = 12.9 mm is the internal diameter of the annular channel (outer diameter of the heating element), d_2 = 16.3 mm is the outer diameter of the annular channel and *d* is the diameter of the median line of the swirling flow, determined by the ratio:

$$\bar{d} = \frac{d_2 + d_1}{2} + h.$$

Table 1 shows the values of the geometric characteristics of the intensifiers used. The "+" sign in the table indicates that this step was used for the represented value of $\dot{H}$.

Table 1. Geometric characteristics of intensifiers

Parameter	Value					
h, mm	1.00	0.90	0.75	0.60	0.50	0.40
d, mm	0.50	0.60	0.75	0.90	1.00	1.10
$\dot{H}$	0.59	0.53	0.44	0.35	0.29	0.23

t=40 mm	+	+	+	+		+
t=50 mm	+	+		+	+	
t=60 mm	+	+	+	+	+	+
t=100 mm		+	+	+		+

3. Effect of an intensifier on heat exchange and hydrodynamics in the single-phase convection

To calculate the heat transfer coefficient in a smooth annular channel, the modified Isachenko-Galin formula is used:

$$Nu = 0.017\,\mathrm{Re}^{0.8}\,\mathrm{Pr}_{жс}^{\;0.4}\left(\frac{\mathrm{Pr}_{жс}}{\mathrm{Pr}_{cm}}\right)^{0.25}\left(\frac{d_2}{d_1}\right)^{0.18}.$$

The measurement results are presented as Nu/A = f(Re), where $A = \mathrm{Pr}_{жс}^{\;0,4}(\frac{\mathrm{Pr}_{жс}}{\mathrm{Pr}_{cm}})^{0,25}(\frac{d_2}{d_1})^{0,18}$.

Parameter A considers the properties change (kinematic viscosity and coefficient of thermal diffusivity) at various temperatures of the coolant in the working area, as well as the geometric dimensions of the annular channel.

The influence of the geometric parameters of the intensifier on the heat transfer in the form of the dependences $Nu/A = f(Re)$ and $Nu/A = f(\dot{H})$ is shown on Fig. 2.

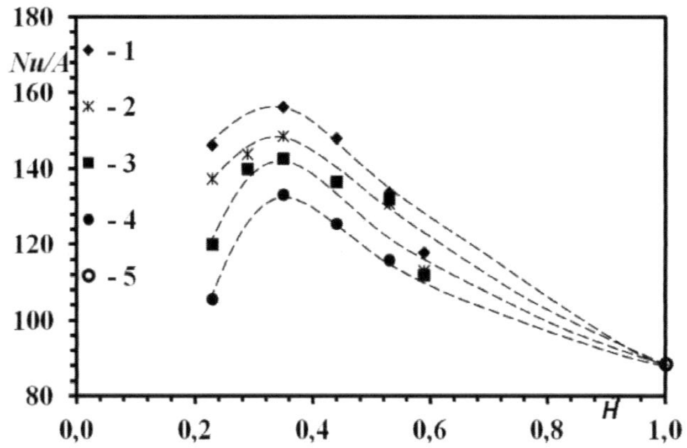

Fig. 2. The dependence of Nu/A on the parameter $\dot{H}$ (Re = 45000):
1 – t = 40·mm.
2 – t = 50·mm.
3 – t= 60·mm.
4 – t = 100·mm.
5 – smooth annular channel (experiment).

As follows from the analysis of Fig. 2 the dependence $Nu/A=f(\dot{H})$ is not monotonic and has a maximum value at $\dot{H} = 0.35$ for all values of the coiling step t. The same value was obtained in [4]. in which the research was carried out only for a coiling step t = 50 mm.

One of the conditions for the efficiency of any heat exchange intensifier is exceedance the heat transfer coefficient against the hydraulic resistance coefficient. For this purpose, studies of hydrodynamic resistance were carried out in the entire range of $\dot{H}$ and t. During the experiment, measurements of pressure losses, mass flow and temperatures at the inlet and outlet of the working area were made.

Fig. 3 shows the dependence of the coefficient of hydraulic resistance on the dimensionless parameter $\dot{H}$ for all steps of wire coiling. This dependence is not monotonic, and the maximum ξ/ξ_0 is observed at $\dot{H} = 0.23$ for all values of t.

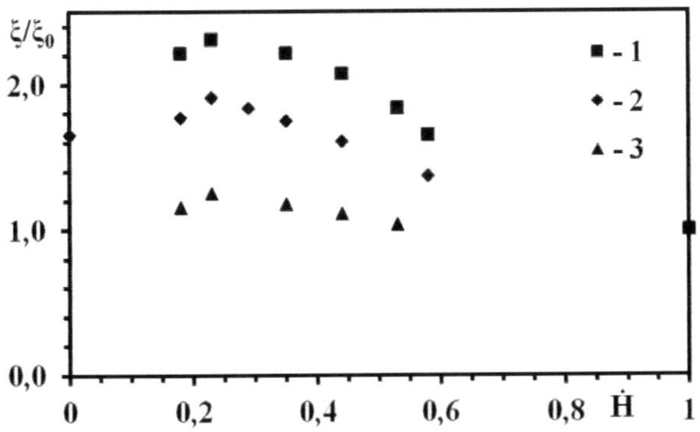

Fig. 3. The dependence ξ/ξ_0 on the parameter $\dot{H}$ (Re = 60000):
1 – t = 40·mm;
2 – t = 50·mm;
3 – t = 100·mm.

4. Calculated ratio

The obtained array of experimental data allows getting equations for hydraulic resistance and coefficient with the given type of intensifier.

The structure of the equations has the following form:

$$\frac{\xi}{\xi_0} = 1 + C_1 k^{m_1}\left(\frac{1}{\dot{H}}\right)^{n_1}; \frac{Nu}{Nu_0} = 1 + C_2 k^{m_2}\left(\frac{1}{\dot{H}}\right)^{n_2}.$$

As a result of the experimental data processing, the following equations were obtained:

$$\frac{Nu}{Nu_0} = 1 + 0{,}37k^{0{,}18}\left(\frac{1}{\dot{H}}\right)^{0{,}43}. \quad (1)$$

$$\frac{\xi}{\xi_0} = 1 + 0{,}54k^{0{,}55}\left(\frac{1}{\dot{H}}\right)^{0{,}31}. \quad (2)$$

Here Nu_0 and ξ_0 are the Nusselt number and the hydraulic resistance coefficient of the smooth annular channel, respectively. The values of the exponents m_1. m_2 and the constants C_1 and C_2 in the equations are obtained by averaging the entire array of experimental data, and the deviations of the experimental data from the results of calculations on the obtained correlations do not exceed the experimental data error estimated by the authors at 12%.

5. Results of bubbly boiling experiments

To review the effectiveness of the studied intensifier in the bubbly boiling, experiments were carried out using the technique, described below:

1. A preselected working pressure was set in the circuit. The circulation pump was switched on, and a certain consumption rate of the coolant was selected.
2. At the selected pressure, using the heaters located in the circuit, the value of the water inlet temperature in the working area was set equal to the saturation temperature.
3. In the absence of a load on the working area heater, the indications of the wall thermocouples were recorded in order to test them and to measure the thermal losses in the working area.
4. 4. The load was applied to the main heater of the working area, and the bubbly boiling regime was reached by a gradual increase in the power of the heater. The indications of thermocouples located on the internal adiabatic surface of the heater tube, the temperature of the water at the inlet and outlet of the working area, the pressure in the circuit, the pressure losses in the working area, the current strength and the voltage on the heater were recorded. In the process of entering the new mode, all parameters were recorded by ASNI. The start of bubbly boiling was characterized by a sudden change in the consumption rate of the coolant and pressure losses in the working area, which happened due to the formation of the vapor phase and additional pressure losses to change the velocity of the flow.

Fig. 4 shows the distribution of the temperature field along the length of the working area at a pressure p = 3.0 MPa for the intensifier with parameters $\dot{H}$ = 0.35 and t = 50 mm. Fig. 5 shows the distribution of the liquid temperature in the annular channel at a constant power of the main heater for different consumption rates.

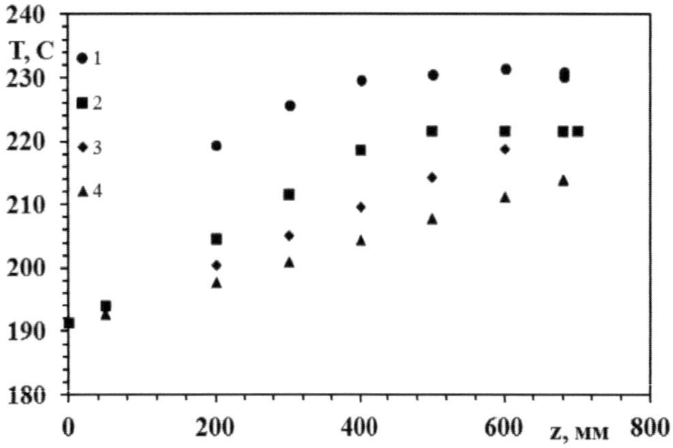

Fig. 4. The distribution of the temperatures of the wall (1) and the liquid (2, 3, 4) along the length of the working area (t = 50 mm. $\dot{H}$ = 0.35 and G = 0.25 kg/s):
1 – N_{el} = 55 kW; 2 – N_{el} = 55 kW;
3 – N_{el} = 50 kW; 4 – N_{el} = 40 kW.

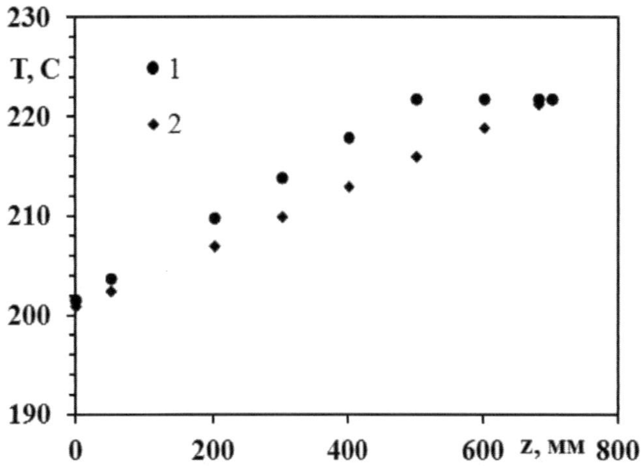

Fig. 5. Distribution of coolant temperature along the length of the working area (N_{el} = 27 kW):
1 – G = 0.189 kg/s;
2 – G = 0.235 kg/s.

The processing of the experimental results enabled calculating the heat transfer coefficient at various regime parameters. The experimental values of the heat transfer coefficient under bubbly boiling conditions were α_{exp} = (110÷180) kW / (m^2·K). The boiling component α_{boil} was isolated using the interpolation formula of S.S. Kutateladze:

$$\alpha_{exp} = \alpha_{con}\sqrt{1 + (\frac{\alpha_{boil}}{\alpha_{con}})^2} = \alpha_{con}{}^2 + \alpha_{boil}{}^2$$

Values α_{con} were selected from the experimental data obtained earlier.

Comparison of experimental values of a_{boii} with known correlations of Borishanskiy V.M., Yagova V.V. and Rassokhina N.G. has showed a noticeable increase in a_{boii} due to the introduction of an intensifier.

The near-critical and critical regimes that determine the boiling crisis appeared at input capacities of more than 50 kW and mass expenditures (0.20 ÷ 0.35) kg/s. The values of the critical thermal loads obtained in 3 working areas are experimentally established. The heat exchange crisis occurred approximately in the same zone along the channel path so quickly that the protection system did not have time to prevent its destruction. A typical example of the destroyed working area is shown in Fig.6.

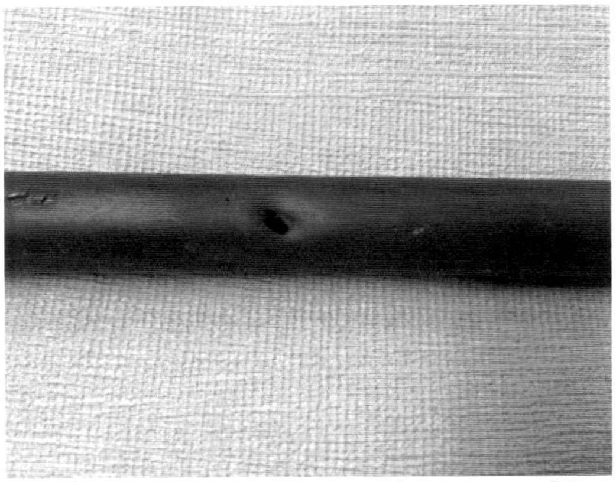

Fig. 6. Consequences of the heat exchange crisis at one of the working areas.

At the same time, the values of critical heat flux density turn out to be equal to q_{cr1} = 1.86 MW/m^2 at ρw = 3079 kg/(m^2 s); q_{cr2} = 1.97 MW/m^2 at ρw = 3720 kg/(m^2 s); and q_{cr3} = 2.23 MW/m^2 at ρw = 4490 kg/(m^2 s). The geometry of the annular channel with intensifier is such that, for a coolant flow with an inlet temperature close to saturation temperature and significant thermal loads, all heat exchange regimes apparently take place: convective, local and developed bubbly boiling, transient boiling, and heat exchange crisis. Calculations carried out using the known correlations [9] have showed that the experimental values obtained by the authors turn out to be lower than the calculated ones. Work in this direction continues.

The work was supported by the Russian Science Foundation under the agreement no. 16-19-10457.

References:

[1] Leontyev A and Olimpiev V 2013 *Thermal engineering* vol 1 (Moscow: MAIK Science/Interperiodika) pp 68-78
[2] Popov I, Yakovlev A, Shchelchkov A, Ryzhkov D and Obukhova L 2011 *Power engineering of Tatarstan Perspective* vol 1 (Kasan: Power engineering of Tatarstan) pp 25-29
[3] Agishev B, Boltenko E, Varava A, Dedov A, Zakharenkov A and Komov A 2015 *New trends in the Russian electric power industry* vol 12 (Moscow: energo-press) pp 6-15
[4] Boltenko E, Varava A Dedov A, Zakharenkov A, Komov A and Malakhovskiy S 2015 *Thermal engineering* vol 3 (Moscow: MAIK Science/Interperiodika) pp 22-28
[5] Boltenko E 1992 Patent 1540426 Russia MKI 3F28F13/12 Heat transfer device 31
[6] Zakharenkov A, Komov A, Varava A et al 2013 *Vestnik MEI* vol 2 (Moscow: Publishing house MEI) pp 39-45.
[7] Komov A, Varava A, Dedov A, Boltenko E, Zakharenkov A and Agishev B 2014 *6th Russian National Heat Exchange Conference* (Moscow: Publishing house MEI) CD
[8] Tsvetkov F F and Grigoryev B A 2005 *Heat exchange: tutorial for high schools : 2nd ed. amended and supplemented* (Moscow: Publishing house MEI) pp 550
[9] Petukhov B, Genin L and Kovalev S 2003 *Heat exchange in nuclear power plants* (Moscow: Publishing house MEI) pp 548

IOP Conf. Series: Journal of Physics: Conf. Series **980** (2018) 012023 doi:10.1088/1742-6596/980/1/012023

Calculating processes of laminar and turbulent heat transfer around the elements of the aircraft

V V Kuzenov[1,2], A O Dobrynina[1] and V V Shumaev[1]

[1] Thermal Physics Department, Bauman Moscow State Technical University, Russia, 105005, Moscow, 2-nd Baumanskaya, 5-1
[2] Institute for Problems in Mechanics of the Russian Academy of Sciences, Russia, 119526, Moscow, Prospect Vernadskogo, 101-1

E-mail: dobryyninaao@student.bmstu.ru, shumaev@student.bmstu.ru

Abstract. The physical-mathematical model and a numerical method for solving the nonstationary two-dimensional one-fluid (one-temperature) Reynolds equations closed with turbulence models, taking into account the diffusion and thermal conductivity processes for bodies of simple geometric forms were described and verified. The effects of the temperature field inside the streamlined body and at changing its shape under the action of temperature stresses were analysed. Temperature fields and thermal stresses in the simple structural elements have been obtained. The corrections to the value of convective heat flux due to non-uniform surface temperatures, the presence of temperature gradients and the shape change may be significant and have to be considered to increase the accuracy of aerothermodynamic modelling.

1. Introduction

A great interest of researchers is attracted to the problems of high-speed aircraft aerothermodynamics. Commercial software does not give a solution of the required accuracy for the case of hypersonic flows. This is confirmed by numerous works in which the LAURA software package is used for these complex purposes. LAURA is not a commercial US code and is not sold by the US government [1]. Therefore, this problem is new and relevant. It is necessary to know the surface temperature of high-speed aircraft for accurate calculation of its external aerothermodynamics (e.g. when there is a need to take into account chemical reactions). Many computational techniques use the approximation of a constant surface temperature and often greatly underestimate it. This can lead to unaccounted heat fluxes, which are proportional to the temperature gradients of the solid and its vicinity [2-9]. The second important effect is the change in the body shape (in particular, radius of curvature of the body) under the influence of thermal stress. Since the thermal flows depend on the form of the streamlined body, the problem has a self-adjoint character in the general case [10-16].

This paper aims to describe, validate and verify the engineering-physical-mathematical model and a numerical method for solving the nonstationary two-dimensional one-fluid (one-temperature) Reynolds equations closed with turbulence models, taking into account the diffusion and thermal conductivity processes for bodies of simple geometric forms. Another task is to study the influence of thermal effects on the accuracy of aerothermodynamic calculations. The mathematical statement of the

Content from this work may be used under the terms of the Creative Commons Attribution 3.0 licence. Any further distribution of this work must maintain attribution to the author(s) and the title of the work, journal citation and DOI.
Published under licence by IOP Publishing Ltd

problem and a brief description of the numerical technique are given. For verification and validation of the developed computer codes, two test problems are solved. The second test is based on the estimates of heat fluxes in a blunt cylinder with a streamlined high-speed air flow. Heat flows that are obtained on the basis of the solution of the problem of external aerothermodynamics under the assumption of a constant surface temperature are known.

2. The description of the mathematical model

A mathematical model of the considered processes is based on multicomponent Reynolds equations. To solve these equations by the finite-difference method a coordinate transformation $r = r(\xi,\eta,\zeta)$, $z = z(\xi,\eta,\zeta)$, $\varphi = \varphi(\xi,\eta,\zeta)$ is introduced. These equations have the following form:

$$\frac{\partial \rho c_i}{\partial t} + Div\left(\rho c_i \vec{V}\right) = -\alpha \frac{\rho c_i u}{r} + Div\left(\rho D_i \nabla c_i\right) + \left(\frac{\partial \rho c_i}{\partial t}\right)_x, \quad \frac{\partial \rho}{\partial t} + Div\left(\rho \vec{V}\right) = -\alpha \frac{\rho u}{r}, \tag{1}$$

$$\frac{\partial \rho u}{\partial t} + Div\left(\rho u \vec{V}\right) = -\xi_r \frac{\partial P}{\partial \xi} - \eta_r \frac{\partial P}{\partial \eta} - \alpha \frac{\rho u^2}{r} + \frac{S_r}{\mathrm{Re}}, \quad \frac{\partial \rho \mathrm{v}}{\partial t} + Div\left(\rho \mathrm{v} \vec{V}\right) = -\xi_z \frac{\partial P}{\partial \xi} - \eta_z \frac{\partial P}{\partial \eta} - \alpha \frac{\rho u \mathrm{v}}{r} + \frac{S_z}{\mathrm{Re}},$$

$$\frac{\partial \rho e}{\partial t} + Div\left(\rho e \vec{V} + \sum \vec{q}_i\right) = -\frac{P}{J} Div\left(\vec{V}\right) - \alpha \frac{Pu}{r} - \alpha \frac{\rho e u}{r} + \frac{S_e}{\mathrm{Re}} + D_x, \quad S_e = \mu_\Sigma D + \frac{\gamma}{\mathrm{Pr}} div\left(\lambda_\Sigma grad T\right),$$

$$D_x = \sum_i h_i \frac{t_*}{e_*} Div\left(\rho D_i \nabla c_i\right), \quad Div(\ ) = \frac{1}{J}\frac{\partial (J\ )}{\partial \xi} + \frac{1}{J}\frac{\partial (J\ )}{\partial \eta},$$

where S_r, S_z are the forces [6, 17], arising due to the presence of viscous friction forces, S_e is the volumetric (specific) energy release, Re is the Reynolds number, Pr is the Prandtl number, $u(r,z,t), \mathrm{v}(r,z,t)$ is the projection of the velocity vector $\vec{V}(r,z,t)$ on the axis *R* and *Z*, e is the specific internal energy of gas, $J = \partial(r,z)/\partial(\xi,\eta)$ is the Jacobian of transition from a cylindrical coordinate system r,z to a curvilinear coordinate system ξ,η, $V_\xi = \xi_r u + \xi_z \mathrm{v}$, $V_\eta = \eta_r u + \eta_z \mathrm{v}$ are the contravariant components of the velocity vector $\vec{V}(r,z,t)$ in the curvilinear coordinate system ξ,η, ρ, P are the density and pressure of gas, $\sum_i q_{i\xi}, \sum_i q_{i\eta}$ are the projections of the vector of the flux density of radiant energy $\vec{q}$ on the axis of the curvilinear coordinate system ξ and η, $\alpha = 0$ corresponds to the flat geometry, and $\alpha = 1$ corresponds to the one-dimensional cylindrical geometry.

The transfer of broadband radiation is considered using a multigroup diffusion approximation, the equations of which have the following form:

$$\frac{1}{J}\frac{\partial\left(J q_{i\xi}\right)}{\partial \xi} + \frac{1}{J}\frac{\partial\left(J q_{i\eta}\right)}{\partial \eta} + \chi_i c U_i = 4\chi_i \sigma_i T^4, \quad \frac{c}{3}\frac{\partial U_i}{\partial \xi} + \chi_i q_{i\xi} = 0, \quad \frac{c}{3}\frac{\partial U_i}{\partial \eta} + \chi_i q_{i\eta} = 0, \tag{2}$$

where $U_i(y,z,t)$ is the bulk density of broadband radiation for *i*–th spectral group, and χ_i is the spectral absorption coefficient.

The optical parameters $\chi_i(T, \rho)$ were calculated using a computer system ASTEROID [7]. The turbulent coefficients of viscosity μ_Σ and thermal conductivity λ_Σ were calculated with the help of the $q-\omega$ Coakley model in a curvilinear coordinate system ξ,η.

The solution to two-dimensional non-stationary equations of viscous one-temperature radiation gas dynamics, written in vector semi-divergence form in a curvilinear coordinate is constructed using the method of splitting into physical processes and directions. These equations are:

$$\frac{\partial \vec{U}}{\partial t} + \frac{\partial \vec{F}}{\partial \xi} + \frac{\partial \vec{G}}{\partial \eta} + \vec{S} = \frac{1}{\mathrm{Re}}\left[\frac{\partial \vec{F}_v}{\partial \xi} + \frac{\partial \vec{G}_v}{\partial \eta} + \vec{S}_v\right]. \tag{3}$$

The type of vectors included in the system of equations is specified in [6, 17]. This system of equations (the time *t* is a variable) includes ordinary differential equations of the first order, which can be solved using the multi-stage Runge-Kutta method (this work applies the three-step variant of this method).

In the first fractional time step $t \in [t, t+\Delta t/3]$, we use a nonlinear quasimonotone compact-polynomial finite-difference scheme of high order of accuracy, developed in [6, 17]:

$$\frac{\partial \vec{U}_{i,j}}{\partial t} + \frac{\vec{F}_{i+1/2,j} - \vec{F}_{i-1/2,j}}{\Delta \xi} + \frac{\vec{G}_{i,j+1/2} - \vec{G}_{i,j-1/2}}{\Delta \eta} + \vec{S}_{i,j} = 0 . \quad (4)$$

On the second fractional time step $t \in [t+\Delta t/3, t+2\Delta t/3]$ the "parabolic" ("viscous") part of the partial differential equation system was solved by the explicit method with adhesion boundary conditions. The first and second spatial derivatives included in this system of equations on the second time step, were found using quasimonotone compact-polynomial finite-difference schemes of high order of accuracy.

At the third fractional time step $t \in [t+2\Delta t/3, t+\Delta t]$ the "hard" part of the $q-\omega$ model system of equations is calculated using an implicit Rosenbrock method. The time step Δt required to integrate this differential scheme is selected taking into account the criterion of Courant – Friedrichs – Levy.

A modified alternating-triangular method using a three-layer iteration scheme is applied for the solution of radiative transfer equations. In this iteration scheme the "time" step is found using the method of conjugate directions [16].

3. Verification and results of numerical simulation

The validation and verification process for the above engineering-physical-mathematical model of heat exchange near the surface of the aircraft can be divided into several stages:

- 2D testing of the developed engineering-physical-mathematical model and a numerical method for solving the Euler equations using bodies of simple geometric forms;
- 3D validation and verification by methods, using unstructured grids for the sample of a complex geometric shape.

For 2D testing, developed with the use of the engineering-physical and mathematical methods of the aircraft, the following set of bodies of simple geometric shapes (separate elements of the HA construction) is used here:

- a wedge interfaced with a plate;
- a cone conjugate to a cylinder;
- a cylinder coupled with a plate;
- a cone blunt in the sphere and conjugated to a cylinder;
- a hypersonic flow around the blunted axisymmetric bodies (in this case, we can use the calculated and experimental data [14]).

In 2D testing, the search for a solution corresponding to the high-temperature heterogeneous gas flow near the aircraft surface may be based on the solution of nonstationary two-dimensional one-fluid (one-temperature) Reynolds equations closed with $q-\omega$ turbulence models, taking into account the diffusion and thermal conductivity processes [18-23]. A special case of such a system of equations is the system of Euler equations. The computational domain includes the flow field in the undisturbed flow behind the front of the shock wave, in the wake of the streamlined body. The procedure for the numerical solution of the Euler equations with the help of a nonlinear quasimonotonic compact polynomial difference scheme of higher order of accuracy consists of three stages:

- integration of equations with respect to the time variable t and transition to a new time layer $t+\Delta t$;
- reconstruction of the values of variables on the faces of the calculated cells from the values in the cell centers;

• calculation of numerical flows through the faces of computed cells by means of an exact or approximate solution of the problem of disintegration of a discontinuity (or its analogues) between values on two (opposite) sides of each face.

For 2D testing of the numerical method for solving the Euler equations, the following parameters of the air flow running onto a body of simple geometric shape were used:

Pressure in the flowing stream: $P = 2060$ Pa;

Speed in the flowing stream: $V = 1860$ m / s;

Temperature in the flowing stream: $T = 223$ K;

Mach number in the flowing stream: $M = 6$;

Gas of the flowing stream: Air;

Height from the surface of the Earth: $H = 25$ km;

All gas-dynamic characteristics behind a flat or conical shock wave (the wedge flow past the plate is considered) are determined [2] (for a given Mach number of the unperturbed gas flow M_∞) by the angle of the wedge or cone β solution, streamlined by supersonic gas flow, and angle of inclination α of the front of the shock wave (figure 1).

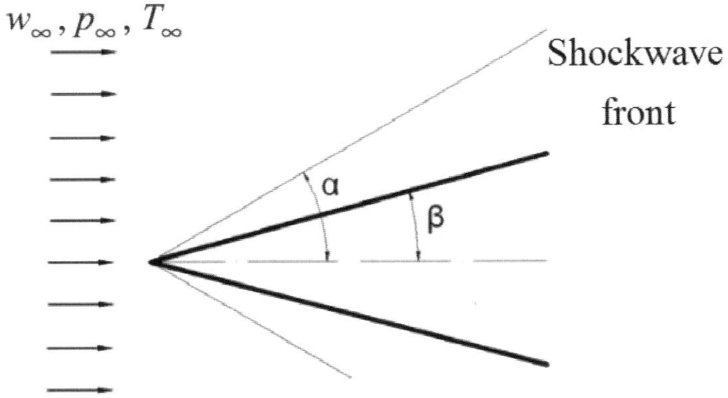

Figure 1. Scheme of the shockwave front location relative to the wedge or cone in the case of supersonic gas flow at zero angle of attack.

In the planar case [2], knowing the angle α (figure 1), we can determine the Mach number $M_{\infty,n}$, which is calculated from the normal velocity component (to the front of the shock wave). This Mach number $M_{\infty,n}$ allows finding the values of all gas-dynamical variables behind the front of the shock wave.

In the case of supersonic flow around the cone for 2D testing of the numerical method for solving the Euler equations, one can use the graphical dependence of the slope angle β of the shock wave on the angle of deflection of the supersonic gas flow for different values of the M_∞ number. It should be also noted here that, since the thickness of the shock wave is very small, the formulas for calculating the gasdynamic parameters behind the plane of the plane-parallel oblique shock ($P_2/P_\infty, \rho_2/\rho_\infty, w_2/w_\infty$ and e.t.c.) are applicable to the calculation of the parameters behind an axisymmetric shock wave (the spatial effects across the shock front are small).

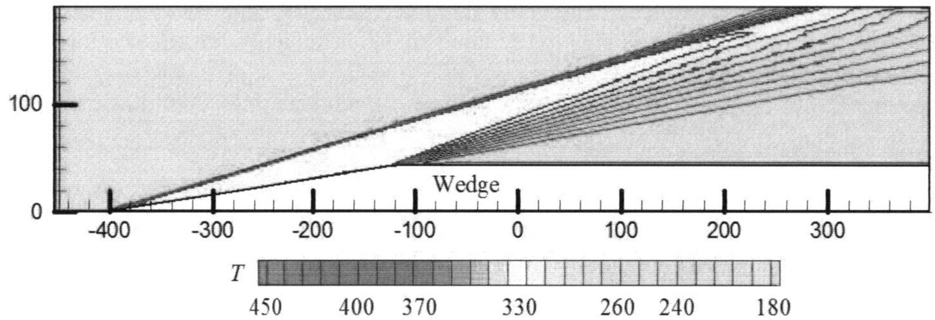

Figure 2. Temperature distribution in flow around an acute wedge.

From the analysis of the graphical dependences shown in figure 2 [17], it follows that the error in the deviation of the inclination angle β of the shock wave is a value $\leq 1\%$, and in gas-dynamic quantities it is of the order of 3%.

The following problem was to find the distribution of temperature and displacements in the structural element of the aircraft - in a cylinder blunted over the sphere. In solving this problem, the following geometry of the aircraft element (spherical blunting radius of 1.27 cm, cylinder length of 10 cm) was given, and the convective heat flux incident on the surface of the element under consideration was specified.

Therefore, at high velocities and, respectively, large deceleration temperatures, while the blunting radius decreases at the critical point, the values of the convective and radiation fluxes sharply increase (if the inverse heat flux q_w of the body surface temperature T_w is not taken into account). These formulas also demonstrate a noticeable effect on the convective heat flux q_w of the surface temperature T_w of the aircraft element.

To find the temperature distribution and displacements in a spherically blunted cylinder the following geometry of the aircraft element (spherical blunting radius is 1.3 cm) and the convective heat flux distribution were taken from [4]. The velocity of the oncoming stream corresponds to the Mach number M = 10. We note that these aerothermophysical values correspond to the hypersonic regime of gas flow.

The calculations of this work (figures 3, 4) allow us to estimate the effect of the blunting radius, the geometric shape, the structural material and the surface temperature distribution on the convective heat flux q_w.

It is assumed that the loss of internal energy by an airborne element is realized only through thermal radiation. The element of the aircraft itself is made of tungsten and rigidly embedded. Figures below show the results of calculations based on the above technique for the element of the aircraft under consideration.

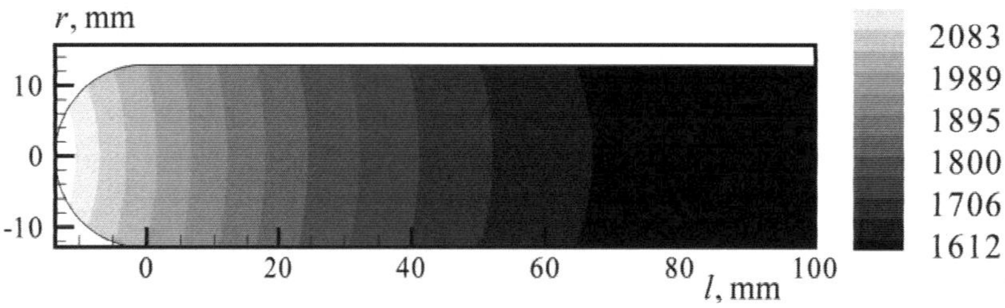

Figure 3. The temperature distribution T [K] in a spherically blunted cylinder.

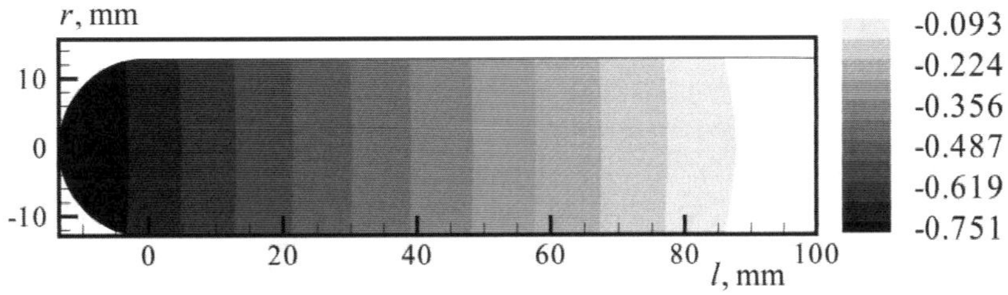

Figure 4. The distribution of displacements U_l [mm] in the direction of a blunted cylinder axis of symmetry.

Figures 3, 4 show, that the temperature distribution of the surface T_w = 1600 - 2100 K, which is streamlined by the external flux of the aircraft element, differs from the constant and significantly exceeds the temperature of the element surface, which was accepted as the boundary condition in the calculation of the aerothermodynamics of the aircraft (this temperature is usually ~ 300 K). The blunting radius changes by 5%. At the same time, the convective heat flux at the front critical point due to the change in the radius of curvature increases by 3%. However, if the substantial growth of the surface temperature is taken into account, then the change will be at the level of 10% near the critical point and 65-70% along the surface of the cylinder.

4. Conclusion

The engineering-physical-mathematical model and a numerical method have been developed and verified by the acute wedge flow test task. The error in the deviation of the inclination angle of the shock wave was a value $\leq 1\%$, and in gas-dynamic quantities it was of the order of 3%.

It has been shown that the temperature of the surface of a cylinder, blunted over a sphere and streamlined by a hypersonic air flow, is not constant and substantially exceeds the surface temperature of the element, usually taken as a boundary condition in the calculation of aerothermodynamics of aircraft. The convective heat flux at the front critical point of the elements under consideration may increase by 10-20% due to a significant increase in the surface temperature and a change in the radius of blunting due to thermal deformation. Corrections to the value of convective heat flux in the problems of external aerodynamics must be taken into account. The consideration of the conjugate problem will increase the accuracy of aerothermodynamic modeling.

Acknowledgment
Author (VVK) is grateful to Prof. S.T. Surzhikov and staff of Laboratory of Radiative Gas Dynamics of Institute for Problems in Mechanics for useful discussions of the problems considered in the study. The presented study was supported by the Russian Science Foundation project № 16-11-10275.

References

[1] Chatwood F M and Gnoffo P A 1996 *User's Manual for the Langley Aerothermodynamic Upwind Algorithm* (LAURA) NASA TM-4674
[2] Lunev V V 2010 *Real gas flows with high velocities* CRC Press
[3] Kuzenov V V and Ryzhkov S V 2017 Numerical simulation of the effect of laser radiation on matter in an external magnetic field *Journal of Physics: Conference Series* **830**(012124)
[4] Surzhikov S T 2013 Convective heating of small-radius spherical blunting for relatively low hypersonic velocities *High Temperature* **51** 231–245
[5] Kuzenov V V, Ryzhkov S V and Shumaev V V 2014 Thermodynamic properties of magnetized plasma evaluated by Thomas-Fermi model *Applied Physics* **3** 22-25

[6] Kuzenov V V and Ryzhkov S V 2017 Approximate method for calculating convective heat flux on the surface of bodies of simple geometric shapes *Journal of Physics: Conference Series* **815**(012024)
[7] Surzhikov S T 2003 Computing System for Solving Radiative Gasdynamic Problems of Entry and Re-Entry Space Vehicles *Proceedings of the 1st International Workshop on Radiation of High Temperature Gases in Atmospheric Entry ESA* vol. 533 pp 111-118
[8] Djadkin A, Beloshitsky A, Shuvalov M and Surzhikov S 2013 Nonequilibrium radiative gasdynamics of segmental-conical space vehicle of large size. *AIAA* 2013-98224
[9] Kuzenov V V and Ryzhkov S V 2016 Developing a procedure for calculating physical processes in combined schemes of plasma magneto–inertial confinement *Bulletin of the Russian Academy of Sciences: Physics* **80** 598-602
[10] Kuzenov V V and Ryzhkov S V 2013 Evaluation of hydrodynamic instabilities in inertial confinement fusion target in a magnetic field *Problems of Atomic Science and Technology* **4**(86) 103-107
[11] Zarubin V S and Kuvyrkin G N 2012 *Mechanics of Solids* **47** 252-260
[12] Kuzenov V V and Ryzhkov S V and Shumaev V V 2015 Numerical thermodynamic analysis of alloys for plasma electronics and advanced technologies *Problems of Atomic Science and Technology* **4**(98) 53-56
[13] Shumaev V V 2016 Calculation of thermodynamic functions of aluminium plasma for high-energy-density systems *Physics of Atomic Nuclei* **79**(9) 1414-1418
[14] Sharikov I V and Surzhikov S T 2005 Experimental Study of Unsteady Supersonic Underexpanded Ablative Plasma Jet Dynamics *AIAA* 2005-4930
[15] Kuzenov V V and Shumaev V V 2015 Description of the thermodynamic properties of plasma in Saha and Thomas-Fermi models *Applied Physics* **2** 32-36
[16] Samarskij A A and Nikolaev E S1989 *Numerical Methods for Grid Equations* vol. 1 and 2 (Birkhäuser)
[17] Kuzenov V V and Ryzhkov S V 2018 Numerical modeling of laser target compression in an external magnetic field *Mathematical Models and Computer Simulations* **10**
[18] Coakley T J 1983 Turbulence modelling methods for the compressible Navier-Stokes equations *AIAA* 1983-1693
[19] Kuzenov V V, Ryzhkov S V and Shumaev V V 2015 Application of Thomas-Fermi model to evaluation of thermodynamic properties of magnetized plasma *Problems Atomic Science Technology* **1**(95) 97-99
[20] Kuzenov V V, Ryzhkov S V and Shumaev V V 2014 Computer Analysis of Transport, Optical, and Thermodynamic Properties of Plasma *High Temperature Material Processes* **18** 99-109
[21] Loktionov E Y, Ovchinnikov A V, Protasov Y Y and Sitnikov D S 2011 Experimental investigation on spectral-energy efficiency of femtosecond laser ablation of metals *Plasma Physics Reports* **37**(13) 1208-1214
[22] Loktionov E Y and Protasov Y Y 2011 Experimental Study of the dynamics and macrostructure of laser-induced high pressure gas-plasma flows *High Temperature* **49**(1) 36-44
[23] Surzhikov S T and Shuvalov M P 2013 Checking computation data on radiative and convectional heating of next generation spacecraft *High Temperature* **51**(3) 408-420

Comparative analysis of film cooling efficiency at coolant supply into a single array of triangular dimples

A A Khalatov[1,2], O O Petliak[1], S D Severin[2] and N A Panchenko[1,2]

[1] National Technical University of Ukraine «Igor Sikorsky Kyiv Polytechnic Institute», Kyiv, Ukraine
[2] Institute of Engineering Thermophysics of NAS of Ukraine, Kyiv, Ukraine

E-mail: artem.khalatov@vortex.org.ua

Abstract. The purpose of this work is a comparative study of the physical structure and film cooling efficiency of the single array of inclined holes, placed in triangular dimples and in a trench. The software package ANSYS CFX 17.0 was used along with RANS SST turbulence model. Calculations were made in a wide range of the blowing ratio ranging from 0.5 to 2.0. Results of modeling have shown high efficiency of triangular film cooling configuration. At $m \geq 1.5$, the triangular configuration is comparable with the trench configuration in terms of the film cooling efficiency.

1. Introduction

Increase in the flow temperature in front of the gas turbine is the main trend of thermodynamic efficiency growth of the gas turbine engine. In the modern power gas turbine engines the flow temperature at the combustion chamber exit reaches 1500-1800 K, that exceeds the operating temperature of the structural materials of the blade. The turbine blades made of the best heat-resistant materials can operate without cooling at the flow temperature that is not greater than 1000-1100 °C [1]. At the high gas temperature temperatures, the external cooling is used in addition to internal convective cooling to ensure a longer service time of gas turbines. For external cooling, the film cooling is widely used when the coolant from internal blade space is supplied onto external blade surface through a number of discrete inclined round holes with a diameter of 0.3...1.0 mm.

Analysis of the obtained results [2] shows that the traditional scheme of film cooling in the form of a single array of discrete inclined round holes has a few significant disadvantages. Some of them are the low film cooling efficiency at the blowing ratio greater than unity that is caused by the appearance of secondary vortex structures, destroying the coolant film and contributing to the suction of the hot flow to the cooled blade surface, as well as the coolant separation from the cooled surface. Therefore, the search for alternative cooling schemes with a coolant flow rate that is acceptable in terms of thermodynamic efficiency, higher cooling efficiency and relatively simple manufacturing technology is the main challenge of the modern gas turbine engineering. The most promising scheme is a coolant supply into the trench [3], but for gas turbine blades this configuration is not acceptable from the prospective of long service life, as well as reduction in the strength of the thermal-protecting blade coating.

The aim of this paper is to determine the physical structure and film cooling efficiency of the alternative film cooling scheme with a coolant supply into a single array of triangular-shaped dimples [4] (figure 1) and compare it with the trench-shaped scheme that demonstrates the best cooling

Content from this work may be used under the terms of the Creative Commons Attribution 3.0 licence. Any further distribution of this work must maintain attribution to the author(s) and the title of the work, journal citation and DOI.
Published under licence by IOP Publishing Ltd

efficiency. The numerical simulation was used in this study, based on the application of commercial package ANSYS CFX17.0. Investigation was performed at the boundary conditions close to the actual gas turbine.

2. Geometric and computer models

The geometric model of the flat plate film cooling with a coolant supply through a single array of round holes in the triangular dimples with a pitch ratio $t/d = 3$ (figure 1) was performed in the ANSYS Design Manager. All dimensions of the model are shown in figure 2a. To perform CFD simulations the combined calculation grid was constructed using the ANSYS Mesh grid generator (figure 2b).

The solid boundaries borders of the calculated region were defined as adiabatic walls ($q = 0$). The symmetry of boundary conditions was used on lateral surfaces of the computational model. The Reynolds averaged Navier-Stokes correlations were solved for a viscous heat-conducting gas in the steady-state formulation of the problem using the total energy correlations.

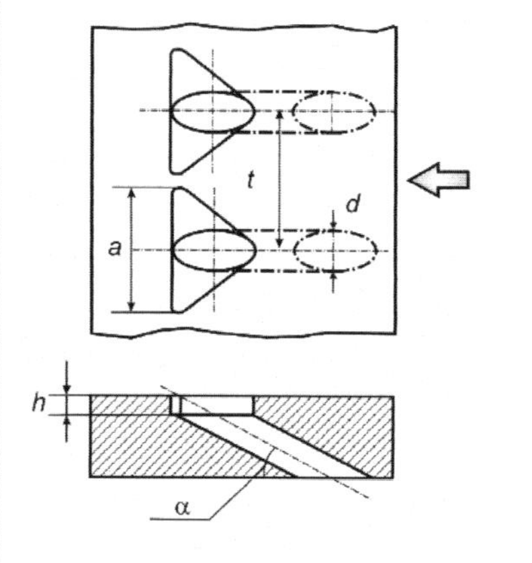

Figure 1. The scheme of triangular dimples

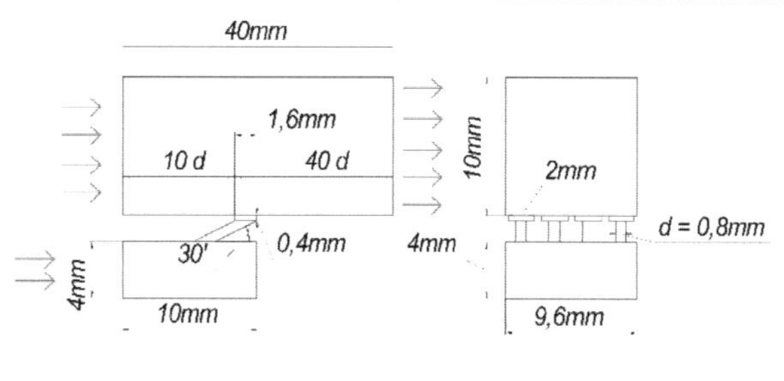

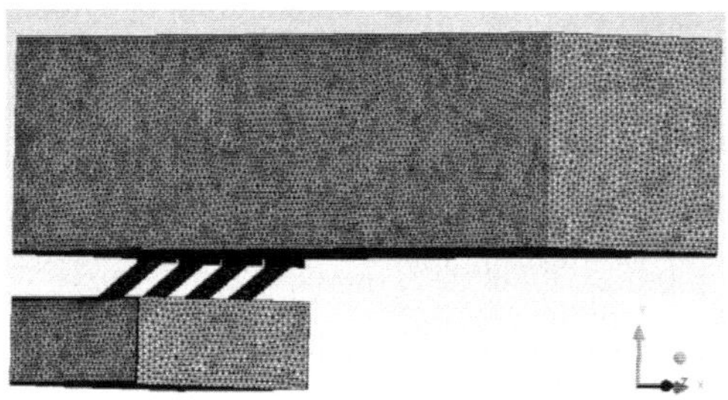

a	*b*

Figure 2. Geometric and grid models of the investigated film cooling configuration: *a* – geometric model, *b* – grid model

The uniform velocity distribution at the inlet for the main flow and the coolant flow rate for the secondary flow were used. assigned. The boundary conditions on two inlet regions (figure 2, regions for inlet No.1, inlet No.2 and outlet were marked by arrows) were set by the corresponding values of a blowing ratio m = 0.5, 1.0, 1.55, 2.0. The boundary conditions are given in table 1.

Table 1. The boundary conditions

Parameters	Inlet No.1	Inlet No.2	Outlet
Average velocity, m/s	37	-	-
Static temperature, °C	20	80	-
Mass flow rate, kg/s ($m = 1.0$)	-	0.000067	-
Mass flow rate, kg/s ($m = 1.5$)	-	0.000107	-
Mass flow rate, kg/s ($m = 2.0$)	-	0.000143	-

Mass flow rate, kg/s (m = 2.5)	-	0.000173	-
Static pressure, Pa	-	-	101315

The adiabatic film cooling efficiency, taking into account that $C_p = C_p(T)$, has the following form (1):

$$\eta = \frac{i_\infty^* - i_{aw}^*}{i_\infty^* - i_1^*}, \tag{1}$$

where $i_\infty^*, i_{aw}^*, i_1^*$ are the enthalpies of the main flow, adiabatic wall (film), and coolant, respectively.

The RANS SST model of turbulence model was used for these studies. This model was well proven at the modeling of film cooling in the previous publication of authors [5,6].

3. Results and discussion

Two well-studied schemes of the film cooling were chosen for comparative analysis, namely the traditional scheme of film cooling with coolant blowing through a single array of round inclined holes [6] and the alternative scheme with coolant supply into the trench [7].

Vortex structures for the traditional scheme and for two alternative schemes (trench and triangular dimples) are shown in figure 3 (hereinafter comparisons are given for the blowing ratio m = 1.5 at x/d = 2.0). It can be seen that scheme with a single array of inclined holes (traditional scheme) demonstrates the greatest intensity of twin vortex structures. Significant decrease in the intensity of twin vortex is observed for a scheme with triangular dimples. Each pair of vortices tightly adhere each other, thus forming a protective cooling film, which prevents the main flow suction towards the cooled surface. Quite a similar effect is also observed for the scheme with a trench. However in this case the vortex structure becomes more complicated by additional vortices of low intensity, generated by the neighbor jets.

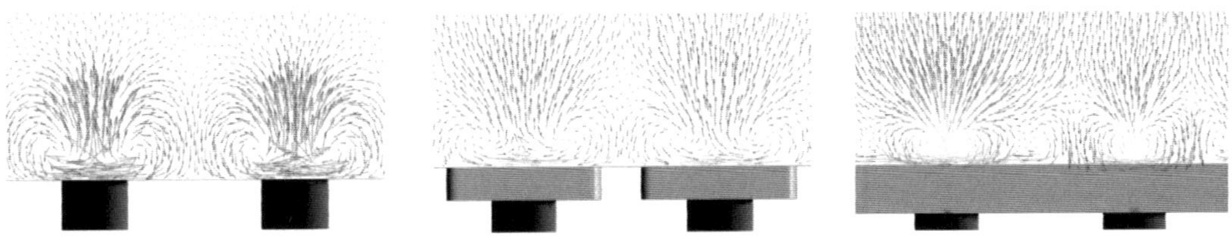

a *b* *c*

Figure. 3. The vortex flow structure at the distance x/d = 2.0 for the blowing ratio m=1.5: *a* – traditional inclined holes configuration; *b* – coolant supply into triangular dimples; *c* – coolant supply into the trench

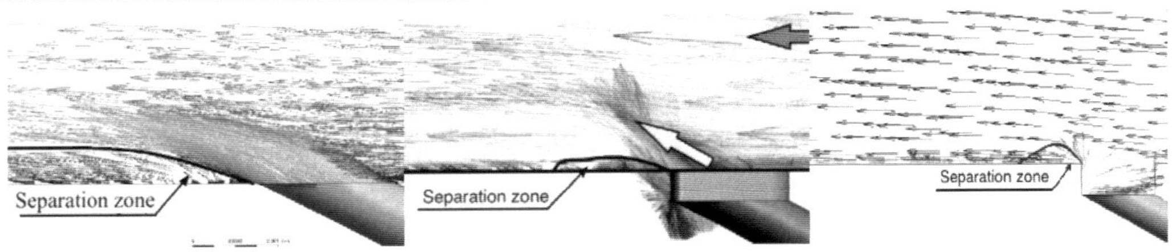

a *b* *c*

Figure. 4. The flow separation zone: *a* – traditional inclined holes configuration; *b* – coolant supply into triangular dimples; *c* – coolant supply into the trench

The decrease in the intensity of twin vortices leads to a decrease in the absolute value of the lift force, which acts on the flow and prevents it from being detached with formation of separation zone in the region of the plate leading edge at large blowing ratio. When triangular dimples are used, the dimensions of the coolant separation zone are significantly reduced in comparison with the traditional scheme (figure 4). The axial length of separation zone for a triangular dimples is about x/d = 1.0.

The above mentioned features influence the adiabatic film cooling efficiency. The distribution of local adiabatic film cooling efficiency in a spanwise cross section is presented in figure 5. For a triangular shape, there should be a significant increase in the uniformity of film cooling in comparison with traditional scheme. When blowing into a trench, the coolant film almost completely covers a flat (protected) surface.

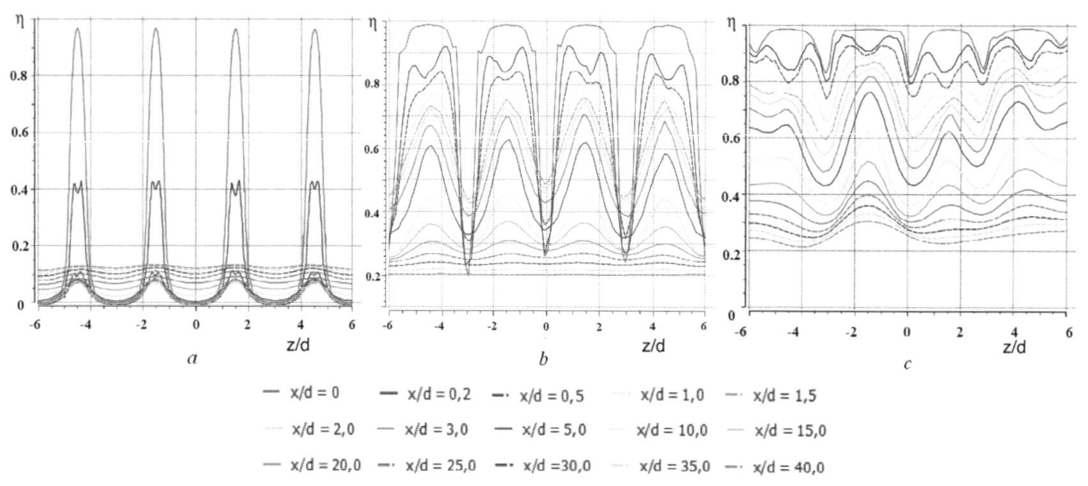

Figure. 5. Adiabatic film cooling efficiency: *a* – traditional inclined holes configuration; *b* – coolant supply into triangular dimples; *c* – coolant supply into the trench

Axial distribution of the laterally averaged adiabatic film cooling efficiency for different blowing ratios is presented in figure 6. For comparison, the calculated data for traditional scheme of inclined holes [6] and experimental data for the trench [7] were used. were taken.

From the analysis of these results it is apparent that for small value of the blowing ratio (m = 0.5), in the region of $x/d \leq 15$ the laterally averaged film cooling efficiency for the triangular configuration is 10-20% higher than this data for the traditional scheme. At m = 1.0, the scheme with coolant supply into triangular dimples has much greater laterally averaged adiabatic film cooling efficiency in comparison with the traditional scheme in the whole area, but it is inferior to the trench scheme data. At m = 1.5 and 2.0, the scheme using triangular dimples is not inferior the scheme with coolant supply into the trench.

Figure 7 shows the dependence of the averaged adiabatic film cooling efficiency on the blowing ratio. The averaged adiabatic film cooling efficiency was calculated from the following correlation:

$$\bar{\bar{\eta}}_{film} = \frac{\sum_N \bar{\eta}_{film}}{N}, \qquad (2)$$

where $\bar{\eta}_{film}$ is the laterally averaged film cooling efficiency for each blowing ratio, and N is the number of sections.

As can be seen from figure 7, the traditional scheme of film cooling is characterized by insignificant decrease in the film cooling efficiency with blowing ratio growth. An important feature of the film cooling scheme with blowing into triangular dimples is a significant increase in the adiabatic film cooling efficiency with an increase in the blowing ratio. The same feature is also inherent in the scheme with the coolant blowing into the trench.

From these results it may be concluded that a single-array of triangular dimples is characterized by high values of the averaged film cooling efficiency, compared to the traditional scheme of inclined round holes, and is of great interest for practical application.

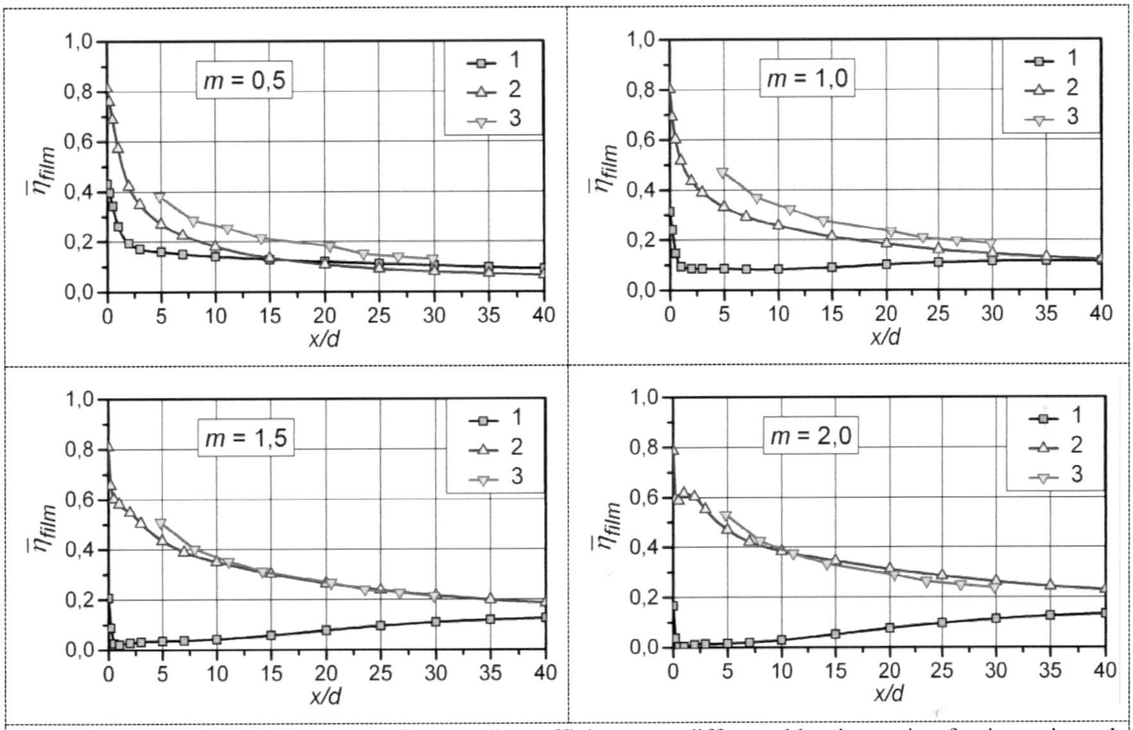

Figure. 6. The laterally averaged film cooling efficiency at different blowing ratios for investigated schemes: 1 – traditional inclined holes configuration [6]; 2 – coolant supply into triangular dimples; 3 – coolant supply into the trench [7]

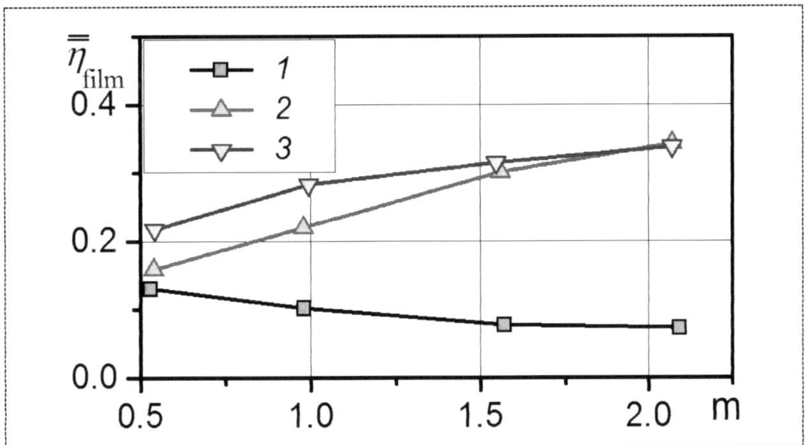

Figure. 7. The averaged flat plate adiabatic film cooling efficiency versus the blowing ratio for the investigated schemes: 1 – traditional inclined holes configuration [6]; 2 – coolant supply into triangular dimples; 3 – coolant supply into the trench [7]

4. Conclusions

The paper theoretically studies the physical mechanism and film cooling efficiency over a flat plate at the coolant supply through a single array of inclined holes in the triangular dimples.

The results of research have proved high adiabatic film cooling efficiency of the scheme using triangular dimples in comparison with traditional scheme of inclined holes. The averaged flat plate adiabatic film cooling efficiency using triangular dimples for the blowing ratio $m = 0.5$ does not differ from the traditional scheme and that using the trench. For $m = 1.0$ and more, the scheme with triangular dimples is better than the traditional scheme by 10...60%. At $m \geq 1.5$, the scheme with triangular dimples is not inferior to the scheme using a trench.

The reasons of such a high efficiency of the alternative scheme of film cooling with triangular dimples are decrease in the intensity of "twin" vortices, as well as reduction of the separation zone length in the region of dimple leading edge. For the scheme with triangular dimples, the separation zone length x/d near dimples is 1.0.

Further research aims at studying the effects of rotation of cooled surface under conditions of temperature factor and external turbulence.

References

[1] Khalatov A A 2008 Heat transfer and hydrodynamics in the fields of centrifugal mass forces
[2] Repukhov V M 1977 *Thermal protection of the wall by blowing gas* (Kiev: Naukova Dumka)
[3] Lu Y, Ekkad V and Bunker R S 2008 Trench Film Cooling: Effect of Trench Downstream Edge and Hole Spacing *ASME paper* **4** 563-569
[4] Khalatov A A, Severin S D, Bezludna M V and Novohatska I V Pat. No. 113452 (State Register of Patents of Ukraine, Kiev, UA), 2017.
[5] Khalatov A A, Bezliudna M V and Dashevsky Yu J 2014 *Aerospace Technic and Technology* **7** 77-82
[6] Khalatov A A, Borisov I I and Severin S D 2012 *Eastern-European Journal of Enterprise Technologies* **3** 4-8
Pedersen D R, Eckert E R and Goldstein R J 1977 Film Cooling With Large Density Differences Between the Mainstream and the Secondary Fluid Measured by the Heat-Mass Transfer Analogy *ASME paper* **99** 620-627

IOP Conf. Series: Journal of Physics: Conf. Series **980** (2018) 012025 doi:10.1088/1742-6596/980/1/012025

Heat transfer under transition and film boiling of liquids at dimpled spheres and cylinders

V. M. Zhukov[1], Yu. A. Kuzma-Kichta[2], A. V. Lavrikov[2], K. I. Belov[1], and V. A . Len'kov[1]

[1] Joint Institute for High Temperatures, RAS, Izhorskaya 13, bld.2, Moscow, 125412, Russia

[2]National Research University (Moscow Power Institute), Krasnokazarmennaya 14, Moscow, 111250

The paper is dedicated to the memory of prominent scientist and Laureate of State Prize Vladimir Zhukov

Abstract. The article presents the results of studies of heat transfer and film and transition boiling mechanism of nitrogen, Refrigerant R-113, and water at spheres and vertical cylinders, which surfaces are covered with spherical dimples.. The data were obtained under the conditions of pool boiling and natural circulation in vertical 1.0 and 2.5 mm wide annular channels. Hemispherical dimples of 3 mm diameter (h/d = 0.17) were made on sample surfaces. The dimples occupied 45% of the sphere surface and 37% of the cylinder surface. In some tests, the dimpled surface was additionally covered with low-conductive coating (10 μm film). Minimal cooling time for the sphere with dimples and low-conductive coating took place under natural circulation in 2.5 mm annular gap and it was almost 2.5 times lower than that for a smooth sphere under pool boiling. It is shown that at pool boiling the presence of dimples and low-conductive coating leads to heat transfer enhancement at transition and film boiling regimes, while at natural circulation such an enhancement occurs at film boiling with high temperature differences. The tests at natural circulation in vertical annular channels of different width showed that in this case an intensity of boiling heat transfer is higher than that at pool boiling. High-speed filming of film boiling process on the surfaces with dimples was conducted.

1. Introduction and experimental method

Creating dimples on a heat transfer surface is one of efficient ways of heat transfer enhancement in channels. In recent years this technology was very popular [1-3]. The dimples provide intense heat transfer between a boundary layer and a flow core. There are many works devoted to heat transfer in single-phase flows using dimples. At the same time, an effect of dimples on boiling heat transfer has been poorly investigated. In this sense, we should point out the work [3], in which boiling crisis was studied under subcooled water boiling in annular channels having dimples, and the work [4], where high-pressure steam-water mixture heat transfer was investigated in a post-crisis zone of the annular steam-generating channel with dimples.

Boiling of liquid film usually appears in any system at an initial phase of cooling, and this phenomenon is just the main mechanism that determines total cooling time. Duration of cooling process strongly depends on an intensity of heat removal at film boiling. The interest in study of

Content from this work may be used under the terms of the Creative Commons Attribution 3.0 licence. Any further distribution of this work must maintain attribution to the author(s) and the title of the work, journal citation and DOI.
Published under licence by IOP Publishing Ltd

nitrogen boiling is due by development of high-temperature superconductors, which are cooled by liquid nitrogen.

In particular, there is a certain interest in studying nitrogen boiling in slot channels under the conditions of natural circulation, which in some cases is accompanied by regimes of "enhanced" as well as "deteriorated" heat transfer [1]. Natural circulation of boiling cryogenic coolant in channels takes place due to buoyancy force, and when this happens, a velocity of the vapor-liquid mixture is determined by an equality of the driving head and the loop hydraulic resistance, and it is a function of a heat flux value and channel dimensions. The objective of the work is study of heat transfer and mechanisms of film and transition boiling regimes of different liquids on spheres and vertical cylinders having dimples on their working surfaces under the conditions of pool boiling and natural circulation.

Two experimental rigs were constructed for boiling heat transfer studying. The experiments in a low-temperature region are conducted in one rig, while the other one - at high medium temperatures. An unsteady-state method is used in all experiments. The samples preliminarily heated above Leidenfrost temperature are immersed into the liquid under study at atmospheric pressure. The tests with refrigerant R-113 and water were conducted at pool boiling [3], while the study of liquid-nitrogen boiling heat transfer - at free convection and natural circulation at the rig described in [4].

To conduct the tests copper spheres and cylinders with smooth and dimpled heating surfaces were fabricated [5]. They are shown in Fig. 1. The dimples on the sample surfaces were fabricated using special punches, at the end of which 5 mm sphere made of steel was fixed. An increase in the heat transfer surface area due to dimpled relief was not more than 5% compared to that of the smooth surface. Two cylinders were made of 18 mm diameter and 81.5 mm long solid copper rod. One of the cylinders had no dimples (it had technical roughness), while the other cylinder had dimples (Fig. 1). The lower parts of the samples had hemispheric shapes to provide better streamlining samples by a steam-water medium. On the cylinder 252 dimples were made in an order (18 rows with 14 dimples in every row, d = 3 mm, h/d = 0.17). The longitudinal pitch between dimple centers was 4 mm, while the peripheral pitch was 4 mm. An increase in the heat-transfer surface area due to dimpled relief was not more than 4%.

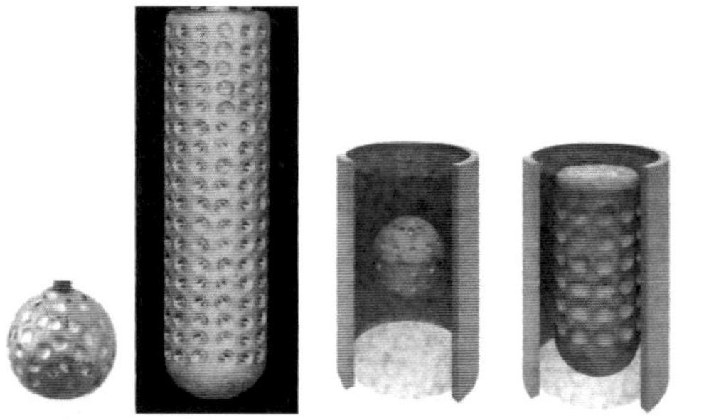

a)

b) c) d)

Fig. 1. a) Copper sphere with dimples D = 20 mm, d = 3mm, h/d = 0.17, n = 45%. b) Copper cylinder (rod) with dimples. D = 18 mm, L= 81.5 mm, h = 0.5 mm, d = 3mm, h/d = 0.17, n = 37%. c) 3-D view of the annular channel. The sphere and the cylinder with dimpled relief in 23 mm cylindrical channel. Minimal gap between the tube and the sphere is 1.5 mm.

Using a thin-wall tube the samples were lifted up into the heating zone. Then, they were kept in this zone at a given temperature and immersed into the vessel with a liquid under study (R-113 or water). The vessel had windows for video filming. When using liquid nitrogen, the glass cryostat was applied. Synchronously with sample immersion the data acquisition and processing system was switched by thermocouples installed in samples, in liquid and in the heater.

The Chromel-Alumel or Copper-Constantan thermocouples were used to measure the sample temperature. The thermocouple wire diameter was 0.2 mm. Using VS 10T high-temperature glue hot junctions of these thermocouples were fitted in dead holes at a depth of 10 mm in spheres and 10 and 30 mm in cylinders at distance equal 2 mm from the surface. Thermocouple signals were transmitted to E-24 L-Card analog-digital converter with 20 Hz interrogation frequency and then to personal computer. Special program was used for data processing. In addition, liquid and heater temperatures were measured. More thoroughly the measurement technique is described in [5].

The temperature and heat flux at the surfaces of spheres and cylinders were obtained by solving the inverse heat conduction problems using asymptotic methods developed in [2]. The final expressions for the temperature and heat flux at the surface of sphere are as follows

$$T(R,t)=T(0,t)+\frac{2}{3}\frac{R^2}{a}\frac{\partial T(0,t)}{\partial t}\sum_{k=1}^{\infty}\frac{1}{\mu_k^2}$$

$$q=-\frac{\lambda R}{3a}\frac{\partial T(0,t)}{\partial t}$$

and that for the surface of cylinder

$$q=-\frac{R\lambda}{2a}\frac{\partial T(r,t)}{\partial t},$$

$$T(R,t)=T(r,t)+\frac{R^2}{8a}\frac{\partial T(r,t)}{\partial t}.$$

Special attention was paid to investigate of dimpled relief effect of the heat transfer characteristics in transition and film nitrogen boiling regimes.

To increase a two-phase flow velocity around the test sample, a tube was mounted coaxially outside. Thus, an annular channel was formed between the sample and the tube. This tube was made of stainless steel, its length was 110 mm. The inner diameters of the tube were 20 or 23 mm. The width of the annular gap was 1.0 and 2.5 mm. In tests with natural circulation of the coolant, three alternatives of the sample surface were applied: smooth surface without dimples, the same with dimples, and the dimpled surface coated by 15 μm low-conductive polymeric film. As it was shown in several works, in particular, in [2], a thin low-conductive coating on the heat transfer surface leads to a decrease in a local wall temperature, cold spot formation and early ceasing of the film boiling regime, thus heat removal intensity increase. In tests, an assembly comprising a tube and a sphere or cylinder mounted on it were heated in air and immersed into the bath with liquid nitrogen. At initial instant a film boiling regime developed on the sphere or cylinder, as well on both sides of the tube. Due to a difference in liquid nitrogen density in the cryostat bath and in vapor-liquid mixture in the tube a driving head provided two-phase medium flow. By low heat capacity of the stainless tube and its large cooling surface the boiling on it stopped in some time, while all boiling regimes took place in succession on the sample surface.

In the tests, high-speed 8-bite monochromic MOCAM-4000 video camera was used for filming a boiling process. At a moment of camera stop a synchronizing signal came to the analog-digital converter. The video frames are correlated with thermocouple readings. Filming was made in passing light, with lightning from the side opposite. A frequency of filming was 2000 Hz with an exposure time equal to 1/50000.

2. Experimental results and data analysis

Boiling of nitrogen

To check a reliability of the experimental technique tests were conducted with nitrogen boiling at the smooth surface without dimples. The data on heat transfer intensity well agreeы with the results published in literature [7, 8]. Video filming of nitrogen film boiling on the sphere and cylinder with dimples under conditions of pool boiling showed that waves of different amplitude and shape with flattened leading and sharp rear fronts were observed at the vapor-liquid interface.

The time temperature changes are shown for the spheres and cylinders without dimples and corresponding dimpled samples under pool boiling condition on Figure 2. We see that the cooling time of dimpled samples decreases from 140 to 100 s as compared to smooth surfaces without dimples.

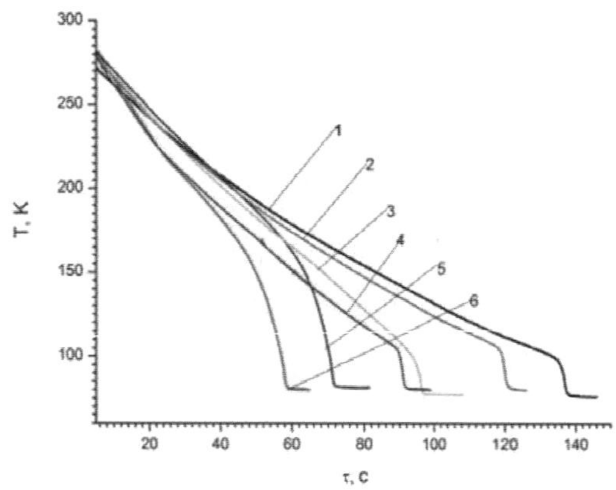

Fig. 2. Sphere cooling in liquid nitrogen. 1 – sphere without dimples, pool boiling; 2 - spheres without dimples, natural circulation; 3 – sphere with dimples, pool boiling; 4 – sphere with dimples, natural circulation; 5 – sphere with dimples and coating, pool boiling; 6 – sphere with dimples and coating, natural circulation.

To additionally enhance heat transfer the dimples were covered by a thin layer of low-conductive coating. In this case, the samples were cooled at pool boiling and under natural circulation conditions. Figure 2 shows the curves of sample temperature change in time from 280K down to the saturation temperature in nitrogen. From the figure we see that the combination of the dimpled relief and low-conductive coating at natural convection reduces a cooling time of spheres. For example, the cooling time for smooth spheres, which was equal to approximately 140 s (curve 1), was down to 60 s for the dimpled sphere with coating (curve 6), i.e. approximately by 2.5 times.

The data on pool boiling heat transfer on spheres with and without dimples (curves 1 and 2) show that the effect of dimpled relief is greater in the region of ceasing film boiling and in the zone of transition boiling. At natural circulation (curve 3), up to temperature difference $\Delta T \sim 140$ K, an intensity of heat removal increases. Boiling process is accompanied by oscillations of the liquid flow rate and sample wall temperature. The vapor-liquid mixture was ejected from both ends of the cylindrical channel. Two-phase flow rate oscillations sharply weakened when the outer tube was cooled down to nitrogen saturation temperature and the following heat input was only from the dimpled sphere. The heat flux at film boiling regime increased approximately by two times as compared to the case of spheres without dimples (curve 1). The considerable increase in heat transfer coefficient values was observed in the range of temperature differences, which corresponded to a transition boiling regime.

The data in Fig. 3 demonstrates heat transfer on spheres with dimples with low-conductive coating and without coating under the conditions of natural circulation. The greatest increase in heat transfer coefficient values took place when the spherical dimpled surface was used in combination with this coating (curve 3). Such a situation was observed over the entire range of temperature differences typical for transition and film boiling zones of liquid nitrogen boiling.

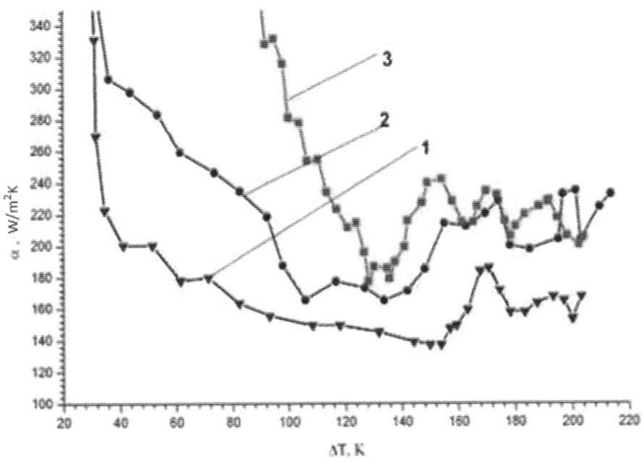

Fig. 3. Dependence of heat transfer coefficient on temperature difference at nitrogen boiling on spheres at natural circulation. 1 – sphere without dimples, 2 – sphere with dimples, 3 – sphere with dimples and low-conductive coating.

Cylinder. Nitrogen boiling heat transfer at natural circulation

As it was pointed out above, the natural circulation of two-phase medium arises in annular channels and heat transfer grows. In commonly applied methods of studying a heat transfer in annular channels, at the experiment beginning, liquid wets annular channel walls and with an increase in heat flux a bubble flow regime originates, further transfers into slug flow. In the experimental technique used in the present study the channel is cooled after heating above Leidenfrost temperature. Therefore, a film boiling regime arises and a liquid is separated from the channel wall by a vapor film. Such a boiling regime is accompanied by flow rate oscillations and liquid-vapor medium ejection from both ends of the channel. Flow rate and temperature oscillations increased when critical heat flux point became closer.

With a decrease in wall temperature transition and bubble nitrogen boiling regimes appeared in succession. A sample temperature was practically constant along the tube length because of high heat conductivity of copper. The curves in Fig. 4 demonstrate this fact.

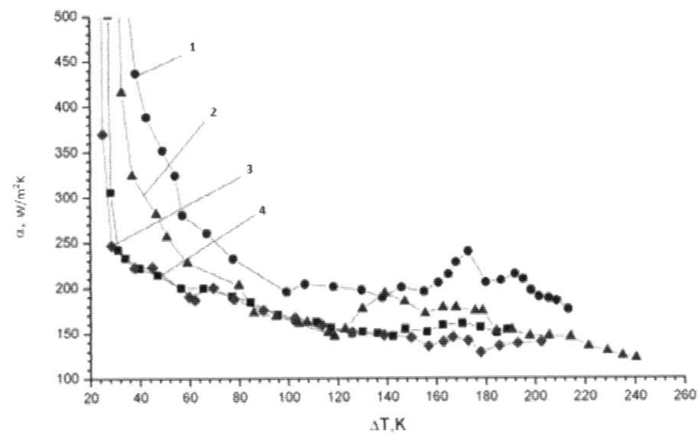

Fig. 4 Heat transfer at nitrogen boiling on smooth and dimpled cylinders under pool boiling conditions and natural circulation.
1-dimpled cylinder with annual channel 2.5 m.
2-dimpled cylinder with annual channel 1 mm.
3-dimpled cylinder pool boiling. 4-cylinder without dimples.

The tests conducted with annular channels showed that the cooling velocity of cylinders, both without dimples and those with dimpled relief, is noticeably greater than that under pool boiling conditions. From Fig 4, where the changes of cylinder temperature in time are shown for annular channels which have a width 1.0 and 2.5 mm one can see that dimples increase cooling velocity. In addition, these results demonstrate an increase in cooling velocity with an increase in annular channel width.

The data on heat transfer intensity from Fig. 4 for cylinders with dimples in the 1 mm wide annular channel demonstrate an efficiency of the dimpled relief under conditions of natural circulation.

Heat transfer intensity at the cylinder with dimples in annular channels, especially in 2.5 mm wide channel, is also higher than that during pool boiling. The difference is equal to 50% at film boiling and up to 100% in transition boiling.

When approaching the maximal heat flux point, a growth of the net vapor content in the channel sharply increases. This leads to two-phase flow chocking and deteriorated heat transfer regime with lowered intensity of heat transfer removal from the sample surface. This regime is unstable through ject of vapor bubbles to the lower end of the annular channel. Two-phase mixture ejection is described in literature as liquid nitrogen boiling in a channel. Assessments based on the data on nitrogen boiling in annular channels [9] show that the void fraction value reaches 0.9-1.0 while a mass-averaged flow velocity tends to 0.5-0.6 m/s. An increase in channel transversal dimension leads to a decrease in friction losses, an increase in two-phase flow velocity and in enhancement of heat removal intensity. Peaks that correspond to maximal velocity of the flow in the channel core are seen on boiling curves.

Refrigerant R-113. Sphere

The dependences of heat transfer coefficient on temperature difference for spheres with dimpled relief are shown on Fig. 5. The presented curves show that the effect of dimples is especially strong in the range of small temperature differences and decreasing vapor film thickness, i.e. near the zone of a ceasing film boiling regime. The heat transfer intensity for this dimpled relief geometry is approximately 15-20% higher than those for the smooth sphere without dimples. Visual observations showed that on the dimpled surface, a vapor film coming-off sometimes arises at the lower part of the sphere, in contrast to the case of using smooth spheres without dimples, when such film coming-off always in upper part of sphere. In the range of nucleate boiling, the presence of dimples on the sphere surface has no effect on heat transfer intensity because adding a limited amount of dimples does not considerably increase a number of active nucleation sites.

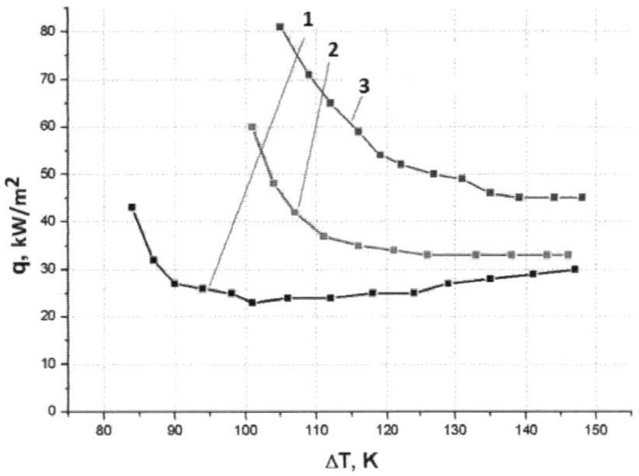

Fig. 5. Heat transfer R-113 film boiling at spheres with different surface treatment
Key: 1- smooth sphere; 2- dimpled sphere; 3 – sphere with porous coating;

The strongest effect of dimples is the change of film boiling ceasing boundary , and this is especially so when liquid is subcooled relative to the saturation temperature. With an increase in subcooling value a vapor film thickness decreases and this leads to film boiling ceasing at larger $\Delta Tcr2$ and qcr2 values. The dependences of critical heat flux qcr2 on subcooling $\Delta Tsub$ both for dimpled spheres and those without dimples. They demonstrate the stronger effect of dimples on film boiling ceasing at larger subcooling values.

If we compare film boiling heat transfer obtained at dimpled spheres, those without dimples, and spheres with porous copper coating [5], we recognize that the dimpled spheres occupy intermediate position (Fig. 5). We should note that at all spheres with different surface treatment, heat transfer intensity sharply increases in the range of small temperature differences, i.e. when approaching the zone of film boiling ceasing. The relatively weak effect of the dimpled relief on heat transfer at film

IOP Conf. Series: Journal of Physics: Conf. Series **980** (2018) 012025 doi:10.1088/1742-6596/980/1/012025

boiling regime, as compared to that in the case of porous coating, can be explained by a low spatial density of dimples (~12%) on the heat transfer surface. In addition, the dimple configuration being studied (h/d = 0.164 < 0.2) belongs to a class of small dimples, at which vapor flow occurs without separation from the heating surface, and heat transfer enhancing effect is lower than that for deep dimples (h/d > 0.2) [2].

Figure 6 shows the heat transfer coefficient under film boiling of refrigerant R-113 on the dimpled sphere. It recognized that heat transfer considerably grows with an increase of subcooling.

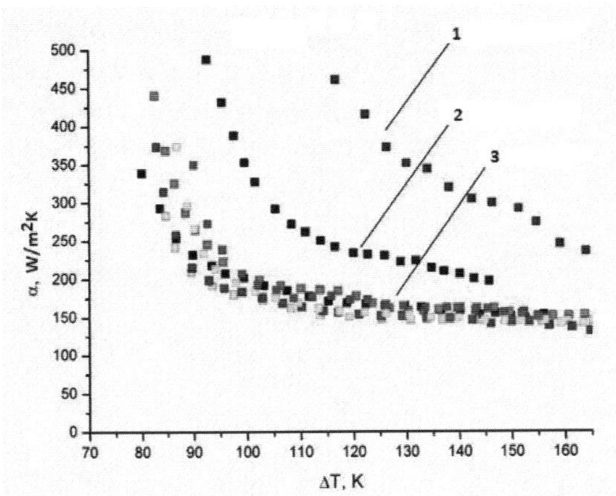

Fig 6. Effect of subcooling on R-113 film boiling on dimpled spheres. 48 dimples, D= 20 mm, h = 0.5 mm, d = 3mm, h/d = 0.17, n = 27%. Key: 1 – subcooling 26 K, 2 – subcooling 13 K, 3 – saturation.

Cylinder

The heat transfer studies of Refrigerant R-113 were conducted on cylindrical surfaces with dimpled relief and without dimples within the temperature difference range up to 170 K, which included all boiling regimes at the liquid saturation temperature.

Figure 7 shows the video filming of the film and transition boiling of the saturated Refrigerant R-113 on the vertical cylinder with dimples. Typical profiles of waves on the interface are shown as well.

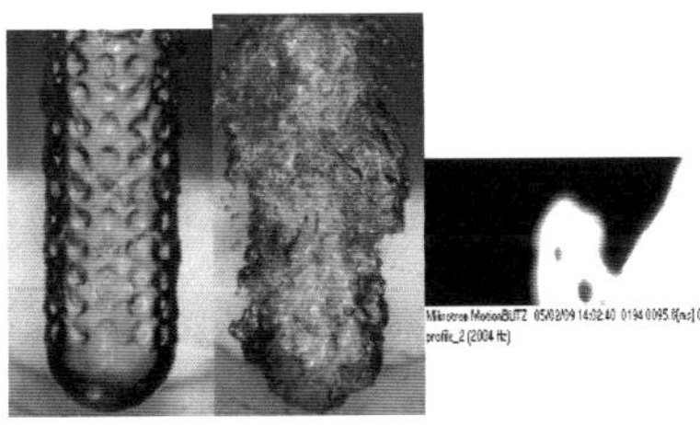

Fig. 7 R-113 film and transition boiling on vertical dimpled cylinder and typical wave profile on the interface

The analysis of the temperature change in time for cylinders without dimples and those with dimpled relief shows that the cooling time of the dimpled sample in R-113 is lower as compared to that for cylinders without dimples.

The data on the wavy flow of the vapor film are similar to the results obtained by Borishanskii and Fokin in the studies of hydrocarbons and water boiling on a vertical tube [10]. In their work the wave amplitude reached several millimeters. With an increase in heat flux density, while reaching the point of maximal critical heat flux, an amount of heat sharply increases. A decrease in heat flux density reduced vapor void fraction and formation of single vapor bubbles.

In the case the sphere with dimpled relief described earlier, nonlinear three-dimensional waves with amplitude alternating in time were noted on the vapor-liquid interface. These amplitudes were comparable in size or even greater than the mean thickness of the vapor film. From typical photos, presented in Fig. 7 for cylinders with smooth surface and dimpled ones, one can see the existence of a complex system of waves on the liquid-vapor interface [9]. At the frontal spherical part of the samples the smooth interface and laminar vapor film flow, which transfers into turbulent one at the cylindrical part of the samples arises. The wavy flow on the cylindrical part of the dimpled sample is more complex. To a certain extent, it repeats the dimpled relief.

The dependences of q on ΔT for cylinders with dimples is shown on figure 8 in the range of ΔT from 50 to 170 K that corresponds to the transition and film boiling. They show a considerable effect of the dimpled relief on ΔT_{cr2} and q_{cr2} values as well as on heat transfer intensity in transition and film boiling regimes. At the dimpled cylinder the crisis occurs at ΔT_{cr2} values equal to 80 K, while at the cylinder without dimples the crisis occurs at ΔT_{cr2} equal 65 K. For cylinders with dimpled relief the value q_{cr2} increases by 1.5 times. In addition, the data presented in this figure show that the effect of the dimpled relief on heat transfer intensity is more pronounced in the range of small temperature differences near the film boiling crisis. This is occur through a decrease in the vapor film thickness. For example, at ΔT= 170 K the heat flux density increases by 20%, while at ΔT = 72 K, by 120%, as compared to the case without dimples.

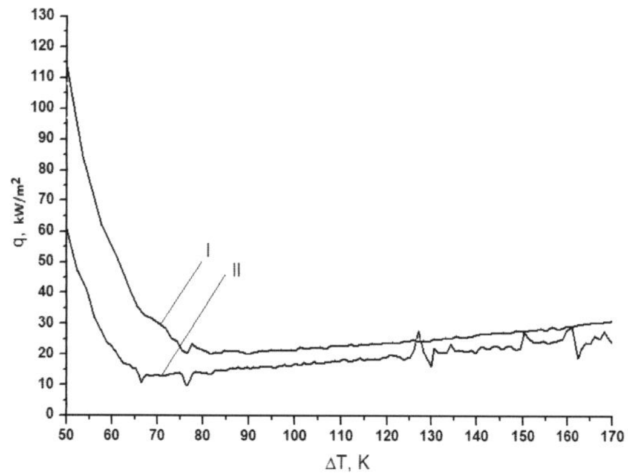

Fig. 8 Heat transfer at transition and film boiling of Freon R-113 on the dimpled cylinder (1) and on cylinder without dimples (2)

Water

The water boiling heat transfer on surfaces with dimpled relief and without dimples is investigated in the temperature difference range that embraces all boiling regimes.

Figure 9 shows the video filming of film and transition boiling of water on the dimpled sphere and cylinder. At the frontal part of the samples a laminar flow of the vapor film is seen. The wavy flow of the vapor film is more complex, and it repeats the dimpled relief.

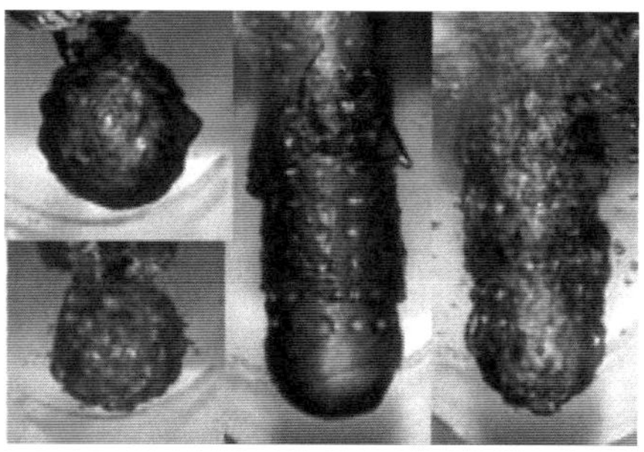

Fig. 9 Subcooled water film and transient boiling at dimpled sphere and cylinder

The data presented in Fig. 10 show that the dimpled relief provides an increase in critical heat flux q_{cr2} value and increases heat transfer intensity in transition and film boiling regime by three times.

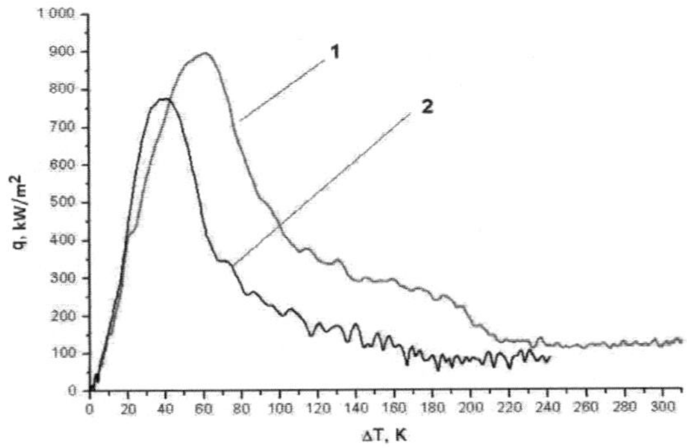

Fig.10. Effect of dimpled relief on water boiling curve. Ts = 373 K
Key: 1 – 31 dimples, d = 2.2 mm, h = 0.28 mm. h/d = 0.138, s = 5.2 mm, 2 – smooth surface.

3. Conclusions

The results of studying heat transfer at film and transition boiling regimes of nitrogen, Refrigerant R-133, and water on dimpled spheres and vertical cylinders are presented under conditions of pool boiling and natural circulation in vertical annular channels with a gap width by 1.0 and 2.5 mm. Semispherical dimples of 3 mm diameter and h/d = 0.17 were made on the sample surfaces. The density of dimples distribution on the surfaces of sphere was 45% and that on the surfaces of cylinder was 37%. In some tests, the low-conductive film of 10 μm thickness was deposited on the dimpled surface.

Under natural circulation of nitrogen in the 2.5 mm wide annular channel the cooling time of the sphere with dimples and low-conductive coating was almost 2.5 times smaller than that under pool boiling on the sphere without dimples.

At film and transition water pool boiling the presence of the dimpled relief provides increase in heat transfer intensity.

Acknowledgments

The work was supported by the Russian Foundation of Basic Research, Grant N 15- 08-00775.
The authors are thankful to Professor Yuri Zeigarnik for his help in preparing of the paper.

References

[1] Kiknadze G I, Krasnov Y K, Podymaka N F and Khabenskii V B 1986 Self- organization of vortex structures at water flow upon semispherical dimples *Doklady of the USSR Academy of Sciencie* vol. 291 no. 6 pp 1315-1318

[2] Shchukin A V, Kozlov A P, Agachev R S and Chudnovskii Y P 2003 *Heat transfer enhancement by spherical dimples under impact of disturbing factors* (Ed. by V.E. Alemasov. Kazan. KGTU Publishing) p 143 (in Russian)

[3] Dzyubenko B V, Kuzma-Kichta Y A, Leont'ev A I et al. 2008 *Heat transfer enhancement on macro, micro, and nanoscales* (Ed. by Yu.A. Kuzma-Kichta. Moscow. FGUP TsNIIAtominform) p 532 (in Russian)

[4] Kiknadze G I, Kryuchkov I I and Chushkin Y V 1989 *Self-organization of tornado-like structures in a coolant flow. Preprint IAE-4841/1* (Moscow, TsNIIAtominform) p 29 (in Russian)

[5] Mostinskii I L, Geshele V D, Goryainov D A and Raskatov I P 2001 Heat transfer from a surface with spherical dimples under water boiling and steam-water mixture flow in post crisis region *IFZh* **74**(3) 13-19 (in Russian)

[6] Kuzma-Kichta Y A, Zhukov V M and Agal'tsov A M 2008 *Heat transfer at cooling dimpled spheres in boiling liquids. VI Minsk International Forum. Theses of reports* (ITMO Publishing, NANB) vol. 1 pp 111-113 (in Russian)

[7] Shumakov N V 1979 *Method succeeded intervals in heat flux metering at unsteady processes* (Moscow, Atomizdat) p 216 (in Russian)

IOP Conf. Series: Journal of Physics: Conf. Series **980** (2018) 012025 doi:10.1088/1742-6596/980/1/012025

[8] Kirichenko Y A, Kozlov S M, Rusanov K Vet al 1992 *Heat transfer at nitrogen boiling and problems of high-temperature superconductors cooling* (Kiev, Naukova dumka) p 278 (in Russian)
[9] Zhukov V M, Kazakov G M, Kovalev S A and Kuzma-Kichta Y A 1974 *Heat transfer under liquid boiling on surfaces with low-conductive coatings In Heat Transfer and Physical Gas Dynamics* (Moscow, Nauka) pp 116-129 (in Russian)
[10] Borishanskii V M, Maslichenko P A and Fokin B S 1964 *Studying of vapor layer at film boiling on vertical surface. In convective heat transfer in single-phase and two-phase flows* (Moscow, Energia) pp 243-248 (in Russian)

IOP Conf. Series: Journal of Physics: Conf. Series **980** (2018) 012026 doi:10.1088/1742-6596/980/1/012026

RANS and LES analysis of a separation region in front of a backward-facing step

T A Baranova[1], V L Zhdanov[13], D A Ivanov[1]and Ja I Smulsky[2]and V I Terekhov[2]

[1] Luikov Heat and Mass transfer Institute, NAS Belarus, 15 P.Brovka, 220072 Minsk, Belarus
[2] Kutateladze Institute of Thermophysics, SB RAS, 1 Lavrentjev Avy., 630090 Novosibirsk, Russia

Abstract. Turbulent characteristics of a reattached flow separated from a rib ahead of a backward-facing step in a flat air channel were numerically analyzed by RANS and LES methods. The separation region ended just before the step edge. The both methods showed low pressure regions in corners, formed by the rib and the channel walls that caused a vorticity tube forming along the rib. Nevertheless in RANS, a separated flow was identified as two-dimensional with a thicker shear layer at the step edge. In LES, the low pressure regions caused both redistributions between velocity components, Reynolds stresses, and a thicker shear layer at the step edge. The flow structure behind the rib has been, thus, identified as three-dimensional.

1. Introduction

Separated flows are common in several engineering applications such as aircraft wings, turbine and compressor blades, diffusers, buildings, pipes with sudden expansion, combustors, etc. These flows drastically change transport of momentum, heat and mass. Due to this fact, interest in separation control has been drawing attention of researchers over decades. A backward-facing step is ideal canonical separated flow geometry, encountered in various technical structures. A flow separated from the step edge due to a favorable pressure gradient reattaches the wall at some distance downstream, forming a separation region. The length of this region in many cases is desirable to be as short as possible. Extensive investigations have revealed parameters that influence the separation region length: expansion ratio, aspect ratio, free stream turbulence intensity, Reynolds number, as well as boundary layer state and its thickness at separation [1]. In earlier studies, the reattachment length of a separated flow was usually a primary parameter of interest. With advances in measuring techniques and discovery of the coherent structure influence on the progress of separation region, main interest has been shifted to flow structures both upstream and downstream of the step. To date, several different passive and active control techniques (fence, rod, tabs, suction and ejection, oscillation [2-7]) have been implemented mainly to examine the transformation of separated flow structures altering the reattachment length.

Originally, experimentation served as a main source of information on transport processes in separated flows, but with progress in computer facilities, numerical simulations have become a powerful tool of research. Nowadays, numerical methods allow obtaining characteristics of a three-dimensional flow with high spatial resolution and, at the same time, visualizing a flow in different directions. Two methods – RANS and LES – are mostly used in investigations. Historically, RANS is the first method extensively applied to study the flow characteristics, including those of separated flows. The method is mostly suitable for studying steady flows while the separated flows are unsteady. RANS showing a good correlation of mean parameters (velocity, reattachment length) with experiment, shades features of flow structural. Being intensely developed over last decades LES

[3] Luikov Heat and Mass transfer Institute, P. Brovka, 15 220072 Minsk, Belarus, e-mail:

valery.zhdanov@hmti.ac.by

Content from this work may be used under the terms of the Creative Commons Attribution 3.0 licence. Any further distribution of this work must maintain attribution to the author(s) and the title of the work, journal citation and DOI.
Published under licence by IOP Publishing Ltd

allows highlighting unsteady flow variations, but at the expense of computer resources and computation time.

The present study aims at showing structural distinctions of the same flow, applying RANS and LES. The reattached flow separated from a rib installed ahead of a backward-facing step is considered. A variable position of a rib relative to a backward-facing step affected a reattachment length behind the step [2]. The experimental measurements have also shown that the reattachment length decreases if the shear layer at the step edge becomes thicker.

RANS and LES simulations of two-dimensional laminar flows above a backward-facing step were compared using OpenFoam [8]. For a lower Reynolds number (Re = 389), the RANS and LES results were in a good correlation with experiment. For a higher Reynolds number (Re = 1000), the RANS simulations were poor but the LES simulations showed a good agreement with experiment.

2. Subject of investigation

Investigations were carried out in a flat air channel with a rib, located in front of a backward-facing step. Channel cross section (0.021 × 0.150 m (h_0 × B)) and step height (H = 0.009 m) were identical to the ones of the experimental setup described elsewhere [2]. The rib of height (Δ) equal to 0.003 m with its thickness of 0.005 m (e) was located across the channel width at different distances from the step edge (S) (figure 1).

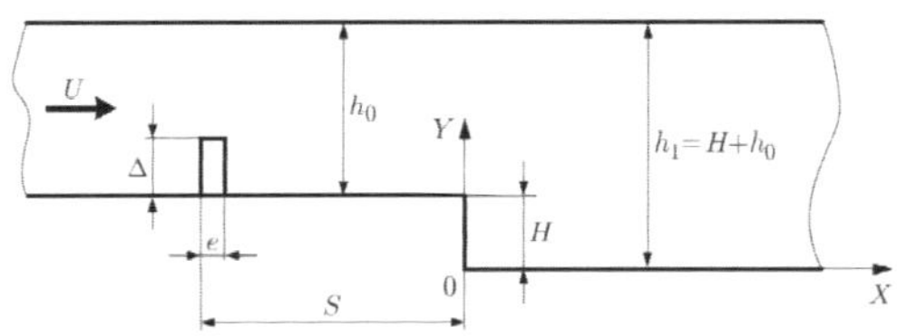

Figure 1. Schematic of a channel with a rib in front of a backward-facing step.

3. Simulation methods

A three-dimensional velocity field was computed by RANS using ANSYST Fluent 18 and two models: k-Ω SST and k - ε with standard wall functions. LES computations were done by OpenFOAM (Open Field Operation and Manipulation) using two eddy viscosity models: Smagorinsky [9] and one-equation (oneEqEddy) [10]. The velocity and pressure fields were linked by a PIMPLE (merged PISO-SIMPLE) algorithm.

In numerical computations, a developed velocity profile (power law of 1/7) was assigned at the channel entrance. The Reynolds number based on step height and bulk velocity (U_0= 25 m/s) was Re_H=15500. A turbulence intensity of 5% was set for each velocity components.

The computational domain included the channel part before the rib as well as the length of the recirculation zone behind the backward-facing step. In RANS, this domain was one meter in length and had a computational grid of 12 000 000 cells. In LES, the domain was shorter – 0.5 m, and the grid had 2209500 cells.

4. Results

The experiment [2] showed that the reattachment length behind the step increased when the shear layer separated from the rib merged with that from the backward-facing step (S = 0.02 m). When the separated layer from the rib attached the channel wall before the step edge (S = 0.04 m), the reattachment length decreased and changed insignificantly with a further increase in S. Therefore turbulent flow characteristics were numerically simulated for two rib positions: S = 0.02 and 0.06 m to correctly select a turbulence model. The boundary of the recirculation zone behind the step was determined by the zero longitudinal velocity coordinate at the channel bottom in the central section. The best agreement of reattachment lengths with experiment was observed for the k-ε model

(RANS) and oneEqEddy (LES) (figure 2). The k-Ω SST model usually used to study a near-wall flow showed a poor agreement with experiment as also noted in [8].

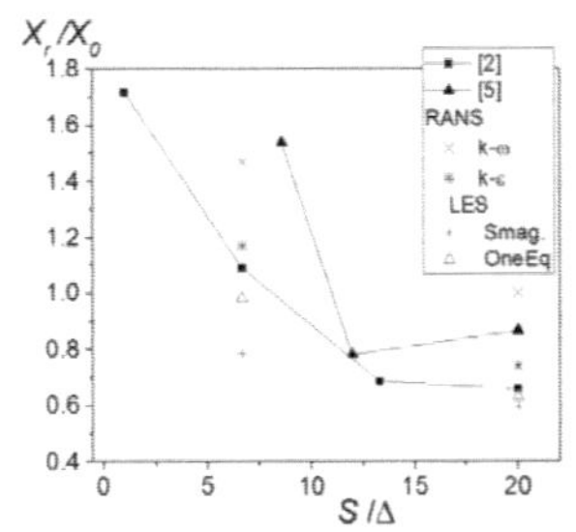

Figure 2. Reattachment length variation with rib position

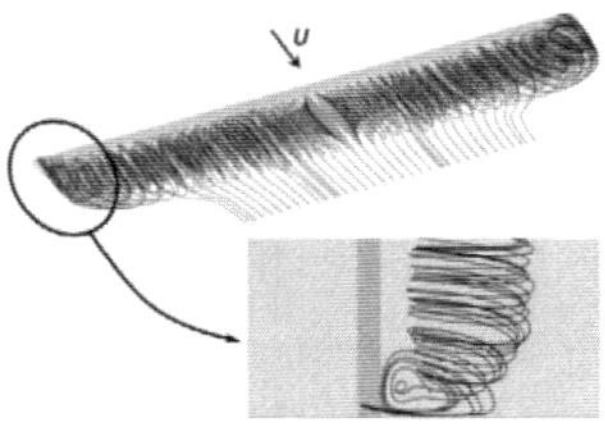

Figure 3. Vorticity tube near a rib.

Due to separation the low pressure regions were generated in the corners, formed by the rib and the channel sidewalls. Fluid was sucked into these regions, which were the sources of a vorticity tube formed along the rib, involved into the tube and flowed in the middle part of the channel (figure 3). This flow structure was experimentally visualized [11].

Since flow parameters before the edge of the backward-facing step influence progress in the recirculation zone behind the step, mean and turbulent flow characteristics were calculated at a distance of 1.5Δ ahead the step edge for two heights 0.2Δ, 0.5Δ and in two cases: a) without rib; b) with rib at S = 0.06 m.

4.1. RANS investigations

A longitudinal velocity profile (U_x) across the channel height was normalized by a maximum velocity (U_{max}) above the edge of the backward-facing step [2]. The numerical profiles differed from the experimental one due to the asymmetry of the latter only near the lower wall. Perhaps, this asymmetry was caused by the influence of the separation region behind the step. In the numerical study, velocity profiles calculated at the step edge were symmetric. Owing to the rib, the velocity decreased near the channel bottom (stronger in experiment) and increased in the channel upper part (figure 4a).

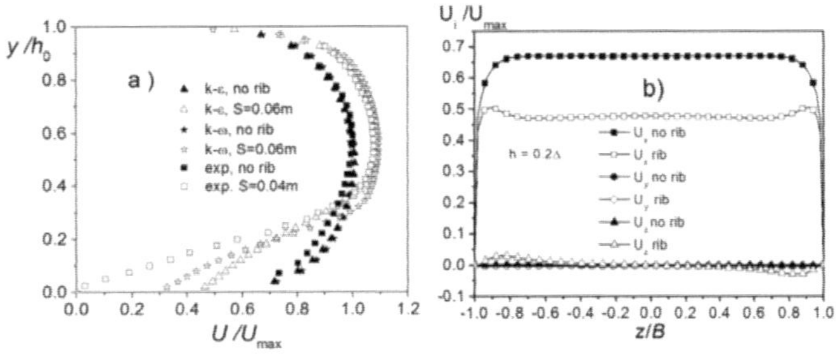

Figure 4. Velocity distributions across the channel at the edge step: a) along the height; b) along the width at h = 0.2Δ.

Distributions of flow turbulent characteristics across the channel were presented only for the k-ε model. The rib effect was shown in small deviations from uniformity for longitudinal and transverse velocity components near the sidewalls (figure 4b).

Reynolds stresses sensitive to structure transformations also showed rather uniform distributions $\overline{u'_x u'_y}$ across the channel width at different heights (figure 5a). The stress $\overline{u'_x u'_z}$ was almost zero along

IOP Conf. Series: Journal of Physics: Conf. Series **980** (2018) 012026 doi:10.1088/1742-6596/980/1/012026

the channel width. The stress $\overline{u_y' u_z'}$ deviated from zero to the sidewalls the stronger the larger was the parameter h (figure 5b, c). But these deviations were more than by an order of magnitude smaller in comparison with the stress $\overline{u_x' u_y'}$.

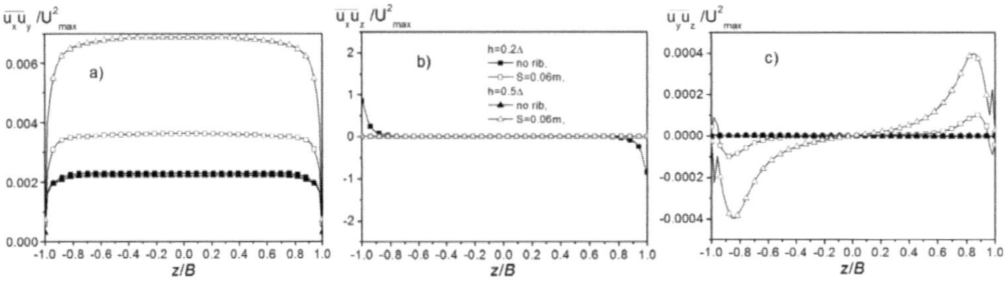

Figure 5. Reynolds stress distributions across the channel width at the step edge.

Thus, the changes in flow parameters due to the rib were mostly shown only in the thickening of a shear layer at the step edge because the velocity decreased as measured in experiment [2] and the reattached flow remained two-dimensional.

4.2. LES investigations

In the channel without rib, the longitudinal velocity profiles computed by the Smagorinsky and oneEqEddy models at the edge of the backward-facing step were more flat in the channel central part and lower at the walls in comparison to the experimental and RANS ones (figure 6a). Apparently, it was a result of a shorter computational domain. The velocity normalization was the same as in RANS.

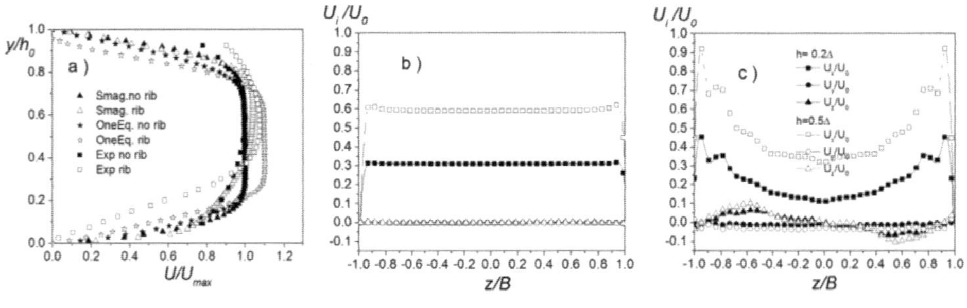

Figure 6. Velocity distributions along the height (a) and width (b, c) of the channel; b) without rib; c) with rib (S = 0.06 m).

The separation region length calculated by Smagorinsky's model was by 4Δ shorter than the length calculated by the oneEqEddy model. Nevertheless, Smagorinsky's model showed a stronger velocity decrease at the step edge than the oneEqEddy model and the experiment [2]. Varying the velocity near the channel wall determines a shear layer thickness at the step edge that strongly influences the development of separation region behind backward-facing step. Therefore flow characteristics along the step edge in the transverse direction were analyzed using only the oneEqEddy model.

In the channel with no rib, the velocity field was two-dimensional: the velocity component U_x uniformly distributed across the channel, and other components were equal to zero (figure 6b). The rib caused the velocity components U_x and U_z to increase strongly near the sidewalls in comparison with their values in the channel center (figure 6c). The normal velocity (U_y) slightly decreased with increasing parameter h.

Small Reynolds stresses were distributed rather uniformly in the channel with no rib, which is typical for a two-dimensional flow (figure 7a). These stresses increased due to the rib by two orders of magnitude and became stronger to the sidewalls reflecting a three-dimensional flow structure (figure 7b).

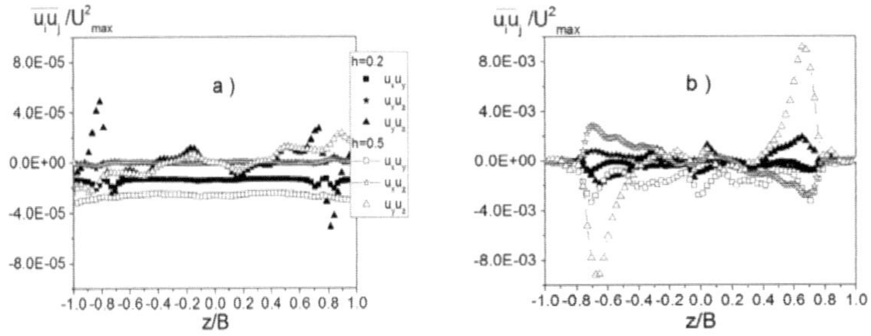

Figure 7. Reynolds stresses distributions along the channel width at the backward-facing step; a) without rib; b) with rib.

The increase in stresses $\overline{u'_x u'_z}$ and $\overline{u'_y u'_z}$ showed the growth of transverse and normal interactions in the flow.

The vorticity $\omega_x = (\partial U_z/\partial y + \partial U_y/\partial z)$ was generated mostly near the walls being almost zero across the channel without rib (figure 8a). The rib caused pairs of opposite-sign vortices to be formed near the channel bottom. The vorticity intensity of these pairs was by an order of magnitude higher than the maximum vorticity in the channel without rib (figure 8b). The generated vortices resulted in energy redistribution within the separation region, which was shown in longitudinal velocity decrease in front of the edge of the backward-facing step (figure 6a).

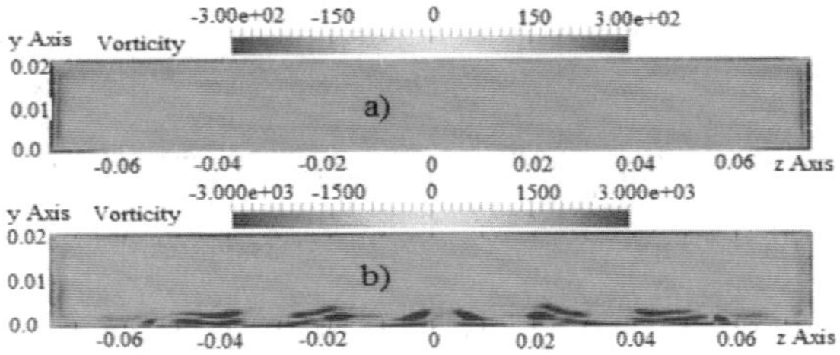

Figure 8. Vorticity field across the channel width at the step edge: a) without rib; b) with rib.

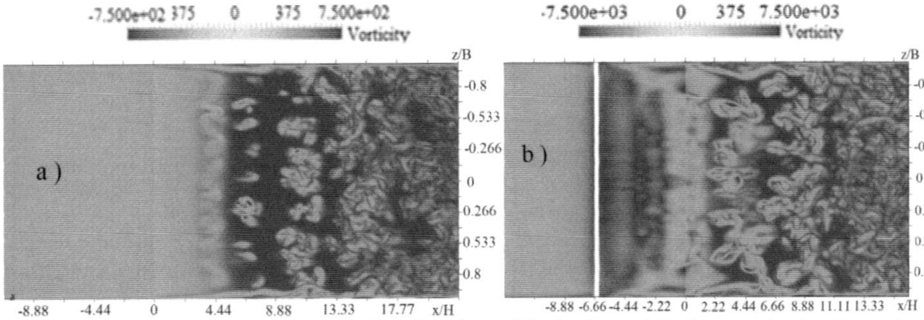

Figure 9. Vorticity maps before and after the edge of the backward-facing step in the channel: a) without rib; b) with rib.

The three-dimensional flow formed ahead of the step resulted in the intense vorticity formation in the recirculation zone behind the step (figure 9). The vorticity maps were presented in the plane over the channel bottom. The rib and step edge coordinates corresponded to numbers -0.06 and zero, respectively.

5. Conclusions

Parameters of a reattached flow separated from a rib ahead of a backward-facing step were documented by two numerical methods: RANS and LES using different turbulence models: k-ε, k-Ω and Smagorinsky and oneEqEddy, respectively. The both methods showed a velocity decrease at the edge of the backward-facing step, but in RANS, the reattached flow was identified as two-dimensional, while in LES – as three-dimensional.

The present study evidences that the k-Ω SST model does not describe the influence of the rib position on the reattachment length behind the backward-facing step when the separation region ends before the step edge.

LES simulations using the oneEqEddy and Smagorinsky models showed a different length of separation regions behind the rib and different velocity profiles at the step edge. The development of the separation region described by Smagorinsky's model is in worse correlation with the natural one known from experiments.

Acknowledgement

The study was made within the framework of Grants T16P -005 (BFFR) and 16-58-00018 (RFBR).

6. References

[1] Eaton J K, Johnston J P 1981 A review of research on subsonic turbulent flow reattachment *AIAA J.* **19** 1093-100
[2] Terekhov V I, Smulsky Ja I, Sharow K A 2016 Experimental study of the separated flow structure behind a backward-facing step under a passive disturbance *J. Applied Mech. and Technical Phys.* **57** 180-87
[3] Isomoto K, Honami S 1989 The effect of inlet turbulence intensity on the reattachment process over a backward-facing step *ASME Trans. J. Fluids Eng.* **111** 87
[4] Park H, Jeon W, Choia H and Yoo J Y 2007 Mixing enhancement behind a backward-facing step using tab *Phys. Fluids* **19** 105103
[5] Miau J J, Lee K C, Chen M H and Chou J H, 1991 Control of separated flow by a two-dimensional oscillating fence *AIAA J.* **29** 1140-48
[6] Chun K B, Sung H J 1996 Control of turbulent separated flow over a backward-facing step by local forcing *Exp. Fluids* **21** 417-26
[7] Yoshioka S, Obi S, and Masuda S 2001 Organized vortex motion in periodically perturbed turbulent separated flow over a backward-facing step *Int. J. Heat Fluid Flow* **22** 301-07.
[8] Gong Y, Tanner F X 2009 Comparison of RANS and LES Models in the laminar limit for a flow over a backward-facing step using OpenFOAM *19th Int. Multidimensional Engine Modeling Meeting* SAE Congress, 19 April (Detroit) Michigan USA.
[9] Pope S B 2000 *Turbulent Flows* (Cambridge: Cambridge University Press).
[10] Schumann U 1975 Subgrid scale model for finite difference simulations of turbulent flows in plane channels and annuli *J. Comput. Phys.* **18** 376-404
[11] Terekhov V I, Bogatko T V, Dyachenko A Ju et al. 2016 *Heat transfer in the subsonic separated flows* (Russian) ed. V V Larichkin, S V Alekseenko (Novosibirsk: Novosibirsk State Technical University Press) p 214

IOP Conf. Series: Journal of Physics: Conf. Series **980** (2018) 012027 doi:10.1088/1742-6596/980/1/012027

Investigating the influence of cells with various geometric parameters on the air flow in the channel

V I Terekhov [1, 2], Ja I Smulsky [1], K A Sharov [1] and A V Zolotukhin [1, 2]

[1] Kutateladze Institute of Thermophysics SB RAS, 630090, Novosibirsk, 1 Ac. Lavrentiev Ave., Russia
[2] Novosibirsk State Technical University, 630073, Novosibirsk, 20 Karl Marks Avenue, Russia

E-mail: terekhov@itp.nsc.ru, sharov@itp.nsc.ru, smul@ngs.ru, zoloav@mail.ru

Abstract. This paper experimentally studies the gas flow structure in a longitudinal flow around a honeycomb surface with hexagonal cells of various sizes and depths. The measurements are carried out using the PIV system for the developed flow in a flat channel with the cross-section of 21×150 mm and the length of 1000 mm. The flow is visualized, and the velocity and turbulence components are measured in a smooth channel and in the presence of honeycomb surfaces. It has been shown that a boundary layer with high gradient of velocity and turbulent pulsations is formed near the honeycomb.

1 Introduction

An important direction in the development of the heat transfer theory and aerohydrodynamics is the search for new methods of heat transfer intensification with a minimal growth of hydraulic losses. In this respect, the surfaces with lunar reliefs are illustrative [1,2]. A number of experimental and numerical studies in this field prove great potential of such methods.

The honeycomb surfaces considered in this work, used in particular for the purposes of thermal protection of spacecraft elements and as the seals for gas turbines, have a number of similarities to the well vortex generators. This is evidenced by the results of the first experimental studies of the longitudinal flow around the surfaces with honeycomb coatings [3,4]. Data on acoustic

Content from this work may be used under the terms of the Creative Commons Attribution 3.0 licence. Any further distribution of this work must maintain attribution to the author(s) and the title of the work, journal citation and DOI.
Published under licence by IOP Publishing Ltd

damping [5], which helps to reduce friction, gives all the grounds for using such vortex generators as the effective methods for controlling the processes of turbulent transfer. However, due to the lack of detailed experimental information on the flow structure, it is not possible to create an integrated pattern of this complex phenomenon, and therefore its deeper investigation is required.

The development of the flow field and turbulent characteristics along the honeycomb surfaces are studied in this paper. It compares data on the averaged and turbulent characteristics of flowing in a channel with honeycombs and with smooth walls.

2 Experimental setup and measurement methods

The working channel of the experimental setup has a rectangular section of 21 × 150 mm, length of 1 m, and height h = 22 mm. It is made of transparent organic glass of 8 mm thickness. The transverse dimensions of the inlet channel are chosen to ensure the two-dimensional flow in the central longitudinal cross-section of the channel. At the distance of 580 mm from the channel inlet, the plates are mounted flush with the following dimensions: the first plate of 40 × 200 mm, and the second and third plates of 58.4 × 182 mm with hexagonal cells. The cells are blocked from outside. Their heights b are 21 mm, 6 mm, and 2.4 mm. The first honeycomb surface is made of textolite with the thickness of cell partitions of 0.2 mm and the diameter of a circle circumscribed around a cell of 5 mm. Honeycombs No. 2 and No. 3 are made of plastic and have the cell diameter of 12 mm and wall thickness of 0.5 mm. The cell diameter-height ratio d/b is 0.24 for the first plate, and for the second and third plates, it is 2 and 5, respectively.

The experimental setup includes all elements of the wind tunnel to form a qualitative flow: prechamber, nozzle and working channel. The detailed description of setup and measurement system is given in [6]. In the present experiments, development of the flow along the entire length of the honeycomb plate is studied at three different velocities: 5, 10 and 20 m/s; in the beginning of the plate with honeycombs, the flow is developed. The measurements are performed using the PIV system.

3 Experimental results and discussion

The paper presents the results of measurements of longitudinal velocity profiles, longitudinal and transverse pulsations, and shear stresses in a mooth channel and channel with honeycombs.

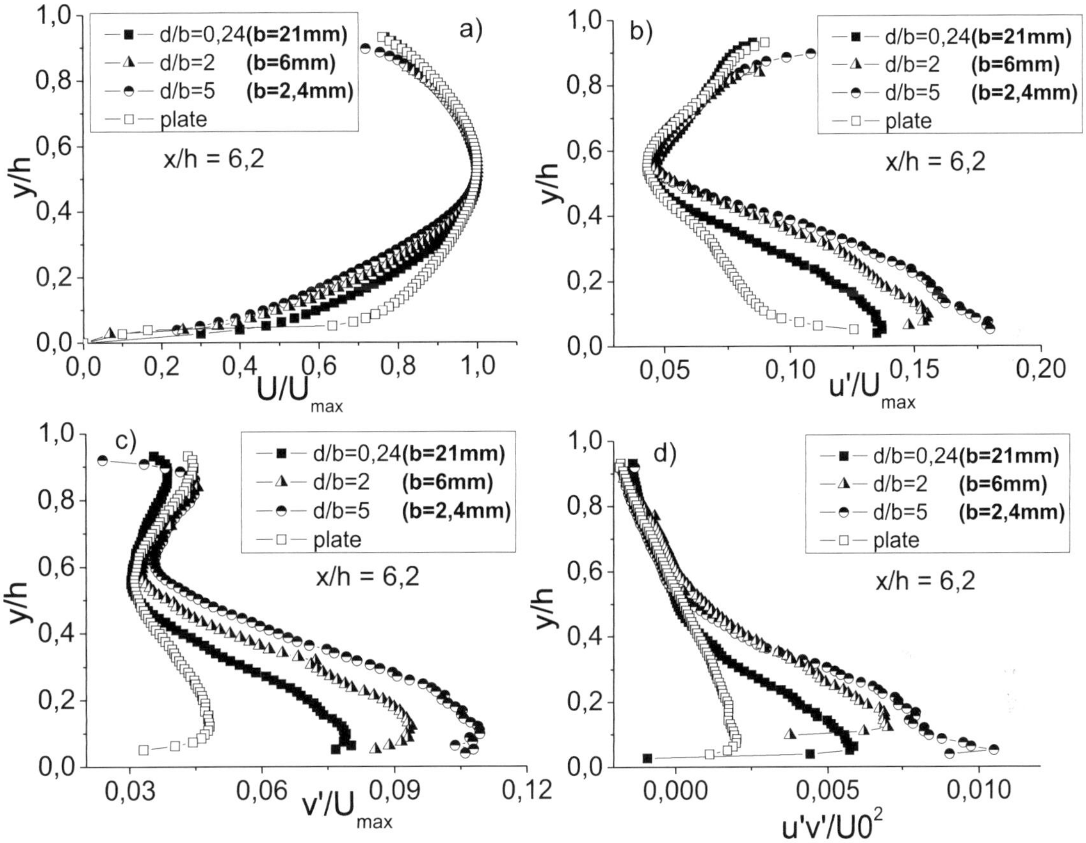

Figure 1. Distribution along the channel height: a) longitudinal velocity, b) longitudinal, c) transverse fluctuations, and d) shear stresses.

All data are obtained for cross-section at x/h = 6.5 from the beginning of the surface with hexagonal cells. According to experiments, the flow in the vicinity of the honeycomb surface is fundamentally different from the flow around a smooth wall. In the near-wall region of the honeycomb surface, the velocity profile is gentler in comparison with the classical distribution on a smooth wall (Fig. 1a). At the same time, longitudinal and transverse velocity pulsations (Figs. 1b and 1c, respectively) near the honeycomb surface are several times greater than on the smooth surface. A similar tendency exists for distribution of shear stresses across the channel (Fig. 1d). If for the channel with smooth walls, as expected, there is the linear distribution of shear stresses along the height, in the channel with cells at distance y/h ~ 0.2, the stresses increase more than 2-3 times. At the same time, a steady tendency to a decrease in transverse velocity pulsations and turbulent shear stresses is observed directly in the near-wall region.

Such a complex scenario of the flow development is caused by interaction of the boundary layer with the honeycomb cells. Indeed, the flow around depressions of various shapes (cylindrical cavity [7], hole [8], transverse trench-groove [9], etc.) on the surface leads to formation of a periodic vortex layer behind them. However, we failed to detect the existence of such structures in the present work.

One of the principal differences between the obtained results and experimental data of [3,4] is the absence of the flow slip along the surface with honeycombs, which, according to this data, could reach about 20-30% of the velocity in the flow core.

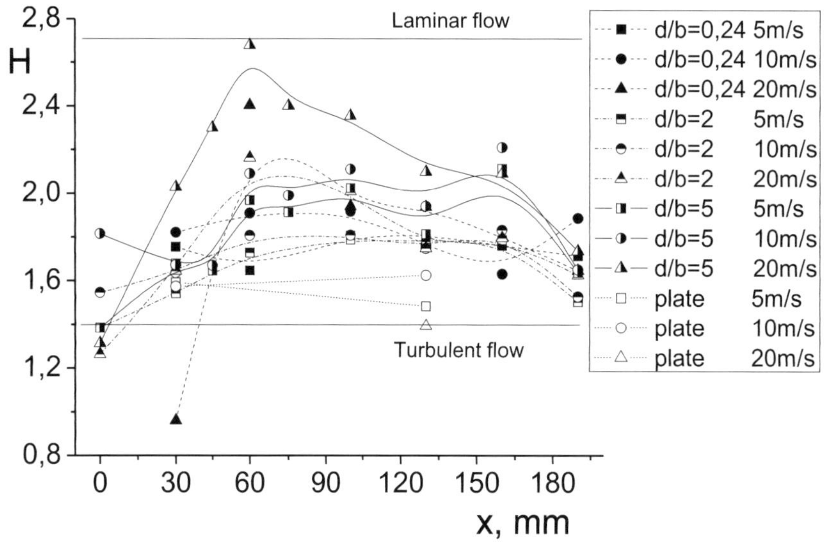

Figure 2. Change in form parameter H along the honeycomb surface.

Distribution of layer form parameter $H = \delta^{*}/\delta^{**}$ along the honeycombs with different ratio d/b is shown in Fig. 2 for three flow velocities (coordinate $x = 0$ corresponds to the beginning of honeycomb region).

In the diagram, value H corresponds to the developed turbulent flow; then, we can observe its increase to the value characteristic of the laminar flow regime ($H \rightarrow 2.7$) and subsequent decrease. This behavior of the form parameter is caused, most likely, by the "flattening" of velocity profile with simultaneous increase in turbulence and tangential stresses.

It is not possible now to formulate clearly the reasons for such a behavior of local and integral characteristics of the turbulent flow around the surface with honeycombs, primarily because of multifactority and complexity of the aerodynamic structure, and this will require further detailed studies.

4 Conclusions

The paper presents the results of experimental study of the averaged flow, normal and tangential stresses in gas, flowing with different velocities around the honeycomb surfaces with characteristic cell size of ~ 5 mm and ~ 12 mm.
The measurements were performed using the two-component PIV system. The flow was organized in a flat channel of 21 × 150 × 1000 mm.
It has been found that the boundary layer on the surface with hexagonal cells is less filled than on a smooth wall. At the same time, the growth of turbulent pulsations and tangential stresses is observed simultaneously. However, there is no effect of flow sliding relative to the cells.

Acknowledgments
The work was financially supported by the grant of the President of the Russian Federation for the leading scientific schools (project NSh-8780.2016.8.).

References

[1] Isaev S A, Kornev N V, Leontiev A I and Hassel E 2010 *Int. J. Heat Mass Transfer* **53** 178
[2] Kovalenko G V, Terekhov V I and Khalatov A A 2010 *J. Appl. Mech. Tech. Phys.* **51** 839
[3] Klimov A A and Trdat'yan S. A 2003 *High Temperature* **41** 801
[4] Trdatyan S A and Klimov A A 2002 *Proc. 12-th Int. Heat Transfer Conf. Grenoble* p 221
[5] Koval'nogov N N 2003 *Energy Problem* **5**(6) 41 (in Russian)
[6] Terekhov V I, Smulsky Ya I, Sharov K A, and Zolotukhin A V 2014 *Thermophysics and Aeromechanics* **21** 6
[7] Hiwada M, Kawamura T, Mabuchi J and Kumada M 1983 *Bull. JSME.* **26** 1774
[8] Terekhov V I, Kalinina S V and Mshvidobadze J M 1992 *Siberian Phys.-Tech. J.* **1** 77
[9] Terekhov V I, Yarygina N I and Shaporin A V 1999 *Industrial Heat Eng.* **21**(2-3) 22

IOP Conf. Series: Journal of Physics: Conf. Series **980** (2018) 012028 doi:10.1088/1742-6596/980/1/012028

Features of round air jet flowing at low Reynolds numbers

V V Lemanov [1], K A Sharov [1], and N V Gorinovich [1, 2]

[1] Kutateladze Institute of Thermophysics SB RAS, 630090, Novosibirsk, 1 Ac. Lavrentiev Ave., Russia
[2] Novosibirsk State Technical University, 630073, Novosibirsk, 20 Karl Marks Ave., Russia

E-mail: lemanov@itp.nsc.ru, sharov@itp.nsc.ru, dina-9494@mail.ru

Abstract. The laminar-turbulent transition in a round jet flowing from a cylindrical channel with the diameter of 3.2 mm was studied experimentally. In experiments, the range of Reynolds numbers determined by the mean-flow velocity was Re = Ud/ν = 700-12000. The measurements were carried out using a PIV system and one-component hot-wire anemometer. The profiles of average velocities and their pulsations in the zone of laminar-turbulent transition were obtained along with axial distributions of longitudinal velocity and pulsations of longitudinal velocity.

1 Introduction

It is known that jets are unstable at low Reynolds numbers Re = 10-30 [1], however, the transition to turbulence occurs at significantly higher values of Re = 30-2000 [2, 3]. The gas jets flowing from the contoured nozzles and at large Reynolds numbers $Re>10^4$ are studied. The initial conditions are known to significantly affect the dynamics of the flow development in the jet. The laminar-turbulent transitions occurring in the jet flowing from the nozzle and cylindrical channel differ. This is due to the difference in the initial velocity profiles. There are almost no publications on the jets from the long channels, although in practice, such cases are rather frequent.
The feature of extended channels is that the laminar, turbulent and transient flows can occur there, and this fundamentally determines characteristics of

Content from this work may be used under the terms of the Creative Commons Attribution 3.0 licence. Any further distribution of this work must maintain attribution to the author(s) and the title of the work, journal citation and DOI.
Published under licence by IOP Publishing Ltd

the jets flowing from these channels. By now, the influence of these factors has been studied insufficiently.

2 Experimental setup and measurement methods

The submerged jet was formed by a round metal channel with diameter d=3.2 mm and length of 100 d. Air from the high-pressure manifold was sent to the channel through a reducer and valve of fine adjustment. The air flow rate was measured by rotameters Krohne DK 800.
The air jet flowed into the flooded air space (a plexiglass flow chamber of 150x150x400 mm). To visualize the flow and measure the velocity fields, a water aerosol with particle diameter of 1-2 μm was added into the flow. The surrounding space in the flow chamber was also filled with a mixture of air and water aerosol. When using the hot-wire anemometer, the compressed air line, feeding the aerosol generator, was turned off. The automated flow controller EL-Flow of Bronkhorst was included into the compressed air line feeding the main flow instead of the DK 800 rotameter.
The experiments were carried out using a PIV system, including digital camera, solid-state pulsed laser, pulse synchronizer and computer. Resolution of matrix of the digital camera was 1024 × 1360 pixels, laser pulse duration was 5 ns, radiation energy was 50 mJ, laser sheet thickness was 0.2 mm, and minimal time between frames was 20 μs. In experiments with PIV, the measuring region was 15 × 20 mm, and the design area corresponded to the size of 0.53 × 0.53 mm.
Measurements with a hot-wire anemometer were carried out by the DISA 55M system consisting of: single-wire sensor 55P11 with wire diameter of 8 μm, measuring bridge, and voltmeter of mean square values. The signal received from the hot-wire anemometer was converted by a linearizer.

3. Results

3.1 Initial conditions

The initial conditions are very important for the jets; they are distributions of average velocities and velocity pulsations at the outlet of the channel, forming a round jet.

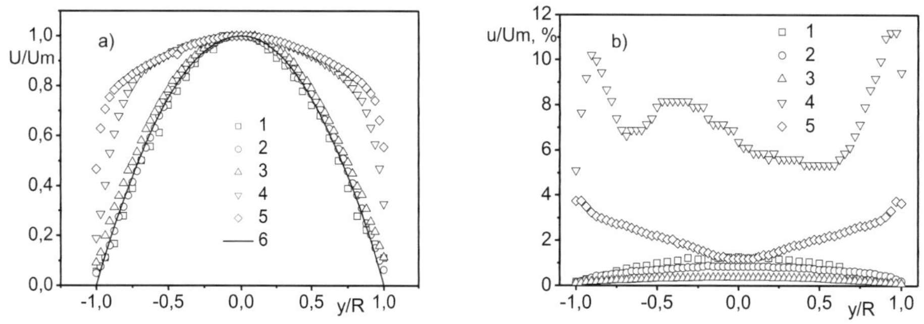

Figure 1. Profiles of average velocities (a) and turbulent pulsations (b) at the edge of a cylindrical channel: *1* – Re = 700; *2* – 1100; *3* – 3990; *4* – 4060; *5* – 12100; line *6* – Poiseuille profile.

The measurements were carried out by a hot-wire anemometer at a distance of 0.5 mm from the tube edge.

Since the length of the channel, which formed a round submerged jet, is sufficiently large, then for Re <2000 in the initial cross-section of the jet, velocity distribution measured along the cross-section must be close to the known Poiseuille profile (Fig. 1a, line 6) for the flow in cylindrical channels. Measurements on the channel edge (Fig. 1a) showed that the average velocity profile is well described by the well-known Poiseuille dependence, under the experimental conditions of up to Re ~ 4000.

This indicates that the flow in the channel was laminar. The corresponding pulsation profiles are shown in Fig. 1b. It can be seen that for the laminar flow the degree of turbulence does not exceed 1.3%. With a further increase in the Re number, the laminar-turbulent transition starts forming in the channel. The velocity profile becomes smoother (Fig. 1a, symbols 4), and pulsations increase significantly (Fig. 1b, symbols 4) near the walls and channel axis. At Re ~ 12000 the fully developed turbulent flow is observed on the channel edge.

3.2 Laminar-turbulent transition in a round jet

Distributions of longitudinal average velocities and longitudinal velocity pulsations on the jet axis measured by the hot-wire anemometer are shown in Fig. 2a and Fig. 2b, respectively. For Re < 1000, the average velocity on the jet axis decreases monotonically along the jet. At that, velocity pulsations here increase insignificantly at large distances from the tube edge. This indicates that the jet remains laminar along the entire length. Starting from

Re = 1000, there is a noticeable increase in velocity pulsations (Fig. 2b, symbols 2) at a distance of ~ 20 x/d from the cylindrical channel outlet. This indicates the beginning of transient processes in the jet.

With further increase in Re, in spite of the fact that the flow in the channel remains laminar, the laminar-turbulent transition is observed in the jet. There is a distinct peak in distribution of axial pulsations (Fig. 2b, symbols 3), then, there is a rapid drop. Distribution of the average velocity on the jet axis also shows a sharp decrease (Fig. 2a, symbols 3). This behavior is caused by a significant increase in the jet half-apex angle after the zone of laminar-turbulent transition, which causes a decrease in the average velocity on the axis.

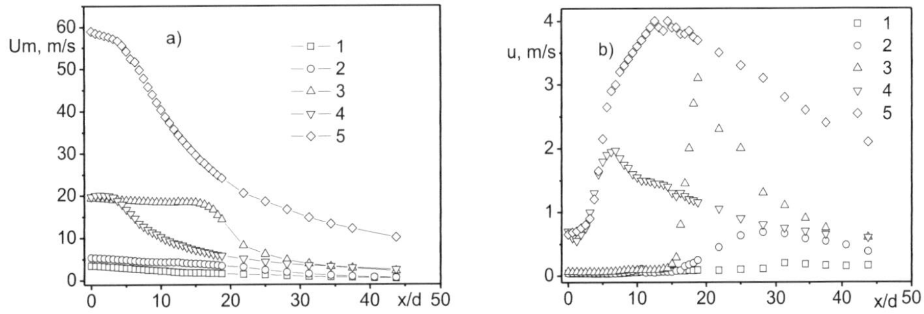

Figure 2. Distributions of average longitudinal velocities (a) and turbulent longitudinal pulsations (b) on the jet axis along its length: *1* – Re = 700; *2* – 1100; *3* – 3990; *4* – 4060; *5* –12100.

In the range of Re ~ 1000-4000, the laminar-turbulent transition was observed in the jet, although the flow in the cylindrical channel was still laminar. The position of the zone of laminar-turbulent transition in the jet depended on the Re number, and with its increase it was shifted to the edge of the channel. When the Re numbers corresponding to the transient processes in the cylindrical channels were achieved, as it can be seen in Fig. 2a, symbols 4, the laminar region of the jet flow became minimal. Fig. 2b, symbols 4, shows attenuation of pulsations in the laminar flow and their sharp increase during transition.

The transition to turbulence in a cylindrical channel leads to an increase in pulsations from the very edge of the tube (Fig. 2b, symbols 5). At some distance, the pulsations reach a maximum, and their recession begins. The average velocity (Fig. 2a, symbols 5) also shows a drop from the very outlet from the cylindrical channel. In this case, this behavior is caused by the turbulent processes.

The profiles of average velocity and pulsations in the round jets were measured by the PIV system. At Re = 1700, which was determined from the mean-flow velocity, the laminar-turbulent transition in the jet started at a distance of about 20 x/d. The transition point determined as for the hot-wire measurements and corresponding to the maximum of turbulent pulsations was located at the distance of 25 *x/d* from the cylindrical channel edge, where the jet flowed out. In experiments, the measurements were carried out at the site, which covered the distance from 20 to 40 *x/d*. Distributions of the average and pulsation velocity components were obtained during the flow transition from the laminar to turbulent regime.
Distributions of average velocities tend to become smoother to the end of the laminar-turbulent transition. This is explained by an increase in the angle of jet half-apex at transition to the turbulent flow. On the jet axis, the turbulence degree is not large and makes up several percents. At transition, pulsations on the axis intensively increase and reach a maximum ; then there is a small decrease, and simultaneously the pulsations profile becomes smoother. Longitudinal and transverse pulsations behave similarly.

4 Conclusions

Distributions of average axial velocities and pulsations along the jet were obtained in a wide range of Reynolds numbers (Re = 700-12000) with attention to initial conditions. The profiles of average velocity, as well as profiles of longitudinal and transverse pulsations in the zone of laminar-turbulent transition in the jet have been obtained. It is shown that the laminar-turbulent transition in a jet is associated with the laminar-turbulent transition in a channel. Data on transverse velocity pulsations and Reynolds stresses in the zone of laminar-turbulent transition may be considered as novel.

Acknowledgments
The work was partially supported by the Russian Foundation for Basic Research (grant 17-08-00958).

References

[1] Abramovich G N 1963 *The Theory of Turbulent Jets* (M I T Press)
[2] Lemanov V V, Terekhov V I, Sharov K A and Shumeiko A A 2013 *Tech. Phys. Letters* **39**(5) 421
[3] Aniskin V M, Lemanov V V, Maslov N A, Mukhin K A, Terekhov V I and Sharov K A 2015 *Techn. Phys. Letters* **41**(1) 46

IOP Conf. Series: Journal of Physics: Conf. Series **980** (2018) 012029 doi:10.1088/1742-6596/980/1/012029

Heat transfer enhancement at increasing water concentration in alcohol in the process of non-stationary film boiling

P K Kanin[1], V A Ryazantsev[1], M A Lexin, A R Zabirov[1] and V V Yagov[1]

[1] NRU "Moscow Power Engineering Institute", Krasnokazarmennaya Street 14, Moscow 111250, Russia

Abstract. New experimental data on heat transfer in pool film boiling of subcooled ethanol-water mixtures at spherical surfaces are considered. The water solutions with ethanol mass fraction from 10 to 91% and temperature of liquid 50°C were examined. All the experiments were conducted under atmospheric pressure, using the stainless steel sphere of 39 mm in diameter as a cooled body. The sphere was heated up to 450-750°C, depending on ethanol concentration, and immersed into the experimental vessel with subcooled mixture. As it is expected, boiling heat transfer intensifies with ethanol concentration decrease, and duration of cooling decreases. It means that stable film boiling duration decreases, and earlier transition to intensive heat transfer regime occurs.

1. Introduction

Basing on our hypothesis presented in [6], the appearance of intensive film boiling mode depends on a liquid latent heat of evaporation (h_{LG}), surface tension (σ) and viscosity (ν). Using liquid mixtures instead of pure components can be considered as a convenient way to investigate the influence of liquid properties. That is why in 2017 we have started the experiments with water/alcohols binary mixtures. The main goal of this work is to reveal the influence of coolant properties (h_{LG}, σ) influence on incipience of high intensive film boiling regime.

It is known that under specific conditions even small admixtures of water in a coolant can lead to high intensity of heat transfer process [7]. The approach of [7] was based on the relative version of the method of controlled pulse heating of a wire probe at constant current mode. As was shown by the authors, in single pulse experiments water addition significantly perturbs the values of both boiling-up temperature and heat transfer intensity. The impact was manifested in a decrease in the spontaneous boiling-up temperature at simultaneous increasing intensity of heat transfer in the watered samples.

The present paper presents new experimental data on heat transfer during film boiling of subcooled aqueous solutions of ethanol. A stainless steel sphere of 39 mm in diameter was used. The experiments were conducted at atmospheric pressure with mixtures of various ethanol concentration (mass fraction of ethanol 10 … 91%, with 10% step). The temperature of mixture was 50°C. To the best of the authors' knowledge, there is no experimental data on nonsteady heat transfer during binary mixtures film boiling. Therefore, the new results described in this article may fill this gap.

2. Experimental facility

The experiments were conducted at the stand, which scheme depicted in Fig. 1. The experimental pattern immersed in liquid through open atmosphere that allow to control the pattern surface's state

Content from this work may be used under the terms of the Creative Commons Attribution 3.0 licence. Any further distribution of this work must maintain attribution to the author(s) and the title of the work, journal citation and DOI.
Published under licence by IOP Publishing Ltd

and cleanness. Besides, such the experimental facility makes it possible conducting large amount of experiments (nearly 90) on various mixtures (mass fraction of ethanol 10 … 91%, with 10% step) with various temperature of liquid (from -10 to +70°C with 20 K step) in short time.

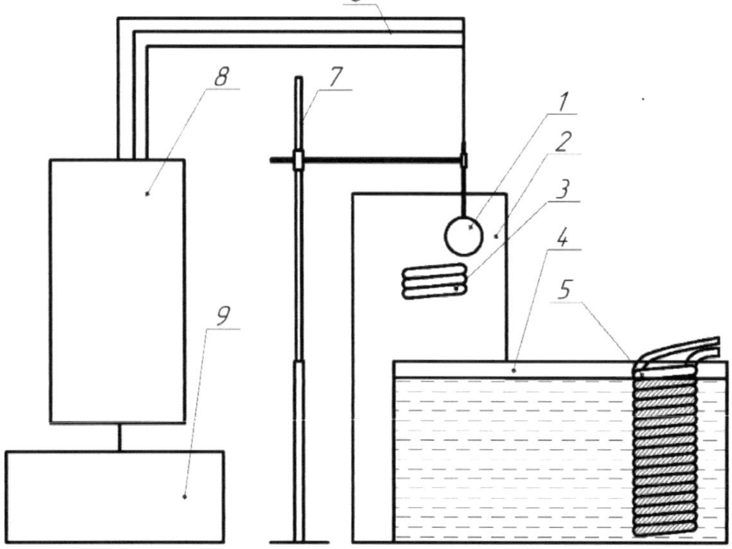

Fig. 1. Scheme of experimental setup: *1* – experimental pattern, *2* – HF inductor, *3* – inductor coil, *4* – experimental vessel, *5* – thermostat coil (connected with the thermostat), *6* – thermocouples wires, *7* – pattern transferring system, *8* – measurement module, *9* – PC.

Depending on ethanol mass fraction in the mixture, the stainless steel ball was heated to temperature 450 – 750°C (the less mass fraction of ethanol the higher initial temperature of the ball). For thermocouples installation 4 through holes were drilled from the top part of the ball to the surface points with polar angles 45, 90, 135 and 180°, and one hole was drilled to the ball centre. The thermocouple junctions (type K) were welded flush with the surface by means of laser welding, their cables passed through the holes and further to the data acquisition system.

At high intensity of heat transfer, the temperature field in the sphere is not uniform and not spherically symmetric. Under such conditions, one has to solve an inverse heat conduction problem (IHCP) in order to calculate heat flux density at the sphere surface. In this study, a numerical solution of direct unsteady heat conduction problem has been found in order to recover the boundary conditions at the surface of the cooled sphere. The problem simplifies, when a cooling process is spherically symmetric; this occurs mainly in film boiling of saturated or weakly subcooled water ($\Delta T_{sub} \leq 20$K). This approximation is also used for rather small spheres even at high intensity of heat transfer at the sphere surface. To restore the surface conditions direct spherically symmetric nonstationary heat conduction problem is solved numerically by means of the RTETA program developed in MPEI [8].

3. Results and discussion

The paper will present only a part of the experimental results obtained. All the experiments were conducted under the conditions, when the stable film boiling regime could be observed. This allows controlling the surface temperature corresponding to vapor film collapse, when the more intensive process starts. All the thermograms at stable film boiling had similar shapes shown in Fig. 2. The curves depict temperature variation in the central (solid line) and the surface points of the sphere during cooling in 91% (ethanol mass fraction) solution with liquid temperature of +50°C. As is seen, the sphere is cooled from 450 to 200°C approximately for 80 s. At low ethanol mass fractions in the solution, quite different type of cooling thermograms is observed. This other type is shown in Fig. 3

that relates to the test pattern cooling in the mixture with 10% ethanol mass fraction and at the same temperature 50°C. In this experimental run the pattern centre is cooled from 630 to 540°C approximately for 1.5 s, while its surface temperature decreases almost by 400 K for the same time. This is obvious that with ethanol mass fraction decrease, the duration of stable film boiling regime decreases.

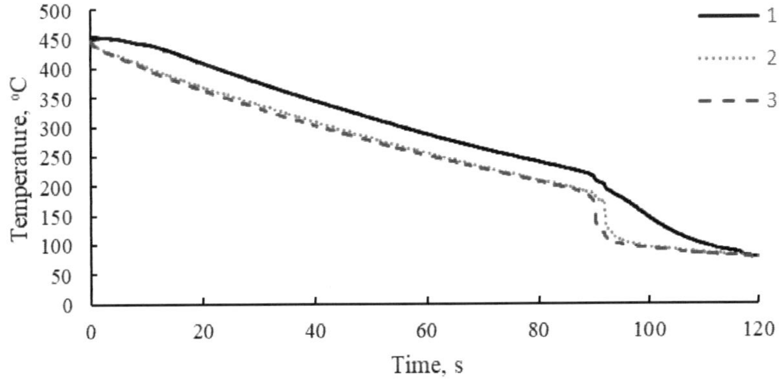

Fig. 2. Thermogram of 39mm stainless steel ball cooling in 91% ethanol solution with liquid temperature of 50°C. Central thermocouple (1) and surface thermocouples at the values of polar angle θ, deg: (2) 45, (3) 180.

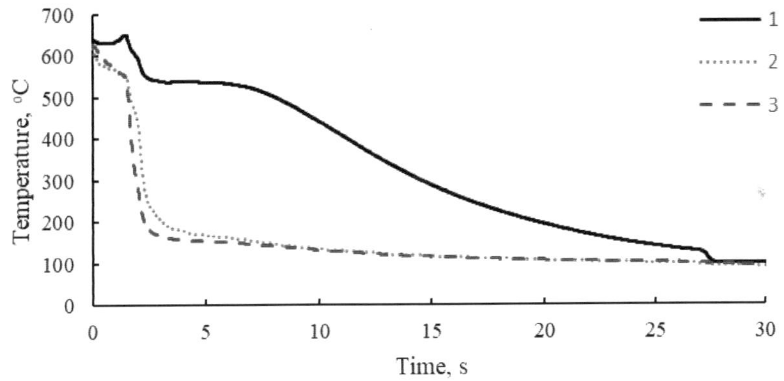

Figure 3. Thermogram of 39mm stainless steel ball cooling in 10% ethanol solution with liquid temperature of 50°C. Central thermocouple (1) and surface thermocouples at the values of polar angle θ, deg: (2) 45, (3) 180.

For testing the hypothesis on the conditions of an incipience of intensive heat transfer regime in film boiling, it is important to determine experimentally the surface temperature values, corresponding to the stable film boiling transition to a more intensive process. Fig. 4 shows how this transition temperature varies with the ethanol mass fraction in the mixture. In this figure the experimental data (dots) present the average temperature of the sphere surface that corresponds to the above transition; a simple linear approximation is also given. It is should be mentioned that all the points in the picture are obtained from a qualitative analysis of the experimental thermograms.

That is why according to our estimations, an uncertainty of this temperature determination is near 50 K. Nevertheless, this dependence seems to be meaningful. As can be seen, the transition temperature dependence on the ethanol mass fraction is close to linear. All the points corresponding to the mass fraction from 91 to 60% are lower than T_{cr} of mixtures and lower than (or near) T_{lim}. For instance, the T_{lim} of the 91% solution is 212°C; as can be seen from Fig.4, the transition temperature is about 165°C. Hence, one can confidently consider the boiling regimes of these mixtures at atmospheric pressure as stable film boiling regimes, similar to our experiments on pure ethanol [5]. As for the solutions with the ethanol mass fraction from 50% and lower, in all the experiments the transition temperature is much higher than T_{lim} or even T_{cr} of the mixtures. For the 50% solution, the average temperature of the sphere surface corresponding to the boiling regimes transition is near 410°C. This is higher than T_{cr} of the mixture with such composition by 90K, to say nothing of T_{lim}. For the mixture with the 10% ethanol mass fraction, intensive heat transfer occurs when the average temperature of the ball surface is near 550°C, while T_{cr} = 364°C. This clearly shows that the cooling process in mixtures with high water mass fraction occurs in micro-bubble boiling regime.

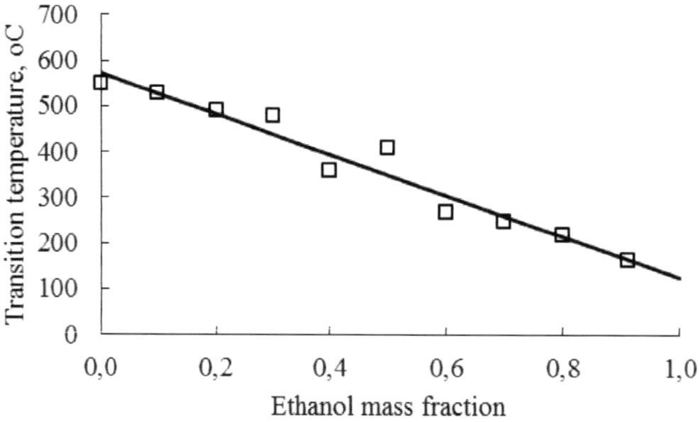

Figure 4. Transition temperature variation with ethanol mass fraction in a mixture

4. Conclusion

Subcooling increase at the fixed water/ethanol mixture composition leads to a decrease in duration of the stable film boiling regime. Cooling process in pure ethanol and in mixtures with its mass fraction ≥ 50% occurs in this regime at rather low heat transfer intensity. At lower ethanol concentration, the vapor film collapses at certain transition wall temperature, and higher intensive process of cooling starts. The cooling process in mixtures with high water mass fraction can be related to micro-bubble boiling regime. On our point of view, the reason of increasing of the transition temperature is in variation of some mixture properties, especially, an increase of latent heat of evaporation and surface tension. This fact is in agreement with our hypothesis of incipience of the intensive cooling regime in subcooled liquids film boiling [6].

The interaction between a hot surface and subcooled liquid may lead to the dangerous and destructive vapor explosion observed in metallurgical and pulp and paper industry and pose a potential hazard in the operation of nuclear reactors [1], in cryogenic systems, and in the transport of liquefied natural gas.

According to the homogeneous nucleation theory [2], if temperature of a hot body surface is much higher than the limiting attainable temperature of a liquid, a vapor film appears around the body almost instantaneously. This means that cooling of high temperature metal bodies in liquids occurs in film boiling regime; under steady conditions, this regime is characterized with low rates of heat transfer. However, in the experiments on cooling the high temperature spheres in water at subcoolings higher than 25 K, the vapor film collapses and huge HTC are observed. Maximum rates of heat flux density can achieve 10 MW/m^2. The English researchers G. Hewitt and D. Kenning with their co-workers first described this unusual intensive process in 1986-1990; they named it micro-bubble boiling [3]. The goal of our study is to reveal regularities and mechanisms of this phenomenon.

Our scientific group has performed many experiments on hot patterns cooling in subcooled water and nonaqueous liquids (ethanol, isopropanol and perfluorohexane) at atmospheric and elevated pressures [4]. As for water, the intensive heat transfer regime occurred in the experiments on the spheres from different metals at pressures of 0.1-1.0 MPa at subcoolings above 25 K [5]. However, among the nonaqueous liquids, the micro-bubble boiling regime was observed only on ethanol under high pressures and high subcoolings [6]. Therefore, the question arises: what water properties, different from those of nonaqueous liquids, cause vapor film collapse and beginning of high intensive heat transfer regime?

Acknowledgments
This study was conducted at the National Research University "Moscow Power Engineering Institute" and financially supported by the Russian Science Foundation, Grant No. 17-79-20402

5. References

[1] Uršic M, Meignen R and Leskovar M 2017 Analysis of film boiling heat transfer during fuel-coolant interaction *Int. J. Heat Mass Transfer* **107** 622–630
[2] Skripov V P 1974 Metastable liquids (Wiley, New York)
[3] Aziz S, Hewitt G F and Kenning D B R 1986 Heat transfer regimes in forced-convection film boiling on spheres, in: *Proc. of the 8th International Heat Transfer Conf* vol. 5 (San Francisco) pp 2149-2154
[4] Yagov V V, Lexin M A, Zabirov A R and Denisov M A 2016 Film boiling of subcooled liquids. Part II: Steady regimes of subcooled liquids film boiling *Int. J. Heat Mass Transf.* **100** 918–926
[5] Yagov V V, Zabirov A R, Kanin P K and Denisov M A 2017 Heat transfer in film boiling of subcooled liquids: new experimental results and computational equations *J. Eng. Phys. Thermophys.* **90**(2) 1466–1475
[6] Zabirov A R, Vagov V V, Kanin P K 2017 Intensive cooling metallic bodies with low thermal conductivity in film boiling of ethanol *Journal of Physics: Conference Series*
[7] Lukynov K V, Starostin A A, Skripov P V 2017 Heat transfer under high-power heating of liquids. The effect of water admixtures on the heat transfer in superheated hydrocarbons *Int. J. Heat Mass Transf.* **106** 657-665
[8] Yagov V V, Lexin M A, Zabirov A R and Kaban'kov O N 2016 Film boiling of subcooled liquids. Part I: Leidenfrost phenomenon and experimental results for subcooled water *Int. J. Heat Mass Transf.* **100** 908–917

IOP Conf. Series: Journal of Physics: Conf. Series **980** (2018) 012030 doi:10.1088/1742-6596/980/1/012030

Numerical Investigation of a Novel Microscale Swirling Jet Reactor for Medical Sensor Applications

G Ogus [1], M Baelmans [2], J Lammertyn[3] and M Vanierschot [1]

[1]Mechanical Engineering Technology TC, campus Group T, KU Leuven, Leuven, Belgium
[2]Department of Mechanical Engineering, KU Leuven, Leuven, Belgium
[3]Division of Mechatronics, Biostatistics and Sensors (MeBioS), KU Leuven, Leuven, Belgium

guven.ogus@kuleuven.be

Abstract. A microscale swirler and corresponding reactor for a recent detection and analysis tool for healthcare applications, Fiber optic-surface plasmon resonance (FO-SPR), is presented in this study. The sensor is a 400 μm diameter needle that works as a detector for certain particles. Currently, the detection process relies on diffusion of particles towards the sensor and hence diagnostic time is rather long. The aim of this study is to decrease that diagnostic time by introducing convective mixing in the reactor by means of a swirling inlet flow. This will increase the particle deposition on the FO-SPR sensor and hence an increase in detection rate, as this rate strongly depends on the aimed particle concentration near the sensor. As the flow rates are rather low and the length scales are small, the flow in such reactors is laminar. In this study, robustly controllable mixing features of a swirling jet flow is used to increase the particle concentration near the sensor. A numerical analysis (CFD) is performed to characterize the flow and a detailed analysis of flow structures depending on the flow rate are reported.

1. Introduction

Fiber optic-surface Plasmon resonance (FO-SPR) is a recent detection and analysis tool with many promising applications in health care diagnostics [1]. The sensor is a 400 μm diameter needle that works as a detector for certain particles. Currently, the detection process relies on diffusion of particles towards the sensor and hence diagnostic time is rather long [1]. As the flow rates are low and the scales are small, the flow in such reactors is laminar. However, a previous study by Ogus et al. [2] shows that it is possible to achieve mixing and turbulent-like characteristics in low Reynolds number regimes by introducing swirl to the flow. This technique will be applied to the microscale reactor in this study.

Annular swirling jets have robust control capability over flows. The flow topologies that can be achieved by changing the inlet swirl are reported thoroughly in literature [2,3]. The challenge for FO-SPR is to increase mixing and particle concentration near the sensor. The swirl has influence on the flow field can meet this challenge. For instance, above a critical value of swirl rate, an isolated recirculation zone downstream of the nozzle appears, called vortex breakdown. Vortex breakdown can be defined as the abrupt widening of the jet core creating a spiral or bubble-like recirculation zone [4,5]. The reviews of Syred et al. [6] and O'Negro et al.[7] are one of the most comprehensive summaries of vortex breakdown and spiral type precessing vortex core studies, mainly related to swirl combustion systems. The nature of the onset of vortex breakdown is not fully understood yet. For a free jet annular or bulk flow jet, previous studies show that the vortex breakdown occurrence is correlated to critical swirl rates [6,8]. In previous studies, it was observed that the vortex breakdown flow state in these type of flows increases mixing [6,8]. There are several types of vortex breakdown. The annular jets with swirl are known to create a spiral type vortex breakdown with a single [10] or a

Content from this work may be used under the terms of the Creative Commons Attribution 3.0 licence. Any further distribution of this work must maintain attribution to the author(s) and the title of the work, journal citation and DOI.
Published under licence by IOP Publishing Ltd

double helix [11]. This type of vortex breakdown, also known as the precessing vortex core (PVC), creates a helical structure inside the jet core. For a Reynolds number in the same order of magnitude, a single helical vortex is observed previously [12].

Enhance mixing in small scale flows, such as bioreactors, is a big challenge as the Reynolds numbers are low and a fully turbulent flow is not feasible to achieve. Using the features of the spiral vortex formation on a swirling jet flow may provide a solution to this problem. The mixing enhancement by vortex breakdown provided by swirl addition is discussed in this paper. By means of direct numerical simulations (DNS), the contribution of vortex breakdown and the PVC to mixing is studied. First, a mixing vessel is designed in the required length scale for the FO-SPR sensor. The vessel is fed by a swirl generator. Then, the fluctuation and precession frequencies are determined for different Reynolds numbers. These frequencies correspond to shedding and laminar instabilities. Turbulent frequencies are not observed. Auto correlation maps for different regions of the flow field are compared. For a further investigation of the flow field, 3D visualization of the vortices are acquired.

2. Swirler and vessel design.

To increase the mixing in small bioreactors, a flow chamber based on a 24 well plate vessel is designed. The design is aimed to work with flow injection methods working with standard syringe pumps for particle detection. The flow is injected from the inlet, shown in Figure 1, coming from the syringe pump. The fluid is transferred to a settling chamber to ensure uniform flow distribution. A microscale swirl generator is connected to this settling chamber and swirl is provided by 6 channels connected to an annular nozzle. The outer nozzle diameter is 3mm, with an inner body diameter of 1.8 mm. The system feeds inside a 15mm diameter to 15mm height vessel of a 24 well plate. Inside this vessel the FO-SPR sensor is planned to be placed for the diagnostics after determining the optimum location for better mixing.

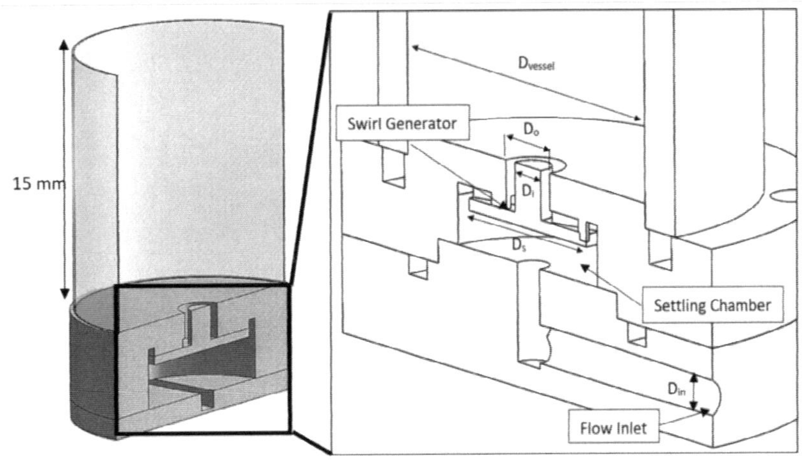

Figure 1. The vessel and swirler design.(Cross section, D_o = 3mm, D_i = 1.8 mm, D_{vessel} = 16 mm, D_{in} = 1.6 mm and D_s = 6.5 mm)

3. Numerical Configuration

The simulations are done using the commercial solver Fluent [13]. Since, the Reynolds number is very low, the Navier – Stokes equations are solved directly using DNS without using any turbulence modelling. The governing equations are

$$\frac{\partial u_i}{\partial x_j} = 0 \quad , \tag{1}$$

$$\frac{\partial \rho u_i}{\partial t} + \frac{\partial \rho u_i u_j}{\partial x_j} = -\frac{\partial p}{\partial x_i} + \frac{\partial \tau_{ij}}{\partial x_j} \text{ and} \tag{2}$$

$$\tau_{ij} = \mu\left(\frac{\partial u_i}{\partial x_j} + \frac{\partial u_j}{\partial x_i}\right) - \frac{2}{3}\mu \frac{\partial u_k}{\partial x_k}\delta_{ij}, \tag{3}$$

where u_i and u_j are velocity components, μ is the viscosity, ρ is the density, p is the pressure, and τ_{ij} is the viscous stress .A structured numerical grid is constructed with a finer midcore region of the flow as high gradients are expected due to shear layers. Figure 2 shows the details of the numerical grid.

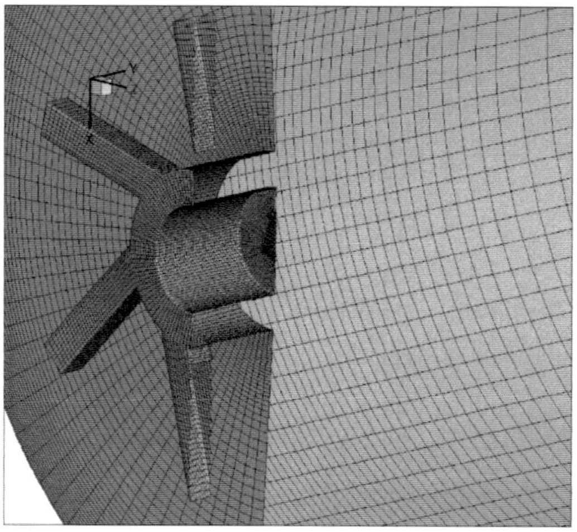

Figure 2. Numerical grid details.

The nature of the flow includes free shear layers that lead to very high gradients. To resolve these gradients, QUICK (Quadratic upstream interpolation for convective kinematics) discretization scheme for momentum has been used [14]. The segregated SIMPLE (Semi-Implicit Method for Pressure Linked Equations) is used for pressure velocity coupling. The discretization of the diffusion term is second order central differencing. A bounded second order implicit time discretization scheme is used for second order accuracy and numerical stability. For a valid DNS study, the spatial and temporal resolution should be taken into account separately. There are two main criteria for the temporal resolution. The first one is that the time step should be as small as to resolve the smallest time scale in the flow. A time step independence simulation is made for 1 ms and 0.5 ms. It is observed that the error between these time steps for second order statistics (root mean square) of centerline velocity is maximally 2.9%, and for the mean velocities around 1.8%. For this study, the time step is chosen to be 0.5 ms for Courant number concerns. The frequency of large scale instabilities such as jet core fluctuation and the vortex precession is found to be around 2-3Hz. Therefore, the chosen time step is small enough to fully resolve these fluctuations.

To validate the DNS study, as well as the temporal resolution, the spatial accuracy should be taken into account. For the spatial resolution, the first criterion checked is the Courant–Friedrich–Lewy condition, $CFL = u\Delta t/\Delta x \approx 1$, which represents that in the flow field an imaginary Lagrangian particle travels less than or equal to one cell size distance. This condition is met inside the vessel areas within the area of interest.

As grid convergence study and to determine the spatial discretization errors, the grid convergence method suggested by Roache [15] is used. At least 3 reference solutions on three grids are required. These grids have approximately 900.000, 1.800.000 and 3.900.000 elements. Then, converged solutions of first and second order statistics on these grids are obtained. The converged solutions are analyzed to determine the grid convergence index , $GCI = F_s e_{21}/(r^p - 1)$, where F_s is the safety factor for the grid convergence, which is 3 for 2 grid studies in 95% confidence interval, r is the grid convergence rate, p is the order of extrapolation (chosen as 2), and $e_{12} = (g_1 - g_2)/g_2$ is the error between the two finest grids, with g being the solved flow parameter. As flow parameter, the maximum RMS values of axial velocity and velocity magnitude, as well as the means of these values are considered. For the maximum RMS values, the convergence index between finest (3.8 million cells) and 2nd (1.8 million cells) grids are 7% and the maximum error (e_{23}) is 2.9%. Therefore the 1.8 million cell grid is used for the rest of the study.

4. Velocity field results

As the aim of this study is to determine the regions and flow states with highest fluctuation values, i.e. regions with intensive mixing, mean flow fields for understanding the flow field and root mean square (RMS) of the axial velocity are reported as it resembles the inner and outer shear layer very clearly. Table 1 shows the mass flow rates, Reynolds and swirl numbers for each setting. The flow rates are limited by the syringe pump that the application requires to be used. The Reynolds number is calculated with the velocity measured on a line on the outlet of the nozzle based on the hydraulic diameter ($D_H = D_o - D_i$) . The Reynolds numbers vary between two extremes, Re= 42, and Re = 168. The swirl rates are determined with the ratio of tangential momentum flux to axial momentum flux as

$$Sw = \frac{\int_{0.5D_o}^{0.5D_i} 2\pi\rho\overline{UW}r^2 dr}{0.5D_o \int_{0.5D_o}^{0.5D_i} 2\pi\left(\rho\overline{U^2}\right) r dr}, \tag{4}$$

where, U and W denote, respectively, instantaneous axial and tangential velocity compo-nents, and the overbar denotes time average of them. The swirl numbers that these flow rates produce also vary with Reynolds number as the swirl generator has a constant geometry.

Table 1. Swirl rates and Reynolds numbers reported in this study.

	Mass Flow Rate (kg/s)	Nozzle Mean Velocity (m/s)	Reynolds number	Swirl rate
Case 1	$1.05\text{x}10^{-4}$	0.036035081	42	0,223
Case 2	$2.63\text{x}10^{-4}$	0.090373696	84	0,608
Case 3	$5.26\text{x}10^{-4}$	0.180747393	168	0,684

Figure 3 shows the mean axial velocity field results for these 3 cases. The first case doesn't provide vortex breakdown. As mentioned before, the vortex breakdown requires a recirculation zone inside the core. However, for the second and third case, this recirculation zone is observed with a negative axial velocity present inside the jet core.

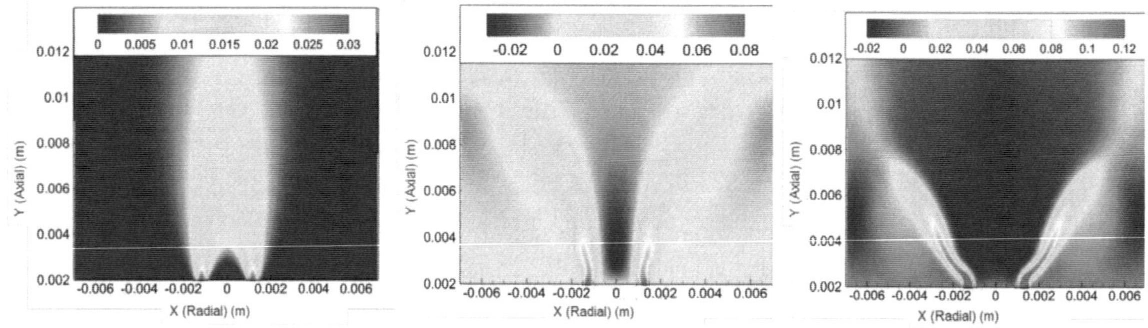

Figure 3. Mean axial velocity fields [m/s]. (Left: Case 1, Middle: Case 2, Right: Case 3)

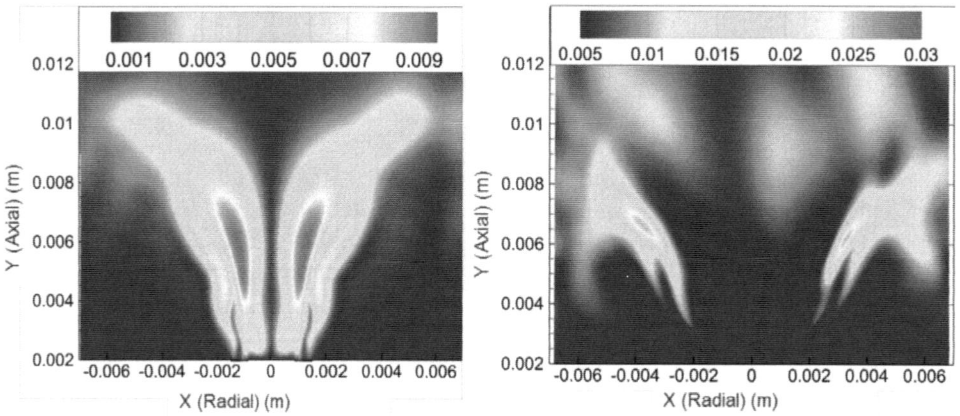

Figure 4. Root mean square of the axial velocities [m/s] (for Case 2 and Case 3)

Figure 4 shows the RMS of axial velocity fields for Case 2 and Case 3. Case 1 is not reported as it does not have significant fluctuations. The root mean square of the flow fields shows that the fluctuation values on these zones are concentrated at the inner and outer shear layer for the Re= 84 and Re = 168 cases (Case 2 and Case 3). Case 2 has a more homogeneous fluctuating velocity region in inner and outer shear layer. On Case 3 with double higher flow rate as the 2nd case, the highest fluctuations are observed in a complete different region compared to Case 2, where the jet loses its momentum. Fluctuations with similar magnitude on inner and outer shear layers, as in Case 2, still exist. The fluctuation magnitudes correspond to 10% of the maximum velocity for the second case as for the inner and outer shear layer of the Case 3. The highest fluctuations observed are around 25% of the maximum velocity for the third case. This suggests that this case might be better for mixing. Application wise, the needle sensor should be aimed to the location of maximal RMS velocity, where increased mixing and particle concentration is to be expected. This optimal location strongly depends on the flow rate. To further understand the mixing behavior, the frequencies of these fluctuations also need to be determined. The frequency of fluctuations are obtained by a fast Fourier transform in a point inside the highest RMS region. Both cases show a very steady fluctuation around 2Hz. The fluctuations have a very periodic nature and turbulent mixing does not exist in these flow configurations. However, the existence of vortices should be proved if any exist. The vortices, with circulation, may trap the particles and enhance the particle concentration. The first analysis is to determine the autocorrelation mapping with 2 - point correlations on a 2D plane with the velocity components as

$$\rho_{ij}(\vec{\xi},t,\vec{x}) = \frac{\langle u_i(\vec{x}+\vec{\xi}/2,\vec{t})u_j(\vec{x}-\vec{\xi}/2,\vec{t})\rangle}{\sqrt{\langle u_i^2\rangle\langle u_j^2\rangle}}, \tag{5}$$

where, u_i and u_j are the velocity deviation from the mean component and ⟨ ⟩ denotes time averaging. This autocorrelation mapping is made in a point where the highest deviation magnitudes are observed for each case. The correlation maps for Case 2 shown on Figure 5 shows a non-isotropic behaviour when the radial (V_x) and the swirling (V_y) velocity components are compared. This behaviour is also reported by Freitag et. al [16] for swirling flows with higher velocity, and is reported as most probably the result of a vortex breakdown spiral. To prove that this spiral type vortex is present in this state, a 3D visualization is required.

The correlation map for Case 3, shown in Figure 6 for the location with the highest RMS magnitude location, shows an isotropic behaviour on the contrary to the previous case. This suggests that the vortex that is present here, is not likely to be a spiral pattern type vortex. Also, the correlation maps for Case 3 are approaching to zero (from red to dark blue) in a slightly smaller distance than Case 2. This suggests that the vortices present in this case are smaller than the previous one, since this distance is a measure of the biggest scale of the vortex structure.

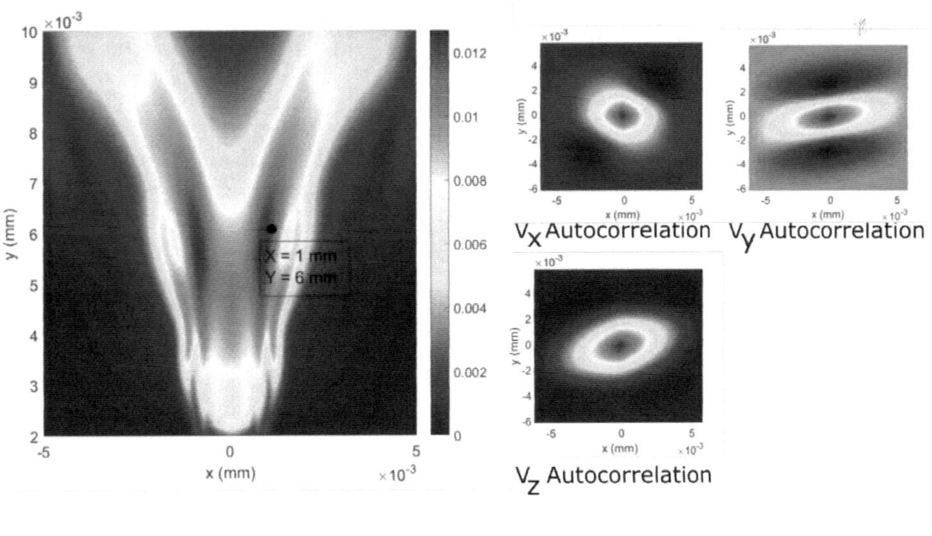

Figure 5. Autocorrelation Map for Case 2.

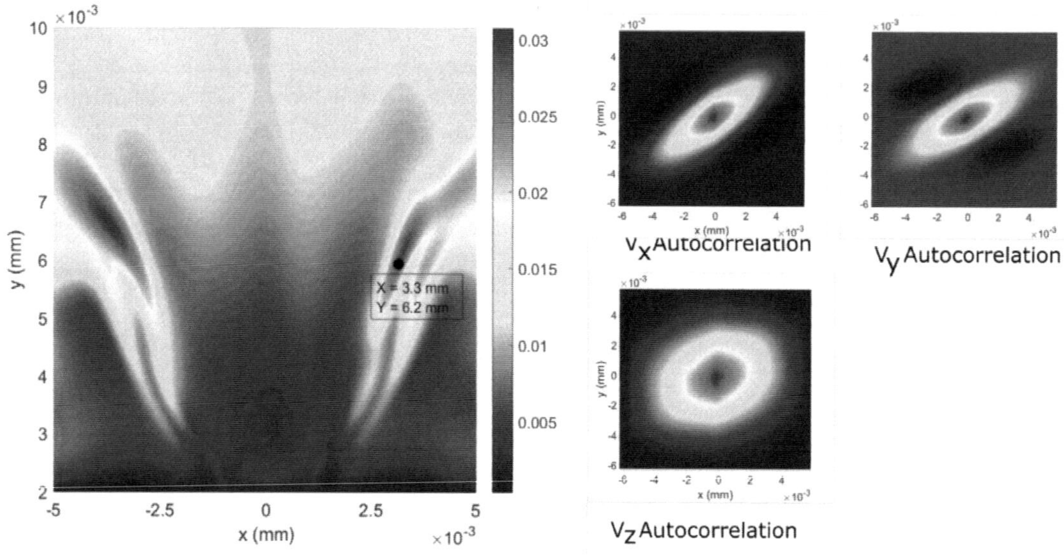

Figure 6. Autocorrelation Map for Case 3.

5. Vortex visualization

A further investigation on the vortex structures is made. This visualization gives more insight, not only for the position of the sensor, but also for the fundamental nature of the precessing vortex core or other flow features. For the visualization, the Q criterion described by Jeong and Hussain [17] is used. It is defined as $Q = 0.5(|\Omega 2| - |S2|) > 0$, where the vorticity tensor (Ω) dominates the strain rate tensor (S) that is derived by partitioning the second order velocity gradient tensor. The visualized cases are discussed in terms of the sensor application and the nature of the vortices .

The first case visualized is Case 1 with Re = 42 and Sw = 0.22. The vortex formation is extremely weak except just downstream the nozzle. Even though the robust location of this circulation bubble provides a good option for the sensor position, fluctuation values lower than 1% of the velocity are considered not to provide enough mixing. The visualization of the second case, with Re = 82 and Sw = 0.6, clearly shows the precessing vortex core or spiral vortex core. The spiral is visible both at the inner shear layer of the core and the outer shear layer. The oscillation frequency of the spirals has been determined as 2 Hz. The configuration provides a solution for mixing and a robust location for the sensor to place in order to obtain a higher particle concentration. The vortex structure shows an irregularity in vortex distribution. These vortices should be regularly shaped in ideal axisymmetric conditions [2]. The irregular shape is caused by the non-ideal conditions of the swirler design. This suggests that the designed swirler is not able to provide fully axisymmetric swirling velocity profiles.

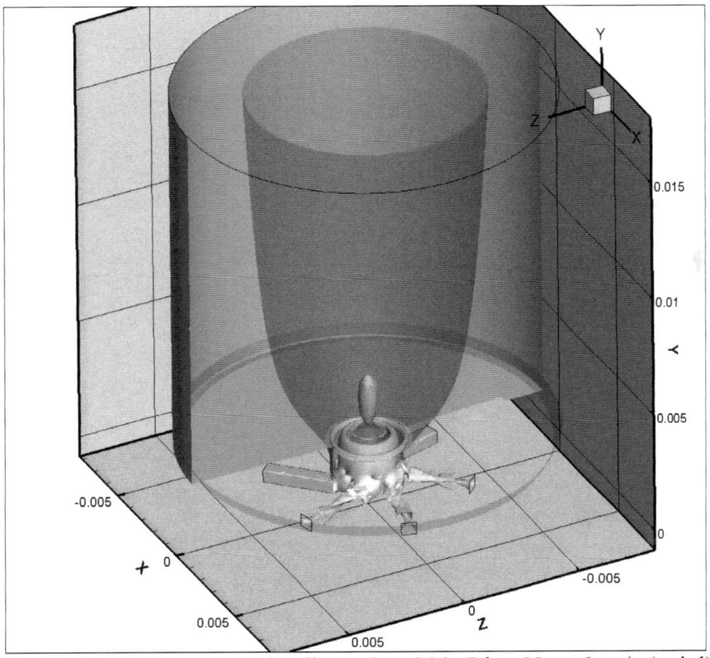

Figure 7. Vortex Visualization for Case 1 (Yellow: Q = 200, Blue Vy = 0 m/s (axial) at an arbitrary instant)

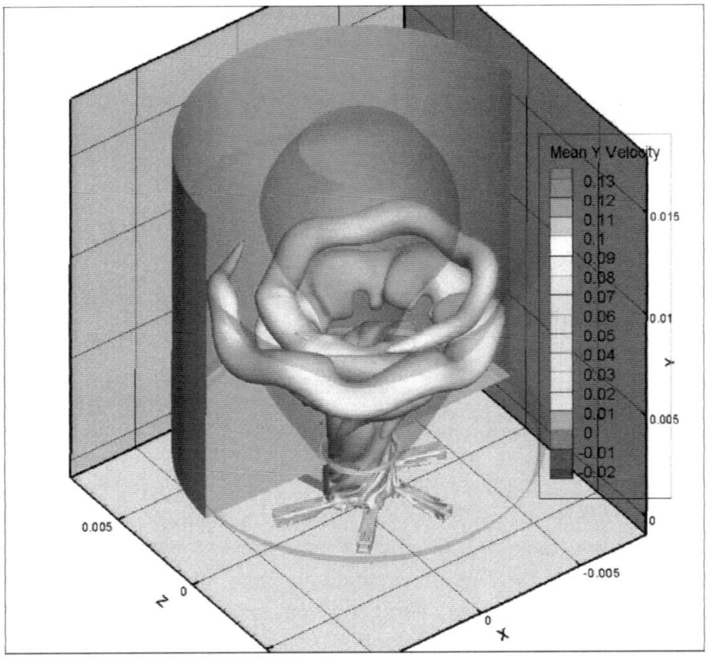

Figure 8. Vortex Visualization for Case 2 (Green: Q = 1000, Blue V_y = 0 m/s (axial) at an arbitrary instant)

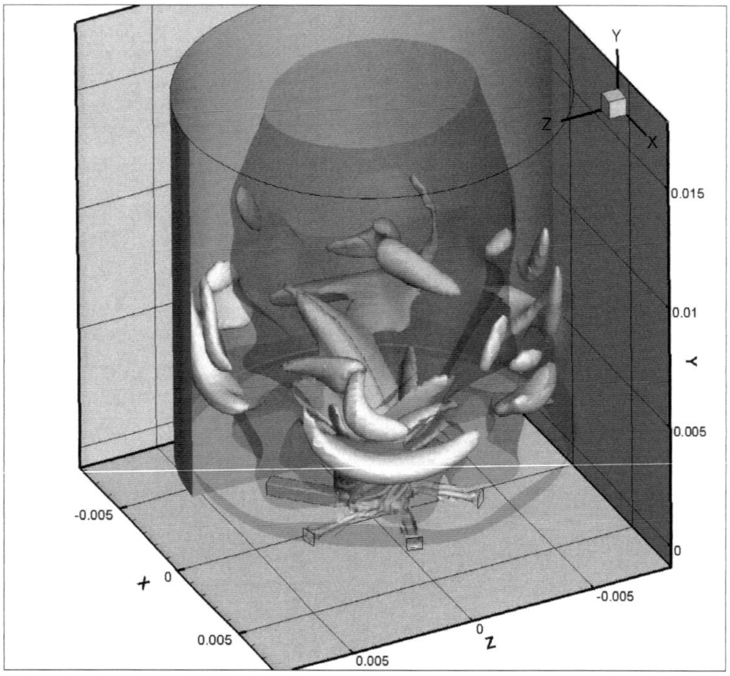

Figure 9. Vortex Visualization for Case 3 (Yellow: Q = 1000, Blue $V_y = 0$ m/s (axial) at an arbitrary instant)

The third case with the highest velocity and swirl rate, the vortex formation is not following the PVC formation pattern. The chaotic positions of the vortices might provide a better mixing resulting in higher fluctuation magnitudes as previously shown. Even though the fluctuation frequency stays the same as the previous case at 2Hz, the higher magnitude of the fluctuations is an advantage. However, if a very precise control of the vortex structures is desired, this flow rate is not sufficient. This result shows that the swirler design could be revised to provide a more regular flow on these Reynolds numbers. A previous study [2] shows for Reynolds number around 180, this type of flow creates robustly controllable flow fields at much higher frequencies. However, the irregularity is an advantage in maintaining higher fluctuation magnitudes.

6. Conclusion

For finding the optimum mixing characteristics for a FO-SPR sensor, a novel mixing tank is designed. The new design consists of a swirl generator feeding a single vessel of a 24 well plate. The swirling flow considered is investigated in term of vortex structures and fluctuation characteristics. An optimum position for the needle–like sensor is tentatively determined by investigating the behaviour of the flow features for different flow rates. The numerical analysis is made by a DNS approach with relatively low computing power as the flow velocities are very low. The resulting flow fields did not provide turbulent mixing. However, a fluctuation of 2 Hz frequency is obtained without using an external mechanical effect, just by controlling the flow. The precessing vortex core obtained, provides an optimum location for the needle like sensor. Also, for higher flow rates, an irregular vortex distribution is obtained that could be beneficial for mixing. The design is open to development by a more optimum swirler design, for a better control of the flow and to obtain higher fluctuation magnitudes at higher flow rates. The future research is to improve the swirler design to provide a more regular flow field, determine the particle concentration ability of the PVC and validate the flow fields experimentally.

References

[1] Lu J, Van Stappen T, Spasic D, Delport F, Vermeire S, Gils A and Lammertyn J 2016 "Fiber optic-SPR platform for fast and sensitive infliximab detection in serum of inflammatory bowel disease patients." *Biosensors and Bioelectronics*, *79*, 173–179.

[2] Ogus G, Baelmans M and Vanierschot M 2016 "On the flow structures and hysteresis of laminar swirling jets." *Physics of Fluids*, *28*(12).

[3] Vanierschot M and Van den Bulck E 2007 "Hysteresis in flow patterns in annular swirling jets." *Experimental Thermal and Fluid Science*, *31*(6), 513–524.

[4] Gupta A and Lilley G, "Swirl Flows," Abacus, Kent, 1984.

[5] Syred N and Beér J M 1974 "Combustion in swirling flows: A review." *Combustion and Flame*, *23*, 143-201

[6] Syred N 2006 "A review of oscillation mechanisms and the role of the precessing vortex core (PVC) in swirl combustion systems." *Progress in Energy and Combustion Science*, *32*(2), 93–161.

[7] Lucca-Negro O and O'Doherty T 2001 "Vortex breakdown: a review." *Progress in Energy and Combustion Science*, *27*(4), 431–481.

[8] Billant P, Chomaz J-M and Huerre P 1998 "Experimental study of vortex breakdown in swirling jets." *Journal of Fluid Mechanics*, *376*, 183–219.

[9] Meunier P and Hourigan K 2013 "Mixing in a vortex breakdown flow". *Journal of Fluid Mechanics*, *731*, 195–222.

[10] Oberleithner K, Paschereit C O, Seele R and Wygnanski I 2012 "Formation of Turbulent Vortex Breakdown: Intermittency, Criticality, and Global Instability." *AIAA Journal*, *50*(7), 1437–1452.

[11] Perçin M, Vanierschot M and van Oudheusden B "Analysis of the pressure fields in a swirling annular jet flow, " accepted for publication in *Experiments in fluids*

[12] Yazdabadi P a, Griffiths a J and Syred N 1994 "Characterization of the PVC phenomena in the exhaust of a cyclone dust separator." *Experiments in Fluids*, *17*(1–2), 84–95.

[13] ANSYS FLUENT 15.0, "User's Guide," 2015.

[14] Leonard B P 1979 "A stable and accurate convective modelling procedure based on quadratic upstream interpolation." *Computer Methods in Applied Mechanics and Engineering*, *19*(1), 59–98.

[15] Roache P J 1994 "Perspective: A Method for Uniform Reporting of Grid Refinement Studies." *Journal of Fluids Engineering*, *116*(3), 405–413.

[16] Freitag M, Klein M, Gregor M, Geyer D, Schneider C, Dreizler A and Janicka J 2006 "Mixing analysis of a swirling recirculating flow using DNS and experimental data," *Int. J. Heat Fluid Flow 27*, 636.

[17] Jeong J. and Hussain F 1995 "On the identification of a vortex", *Journal of Fluid Mechanics 285*, 69–94.

A study of burning processes of fossil fuels in straitened conditions of furnaces in low capacity boilers by an example of natural gas

P V Roslyakov, Y V Proskurin, D A Khokhlov and M N Zaichenko
[1] National Research University "Moscow Power Engineering Institute", 14 Krasnokazarmennaya Street, 111250, Moscow, Russia

RoslyakovPV@mpei.ru

Abstract. The aim of this work is to research operations of modern combined low-emission swirl burner with a capacity of 2.2 MW for fire-tube boiler type KV-GM-2.0, to ensure the effective burning of natural gas, crude oil and diesel fuel. For this purpose, a computer model of the burner and furnace chamber has been developed. The paper presents the results of numerical investigations of the burner operation, using the example of natural gas in a working load range from 40 to 100%. The basic features of processes of fuel burning in the cramped conditions of the flame tube have been identified to fundamentally differ from similar processes in the furnaces of steam boilers. The influence of the design of burners and their operating modes on incomplete combustion of fuel and the formation of nitrogen oxides has been determined.

1. Introduction

Combustion of organic fuel in swirling jets is widely used in a variety of power plants. Fuel burning always occurs in a confined space, but the degree of limitation in different types of units is significantly different. There is a large number of studies, devoted to the burning in combustion chambers of large energy boilers (in condition of small tightness) [1, 2] and to the processes taking place in combustion chambers of gas turbines (very high tightness) [3, 4] . The intermediate position is occupied by a fire-tube and some water-tube boilers with a capacity of 2-5 MW. As practice has shown, the regularities of processes in such units have their own distinctive features in comparison with the two cases cited above. In [5, 6], it is noted that toroidal vortex regions are created in cylindrical furnaces. However, the regularities of their formation and the influence on heat and mass transfer processes have not been sufficiently studied yet. Therefore, the purpose of this work was to study the processes of ignition and burnout of various types of fuel, heat and mass exchange and emission of nitrogen oxides in the constrained conditions of small-sized furnaces of boilers.

2. Main approaches

Preliminary studies have shown that the solution of this problem in isolation from the specific equipment is not correct, therefore the main studies were carried out for a 2 MW fire-tube boiler (Figure 1, a) equipped with a vortex burner. This combination of equipment is widespread in the domestic energy sector.

The burner device (Figure 1, b) used in the work is equipped with one annular channel for supplying secondary air, which is equipped with an axial blending device for its swirling. The outlet part of the

Content from this work may be used under the terms of the Creative Commons Attribution 3.0 licence. Any further distribution of this work must maintain attribution to the author(s) and the title of the work, journal citation and DOI.
Published under licence by IOP Publishing Ltd

burner contains a diffuser-confuser section and a perforated disk that can change its position to influence the characteristics of the flow. In the central channel, which is cooled by a small fraction of air, a fuel injector is installed to spray the liquid fuel into the furnace. Natural gas (Table 1) is fed into the annular air channel after the vortex blades.

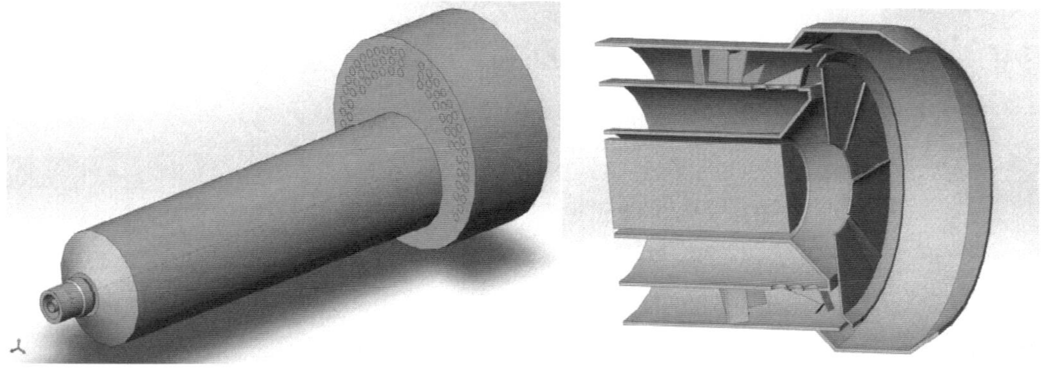

a) b)

Figure 1. Fire-tube of boiler (a) and vortex burner device (b)

ANSYS software was used to simulate the jet flow with fuel combustion. The problem of gas fuel burning was solved using the program ANSYS CFX, and for liquid fuel the ANSYS FLUENT was used. The obtained and processed results were presented in the form of fields of velocity, temperature and concentration variables, mean integral values in different cross and longitudinal sections, etc. In this form they were used for further analysis.

The geometric model was prepared in the SolidWorks program. The ANSYS MESHING software complex with a combination of Delaunay, Sweep and Advancing Front methods were used to build an unstructured calculation grid.

The solution of the known differential equations (continuity, motion, diffusion, energy, etc.) is based on gas dynamics modeling in ANSYS packages and is supplemented by a number of special models. To solve the motion problem in the ANSYS FLUENT program, the RANS method supplemented with the Realizable k-ε model was used [7, 8]. Heat transfer by radiation was calculated using the Discrete Ordinates model [9], and liquid fuel spraying process was simulated using the Pressure Swirl Atomizer model (based on the LISA model) [10] and a number of submodels (Stochastic Collision, Coalescence, TAB Breakup).

Table 1. Composition of natural gas

Components	% vol.
CH_4	98,2
C_2H_6	0,4
C_3H_8	0,1
C_4H_{10}	0,1
N_2	1
CO_2	0,2

The motion of the droplets of liquid fuel, formed as a result of spraying, was described by the Discrete Phase Model, which was supplemented by the Particle Radiation Interaction model. The formation of nitrogen oxides (thermal, fuel, prompt) was described by well-known mechanisms [11]. To describe chemical reactions a diffusion-kinetic model was used [12, 13].

The combustion process was investigated for gas, oil and diesel fuel. Preliminary studies had shown differences in the flame structure for different loads of the boiler, i.e. fuel-air mixture, so the studies were carried out for a range of loads from 40% to 100%. The excess air in the furnace was maintained at 1.17. The inlet air temperature was 30°C.

The realization of a full-scale experiment for verification of the mathematical model was not possible; therefore, to confirm the reliability of the obtained results, the traditional approach - the study of grid convergence - was used. Previously, the problem was solved with a fixed number of elements, which then increased approximately twice. A conclusion about their reliability was made based on the comparison of the obtained results, and further the grid refinement was carried out, if necessary. This approach allowed achieving acceptable results with the reasonable use of mathematical models, included in the modeling process. After analyzing the obtained data and generalizing the values, a conclusion was made about the adequacy of the grid detail with its dimension of about 17 million elements, since most of the studied parameters agreed with the data for a 35 million element grid with an error of less than 2.5%, which is an acceptable indicator.

3. Results

The obtained results for the natural gas combustion indicate that the ignition of the gas-air mixture at maximum loads (80-100% of nominal) occurs due to the creation of two recirculation high-temperature zones in the flame tube (Figure 2). The central return flow zone, which is located not far from the burner section on the axis of the flame tube, brings high-temperature combustion products to the root of the flame and provides a steady ignition in this area. The second torus-shaped recirculation zone is located on the periphery, near the diffuser part of the flame tube (Figure 2). High-temperature combustion products from this zone ensure the fuel-air mixture ignition at the external boundary of the swirling flow in the immediate nearness of the burner section. The maximum temperature region, shaped like a torus, is located approximately one caliber from the burner (that is, just behind the main burning-out zone) and borders directly with the wall of the flame tube, which indicates a decrease in the reliability of its operation. Thus, at maximum loads a stable ignition of the mixture at the outer and inner boundaries of the fuel-air flow is achieved. This is confirmed by the fields of concentrations of oxygen and carbon monoxide, showing that almost complete burn-out of the fuel is provided in the first half of the flame tube at maximum loads.

The efficiency of the processes of ignition and the air-fuel mixture burnup decreases with decreasing load, that is, with a decrease in the swirling flow momentum. If at maximum loads the return flow zones and the ignition borders are practically symmetrical, then when the thermal power decreases, the flame is "stalled" to the wall of the flame tube (Figure 2). The zone of maximum temperatures loses the torus shape and begins to rotate around the circumference near the wall of the flame tube. As a result, the symmetry of all zones and boundaries is violated, and the processes of ignition of the fuel-air mixture are delayed. At the same time, due to the formation of a larger number local active combustion zones with significantly smaller volumes, the maximum temperatures in them noticeably increase to the level of 1950-1990°C. This process, in particular, is significantly affected by the uneven supply of gas fuel through the annular channel of the peripheral air flow.

The calculations have not shown the presence of appreciable concentrations of CO (Table 2) at the exit from the calculated combustion region, even at reduced loads, but the concentration of nitrogen oxides (Figure 3) first decreases, and then increases up to greater values than obtained for the nominal operating mode. This is due to the fact that when the load decreases, the consumption of combustion products decreases. This, on the one hand, reduces the mean integral temperature level at the exit from the calculation area (at the exit from the turnover chamber) and the speed of gas motion inside the flame tube. On the other hand, it increases the time of their stay within the calculated volume. These conflicting trends have different effects on the level of nitrogen oxides emission in the volume of the flame tube.

When the load decreases from 100 to 80% of the nominal value, the maximum temperature level practically does not change (Table 2), while the volume of the zone with a temperature above 1500 °C decreases. This results in an appropriate decrease in the emission of nitrogen oxides. With a further load decrease the symmetry of the aerodynamic organization of combustion was destroyed, which led to the "stalling" of the flame to the wall of the flame tube. Despite occupying smaller volumes in comparison with the maximum loads, high-temperature zones on these regimes were characterized by a rather high level of maximum temperatures T = 1950-1990°C. At the same time, reduced gas velocities led to an increase in the residence time of the combustion products in these high-temperature zones. As a result, both of these factors positively influenced the NO emission, the yield of which in reduced loads decreased by almost 70%, compared to the nominal mode operation, and reached 113-115 ppm.

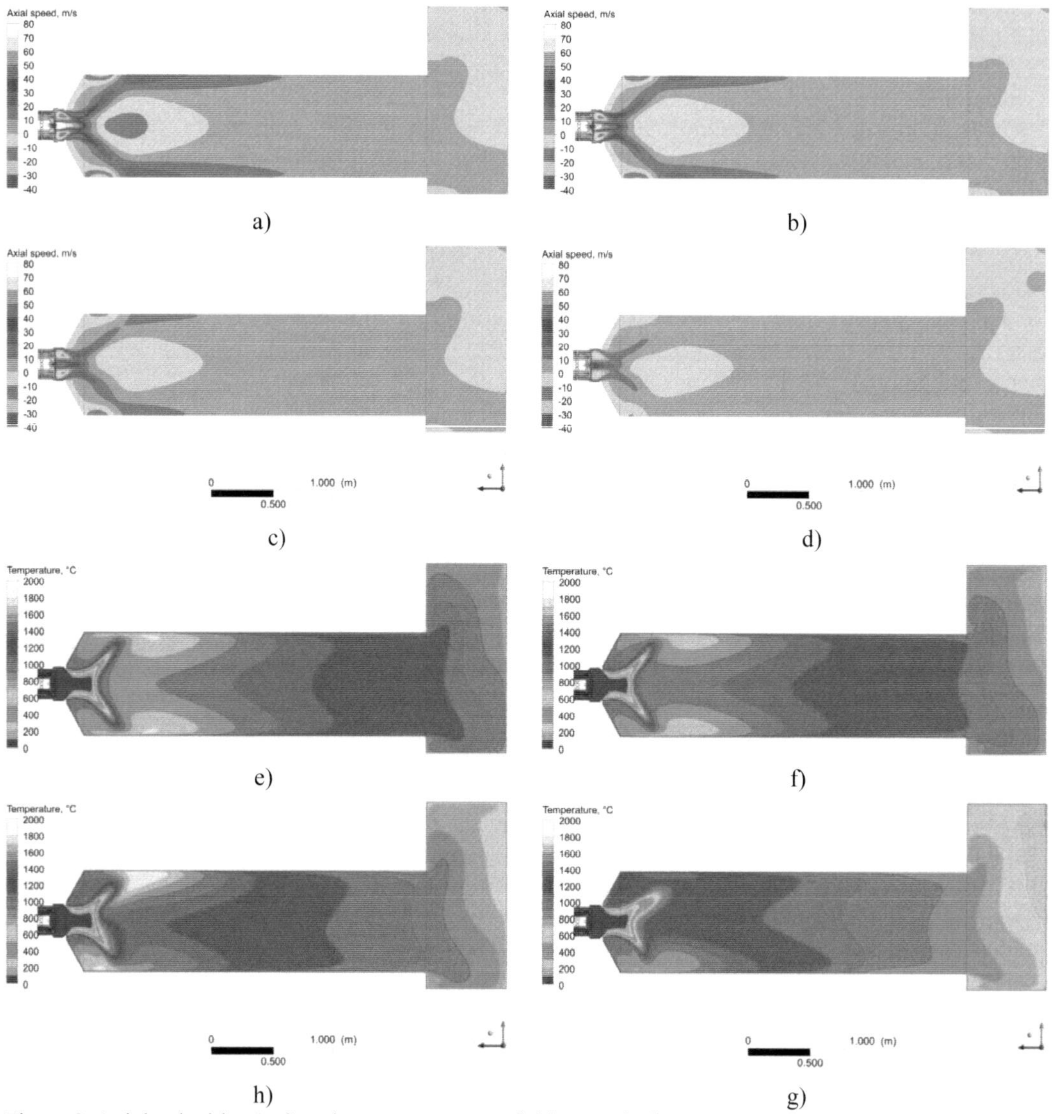

Figure 2. Axial velocities (a-d) and gas temperatures fields (e-g) in fire tube depending on boiler load N: a, e) N = 100%; b, f) N = 80%; c, h) N = 60%; d, g) N = 40%.

4. Discussion

Due to the limited volume of the article, the results of modeling of the combustion processes of oil and diesel fuel are not presented, however, the flow pattern is similar to that for the case of gas combustion, despite the existing differences in the ignition mechanisms [14]. The process of ignition is tightened due to the need for heating and evaporation of the droplets. Combustion of liquid fuel and bottom water is described in detail in [15].

The flow structure in the furnace is regulated by adjusting the burner, or rather the position of the central disk (Figure 1, a). In many respects, the negative effects observed with a reduced load (Figures 2, 3) can be avoided, providing adjustment of the gap at the root of the ring flow. However, in practice this is not used, due to the complexity of implementing such a solution. In this way, the burner is set up once at rated power. That is why it is very important to realize all the subtleties of the furnace operation in the entire range of loads. If the position of the ring at a low load is adjusted, the aerodynamic resistance of the burner at its nominal operating mode will increase substantially, which is unacceptable.

The work of burners of other designs should be studied to obtain more complete data, but most burners for boilers of this type have similar design solutions: one annular channel, a central disk, adjustment of the width of the annular gap between the disc and the outer wall of the channel, preliminary mixing of the gas and spraying of liquid fuel through central nozzle. Therefore, the swirling flow created by such burners should have a very similar configuration.

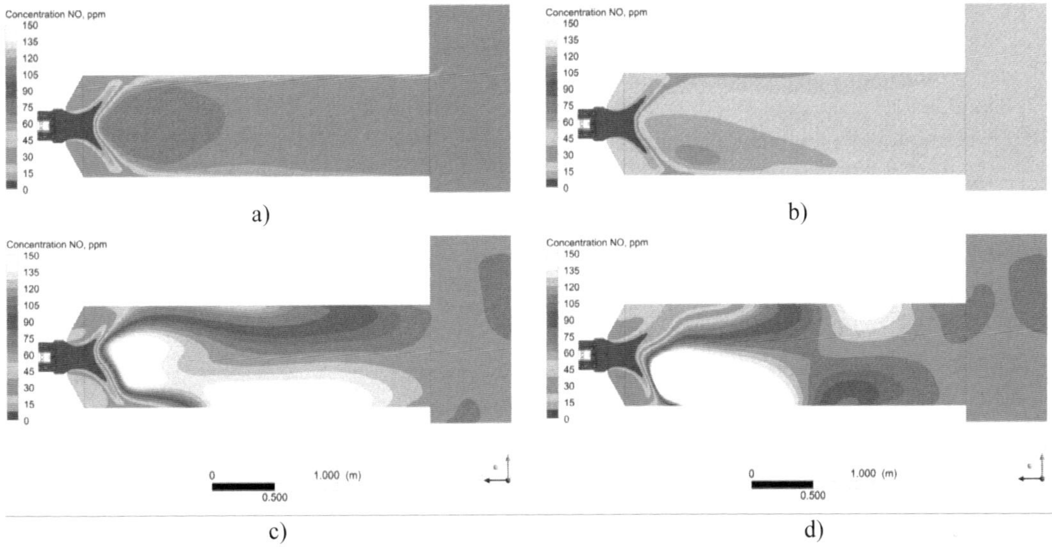

Figure 3. NO concentration fields in fire tube depending on boiler load N:
a) N = 100%; b) N = 80%; c) N = 60%; d) N = 40%.

Table 2. Mean integral gas characteristics in outlet of fire tube

Boiler load, %	Maximum gas temperature, °C	Gas temperature, °C	NO concentration, *ppm*	CO concentration, *ppm*
100	1709	950	67.3	0.39
80	1718	882	47.5	0.43
60	1954	787	114.7	0.54
40	1991	673	113.4	0.74

5. Conclusion

1. The considered flow structure created by a typical swirl burner in the furnace of a flame tube boiler provides almost complete fuel burn-up throughout the all range of workloads from 40 to 100% of the nominal.
2. Combustion of natural gas at reduced boiler loads is accompanied by an increased yield of nitrogen oxides, which exceeds the normative values (GOST 30735-2001), that is associated with a loss of stability of the swirling flow.
3. Avoiding negative effects at low loads may be ensured by constant adjustment of the annular gap, depending on the current load (fuel consumption), while observing the necessary level of excess air (i.e., its flow rate).

References

[1] Dvoinishnikov V A and Khokhlov D A 2015 *Therm. Eng.* **62** 278–85
[2] Roslyakov P V, Morozov I V, Zaychenko M N and Sidorkin V T 2016 *Therm. Eng.* **63** 268–77
[3] Sangl J, Mayer C and Sattelmayer T 2011 *J Eng. for Gas Turbines and Power* **133**(7)
[4] Marosky A, Seidel V, Sattelmayer T, Magni F and Geng W 2013 *J. Eng. for Gas Turbines Power* **135**(11)
[5] Khaustov S A and Zavorin A S 2014 *Energosberezhenie Vodopodgot* **92**(6) 63–67
[6] Alekseenko S V, Kuibin P A and Okulov V L 2003 *Introduction to the Theory of Concentrated Vortices* (Novosibirsk: Inst.Teplofiz. Sib. Otd. Ross. Akad. Nauk) pp 9–68
[7] Launder B E, Spalding D B 1974 *Comput. Meth. Appl. Mech. Eng.* **3** 269–289
[8] Carvalho M G, Farias T and Fontes R 1991 *Fundamentals of Radiation Heat Transfer: Proc.28th National Heat Transfer Conference, ASME HTD* (Minneapolis, MN, July 28–31) vol 160 pp 17–26
[9] Chui E H, Raithby G D and Chui E H 1993 *Numer. Heat Transfer* B **23**(3) 269–288
[10] Schmidt D P, Nouar I, Senecal P K, Rutland C J, Martin J K and Reitz R D 1999 *SAE Technical Paper* No. 1999-01-0496
[11] Roslyakov P V 2007 *Methods of Environmental Protection* (Moscow: Mos. Energ. Inst.) p 336
[12] Spalding D B 1971 *Chem. Eng. Sci* pp 95–107
[13] Magnussen B F, Hjertager B H 1976 *In 16th Symp. (Int'l.) on Combustion*
[14] Roslyakov P V, Proskurin Yu V and Khokhlov D A 2017 *Therm. Eng.* **64** 574–584
[15] Roslyakov P V, Proskurin Yu V and Zaichenko M N 2017 *Therm. Eng.* **64** 644–651

Structure of a swirling jet with vortex breakdown and combustion

D K Sharaborin[1,2], V M Dulin[1,2], D M Markovich[1,2]

[1]Kutateladze Institute of Thermophysics, 630090, 1 Ak. Lavrentyev Avenue, Novosibirsk, Russia
[2]Novosibirsk State University, 2 Pirogov Street, Novosibirsk, 630090, Russia

Email: sharaborin.d@gmail.com

Abstract. An experimental investigation is performed in order to compare the time-averaged spatial structure of low- and high-swirl turbulent premixed lean flames by using the particle image velocimetry and spontaneous Raman scattering techniques. Distributions of the time-average velocity, density and concentration of the main components of the gas mixture are measured for turbulent premixed swirling propane/air flames at atmospheric pressure for the equivalence ratio $\Phi = 0.7$ and Reynolds number Re = 5000 for low- and high-swirl reacting jets. For the low-swirl jet ($S = 0.41$), the local minimum of the axial mean velocity is observed within the jet center. The positive value of the mean axial velocity indicates the absence of a permanent recirculation zone, and no clear vortex breakdown could be determined from the average velocity field. For the high-swirl jet ($S = 1.0$), a pronounced vortex breakdown took place with a bubble-type central recirculation zone. In both cases, the flames are stabilized in the inner mixing layer of the jet around the central wake, containing hot combustion products. O_2 and CO_2 concentrations in the wake of the low-swirl jet are found to be approximately two times smaller and greater than those in the recirculation zone of the high-swirl jet, respectively.

1. Introduction

Swirl is often used in combustors to enhance flame stabilization by creating a low velocity region in high speed flow, or by recirculating hot combustion products and radicals back toward the reactants in the low velocity shear layer where the flame is stabilized [1]. The swirl provides a successful ignition and stable combustion in a compact zone for a wide range of fuel-to-air ratios [2-4]. In particular, it allows stabilizing flames for fuel-lean conditions, which is an effective technology to reduce NO_x emissions [5]. Optimization of the aerodynamics of combustion chambers is also important, because the oxidation of nitrogen to a sufficient extent takes place in hot combustion products behind the flame front [6]. Likewise, low-swirl flame concept (see [7, 8]) has been proved to decrease the amount of NOx in the combustion products in comparison to the high-swirl flame at similar conditions [9].

A detailed information on the flow structure of turbulent reacting flows can be retrieved by using low-intrusive tracer-based optical techniques such as laser Doppler velocimetry (LDV, [10]) and particle image velocimetry (PIV, [9, 11, 12]). PIV provides the spatial velocity distribution for a certain cross-section of the flow. For a more complete description of the processes occurring in the flame, information about the density and temperature distribution is necessary. Currently, there are several methods for measuring the fields of temperature, density and concentration in flames. These methods are based on registration of the local laser-induced fluorescence (LIF), Rayleigh scattering, spontaneous Raman scattering (SRS) and coherent anti-Stokes Raman scattering (CARS) [13-15].

Content from this work may be used under the terms of the Creative Commons Attribution 3.0 licence. Any further distribution of this work must maintain attribution to the author(s) and the title of the work, journal citation and DOI.
Published under licence by IOP Publishing Ltd

LIF-based methods are characterized by high intensity of the detected signal, but the quantitative interpretation of the data is difficult, and sometimes impossible at all, due to the unknown quenching rate of the fluorescence. The planar registration of Rayleigh scattering allows estimating the local density in the flow cross section, but requires a priori information about the resulting scattering cross section of molecules at each measurement point. Interpretation of SRS signal during vibrational transitions of a specific type of molecule at temperature measurements does not require a priori information of the concentration of other molecules and provides information on the relative concentration of the main components of the gas mixture. SRS measurements can be also performed for 1D (along the beam) and 2D (in plane) configurations [16, 17]

The main aim of the present work is an experimental study of the effect of swirl on the spatial distribution of the local velocity, density, and concentration of the main species in low- and high-swirl turbulent lean premixed flames using the PIV and SRS methods.

2. Experimental setup

Sketch of the experimental setup for 2D measurements is shown in figure 1. The measurements were carried out using a combustion rig consisted of a swirl burner, flow seeding device, premixing pipe and section for the air and propane flow rate control. The burner was a contraction nozzle (with the exit diameter $d = 15$ mm) with vane swirler inside. A swirl rate of the flow (the ratio between the angular and axial jet momentum fluxes) was varied by using swirlers with different inclination angles of the blades. The swirl rates were estimated based on the geometric parameters of the swirlers [2]:

$$S = \frac{2}{3}\left(\frac{1-(d_1/d_2)^3}{1-(d_1/d_2)^2}\right)\tan(\psi) \tag{1}$$

Here $d_1 = 7$ mm is the diameter of the centerbody supporting the vanes, $d_2 = 27$ mm is the external diameter of the swirler, and ψ is the vanes inclination angle relative to the axis. Swirlers with $\psi = 30°$and 55° were used to provide the swirl rates of $S = 0.41$ and 1.0, respectively. The Reynolds number Re_{air} (based on d, the bulk velocity of the air flow at the nozzle exit $U_0 = 5$ m/s, and viscosity of the air) was 5 000. The equivalence ratio Φ of the propane-air mixture was 0.7. In order to provide the PIV measurements, the flows were seeded by TiO_2 particles (with the average size of 1 μm). The mass fraction of the particles in the jet was less than 0.03%. The seeding device introduced solid particles to the air flow by using a stirring rod.

The PIV system included a double-pulsed Nd:YLF Pegasus PIV laser and a pair of PCO 1200HS CMOS cameras. The system was running at approximately 770 Hz repetition frequency. The PIV laser sheet, formed by a system of cylindrical and spherical lenses, passed through the central plane of the flow and had a thickness of 0.8 mm in the measurement region. The cameras were equipped with narrow-bandwidth optical filters, transmitting the light from the laser and suppressing the radiation of the flame. The system was operated by a computer with in-house "ActualFlow" software. Stereo calibration was performed using a multi-level calibration target and 3rd-order polynomial transform. For each flow case, 2400 instantaneous three-component velocity fields were measured in three independent runs. The data processing was similar to that in the previous stereo PIV investigation of the swirling flames [18]. The error analysis of the PIV measurements in the reacting flows due to finite spatial and temporal resolution, including those caused by the particle inertia, was reported by [12].

The third (355 nm) or the second (532 nm) harmonic of a pulsed Quanta-Ray Nd:YAG laser was used for the illumination during SRS measurements. The energy of 6 ns laser pulses was monitored by a Coherent LabMax-TOP energy meter. The pulse energy of 355 and 532 nm beams was 200 and 690 mJ, respectively, with 5% RMS fluctuations. During 1D measurements the laser beam was focused with the spot diameter less than 0.8 mm. In case of 2D measurements a collimating optics was used to produce the laser sheet with the width of 45 mm and thickness less than 0.8 mm in the measuring area. The SRS signal was registered by a 16-bit ICCD camera (Princeton instruments PI-MAX-4 with GEN II type of the photocathode). During 1D spectroscopy the scattered light was captured by a Czerny-Turner spectrograph (Newport MS127i 1/8 m), equipped with a UV lens. In the case of 2D SRS spectroscopy, the camera was equipped with a band-pass tunable Lyot optical filter (VariSpec LC), based on liquid crystals. 2D intensity of the Stokes component of the 2D SRS by nitrogen molecules (vibrational-rotational transitions) was recorded in the wavelength range 607.3 ± 5 nm. An additional multi-notch holographic filter was used to block radiation of the Rayleigh scattering, which was not completely blocked by the tunable filter.

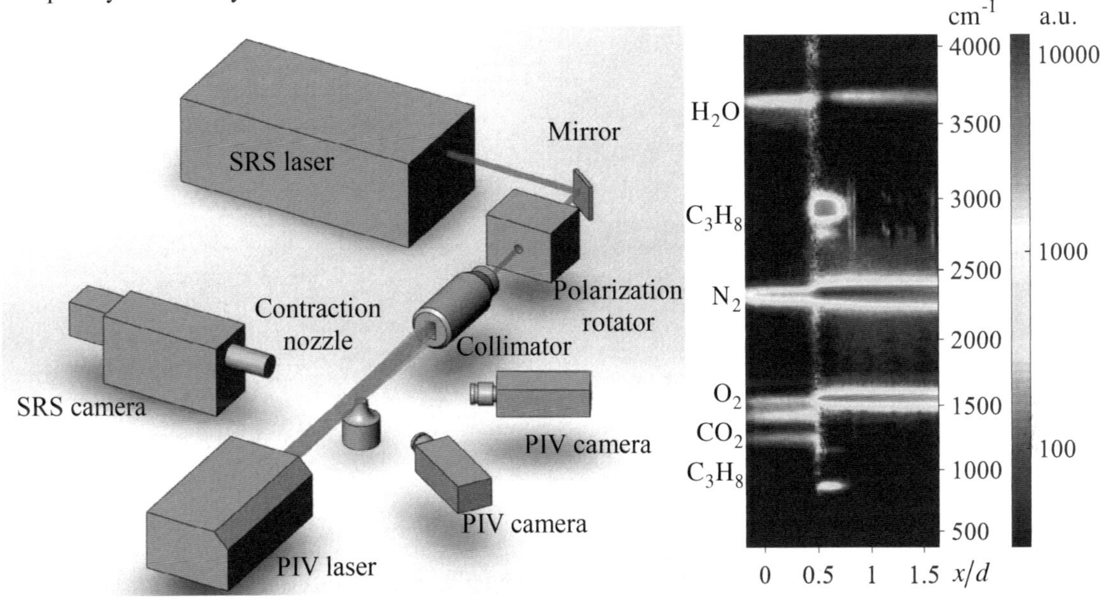

Figure 1. Sketch of the experimental setup for 2D SRS and stereo PIV measurements and example of Stokes SRS spectra along the laser beam intersecting a turbulent high-swirl (S = 1.0) propane/air flame at height of y = 0.5d downstream the nozzle exit

To improve signal-to-noise ratio, each frame captured by the camera was accumulated from SRS signal of 250 laser pulses and was also processed by 8×8 pixels binning. Fifty images were captured and averaged for each measured flame regime to suppress read-out noise of CCD. For further increase of the signal-to-noise ratio, the SRS signal was recorded for two perpendicular linear polarizations of the laser radiation by using a half-wave plate. Signal for the polarization parallel to the scattering plane was subtracted from that for the perpendicular polarization to minimize contribution of the background light, fluorescence and dark current of the photocathode to the SRS data.

Figure 1 shows an example of 1D SRS signal in the cross section of a swirling jet with combustion. There are four different zones. The surrounding air is for $x/d > 0.75$. The annular fuel-air jet surrounds the flame front (a thin layer with remaining broadband fluorescence at approximately 0.5 of x/d) and mixing with the atmospheric air. The reaction zone surrounds the central wake region, containing hot combustion products with temperature above 1800 K, where the signal-to-noise ratio is evaluated as 1:9 and 1:1.7 for the Stokes and anti-Stokes components of the SRS for rotational-vibrational transitions of the N_2. Intensity of the anti-Stokes components of the SRS in the ambient air and reacting flow did not exceed 2% of the Stokes SRS intensity in the air. It was not accounted for during density evaluation because the population of the first vibrational energy level did not exceed 15% for the temperatures below 1800 K.

The data processing for the 1D measurement is similar to that performed by [16]. 2D data processing is based on the ratio between the Stokes components of the rotational-vibrational transitions of the nitrogen in the reacting jet and atmospheric air (when the jet flow was switched off). A more detailed description of the signal processing is given in [19].

3. Results

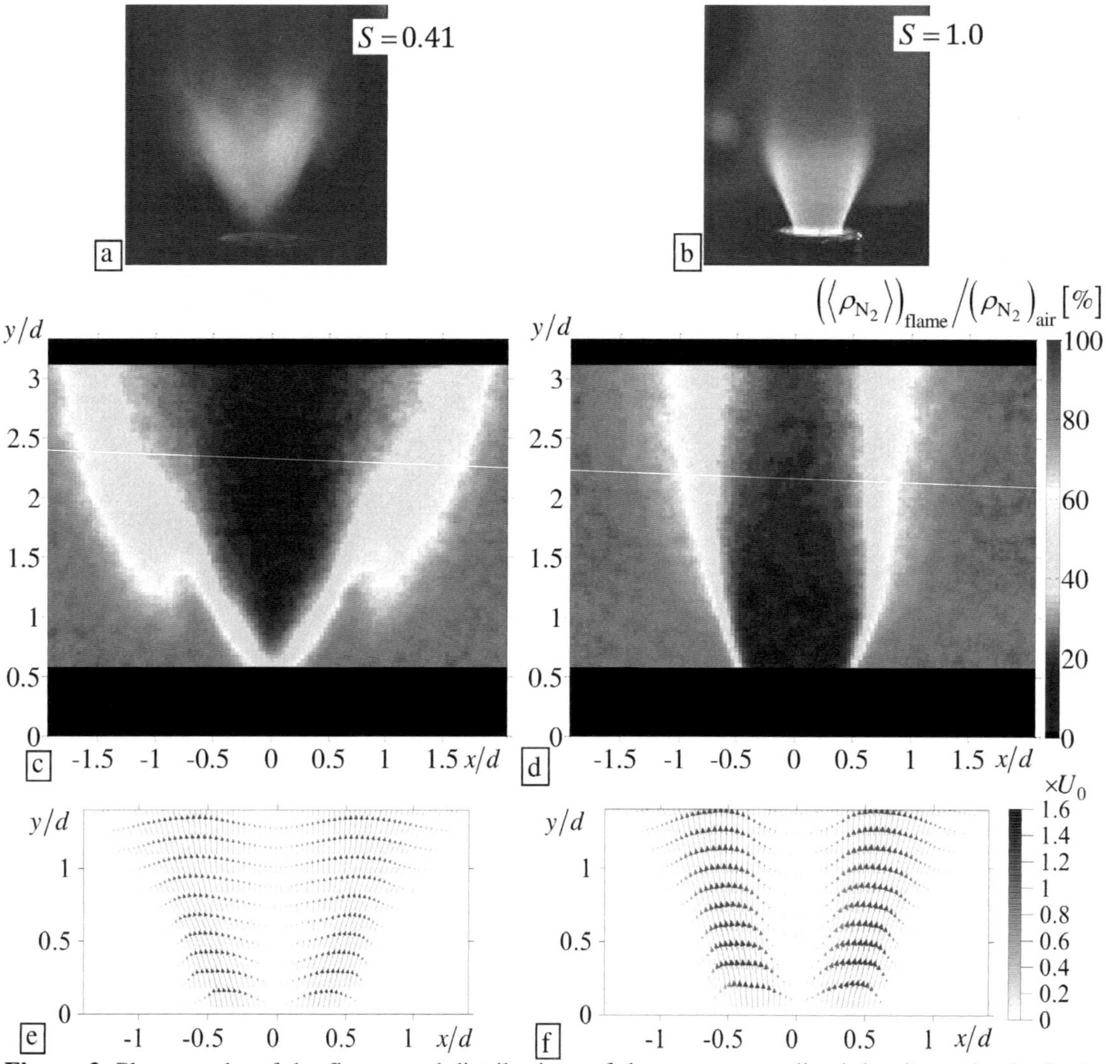

Figure 2. Photographs of the flames and distributions of the mean normalized density and velocity in low-swirl (S = 0.41) and high-swirl (S = 1.0) lean propane/air flames

Figure 2 shows photographs and distributions of the mean velocity and local density of the nitrogen in the low- and high-swirl lean propane/air flames. For both swirl intensities, there are wake regions present at the jet axis, where the local gas density is approximately 6 times lower than that in the surrounding air. For the low-swirl flow the axial velocity reaches minimum near y/d = 0.5 but remains positive, indicating that there is no recirculation zone. The flame front stabilizes around this region. For the high-swirl jet there is a central recirculation zone, partially present in the nozzle. The flame front surrounds it and penetrates into the nozzle. According to the photographs, the flame for the high-swirl flow is less turbulent. This could be due to suppression of turbulent fluctuations by the combustion inside the nozzle.

Figure 3 shows distributions of the temperature and main gas species concentrations for three cross-sections of the reacting jets. There was no sufficient concentration of CO detected in the spectra (see Figure 1). The molar fraction of N_2 is 76±2% for the entire cross-section, indicating that it is a good passive tracer (since it is virtually not consumed during the chemical reactions) for 2D SRS measurements of the local gas density. For y/d = 0.5 and y/d = 1.5 cross-sections of the high-swirl case, the fuel/air jet surrounds the recirculation zone, containing the combustion products with 11% of H_2O, and 7±1% of CO_2 and O_2. The temperature inside the recirculation zone reaches 1845 K. For the low-swirl flame, the hot combustion products also tend to concentrate inside the wake. Remarkable, that for the $y/d = 1.5$ and $y/d = 2.5$ cross-sections the O_2 concentration has minimum of approximately 3% at the axis of the low-swirl jet. CO_2 concentration in the wake is almost two times greater than that for the high-swirl jet.

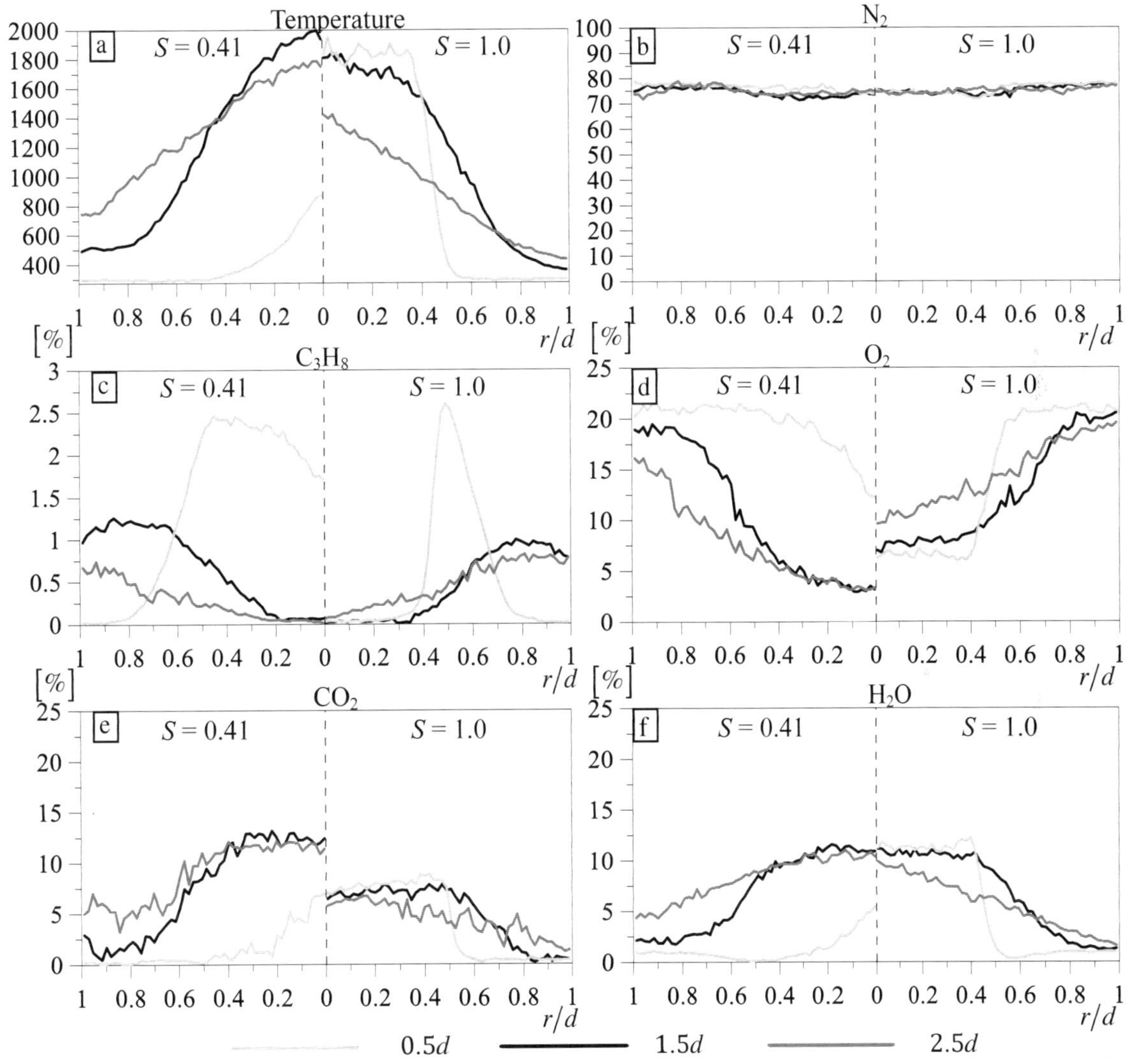

Figure 3. Distributions of the mean temperature and main species concentration in low-swirl ($S = 0.41$) and high-swirl ($S = 1.0$) lean propane/air flames ($\Phi = 0.7$)

4. Conclusion

The time-averaged flow and flame structure of turbulent premixed propane/air lean flames with low- and high-swirl has been studied using the PIV and SRS techniques. In both cases the swirling jet flows from the burner were featured by the presence of a wake zone at the jet axis. The flame stabilized in the inner mixing layer between the wake and the annular fuel/air jet flow. For the high-swirl jet there is a recirculation of the hot combustion products with remaining oxygen. For the low-swirl jet, the oxygen and carbon dioxide concentration in the wake are found to be twice lower and higher, respectively, than those in the recirculation zone of the high-swirl jet.

Acknowledgments
Financial support has been obtained from the Russian Science Foundation (grant № 16-19-10566).

References

[1] Beer J M and Chigier N A 1972 *Combustion Aerodynamics* (Applied Science Publishers, London)
[2] Gupta A K, Lilley D G and Syred N 1984 *Swirl Flows* (Abacus Press, Kent, U.K)
[3] Syred N and Beer J M 1974 Combustion in swirling flows: a review. *Combust. and flame* **23**(2) 143-201
[4] Weber R and Dugué J 1992 Combustion accelerated swirling flows in high confinements. *Progress in Energy and Combustion Science* **18**(4) 349-367
[5] Dunn-Rankin D 2008 *Lean combustion: technology and control* (Academic Press, Elsevier)
[6] Zel'dovich Ya B Sadovnikov P Ya and Frank-Kamenetskiy D A *1947 Okislenie azota pri gorenii [Oxidation of nitrogen during combustion]* (Izd-vo AN SSSR, Moscow–Leningrad)
[7] Yegian D T and Cheng R K 1998 Development of a lean premixed low-swirl burner for low NOx practical applications. *Combust. Sci. Technology* **139**(1) 207-227
[8] Cheng R K 2006 *Low swirl combustion* (The Gas Turbine Handbook)
[9] Johnson M R, Littlejohn D, Nazeer W A, Smith K O and Cheng R K 2005 A comparison of the flowfields and emissions of high-swirl injectors and low-swirl injectors for lean premixed gas turbines. *Proceedings of the Combustion Institute* **30**(2) 2867-2874
[10] Edwards C F, Fornaciari N R Dunsky C M Marx K D and Ashurst W T 1993 Spatial structure of a confined swirling flow using planar elastic scatter imaging and laser Doppler velocimetry. *Fuel* **72**(8) 1151-1159
[11] Stella A, Guj G, Kompenhans J, Raffel M and Richard H 2001 Application of particle image velocimetry to combusting flows: design considerations and uncertainty assessment. *Experiments in Fluids* **30**(2) 167-180
[12] Korobeinichev O P, Shmakov A G, Chernov A A, Markovich D M, Dulin V M and Sharaborin D K 2014 Spatial and temporal resolution of the particle image velocimetry technique in flame speed measurements. *Combust., Explosion, and Shock Waves* **50**(5) 510-517
[13] Kohse-Höinghaus K 1994 Laser techniques for the quantitative detection of reactive intermediates in combustion systems. *Progress in Energy and Combustion Science* **20**(3) 203-279
[14] Hassel E P and Linow S 2000 Laser diagnostics for studies of turbulent combustion. *Meas. Sci. Technol.* **11**(2) R37
[15] Miles R B, Lempert W R and Forkey J N 2001 Laser Rayleigh scattering. *Meas. Sci. Technol.* **12**(5) R33-R51
[16] Rabenstein F and Leipertz A 1998 One-dimensional, time-resolved Raman measurements in a sooting flame made with 355-nm excitation. *Appl. Optics.* **37**(21) 4937-4943
[17] Schefer R W 1997 Three-dimensional structure of lifted, turbulent-jet flames. *Combust. Sci. Technology* **125**(1-6) 371-394
[18] Alekseenko S V, Dulin V M, Kozorezov Y S, Markovich D M, Shtork S I and Tokarev M P 2011. Flow structure of swirling turbulent propane flames. *Flow, turbulence and combustion* **87**(4) 569-595
[19] Sharaborin D K, Dulin V M, Lobasov A S and Markovich D M 2016 Measurements of density field in a swirling flame by 2D spontaneous Raman scattering. *AIP Conf. Proc.* **1770**(1) 030027

IOP Conf. Series: Journal of Physics: Conf. Series **980** (2018) 012033 doi:10.1088/1742-6596/980/1/012033

Investigation of turbulent swirling jet-flames by PIV / OH PLIF / HCHO PLIF

A S Lobasov[1,2], **L M Chikishev**[1,2]

[1]Kutateladze Institute of Thermophysics, 1 Ak. Lavrentyeva Avenue, Novosibirsk, 630090, Russia
[2]Novosibirsk State University, 2 Pirogova Street, Novosibirsk, 630090, Russia

E-mail: Alexey.Lobasov@gmail.com

Abstract. The present paper reports on the investigation of fuel-lean and fuel-rich turbulent combustion in a high-swirl jet. Swirl rate of the flow exceeded a critical value for breakdown of the swirling jet's vortex core and formation of the recirculation zone at the jet axis. The measurements were performed by the stereo PIV, OH PLIF and HCHO PLIF techniques, simultaneously. The Reynolds number based on the flow rate and viscosity of the air was fixed as 5 000 (the bulk velocity was $U_0 = 5$ m/s). Three cases of the equivalence ratio ϕ of the mixture issuing from the nozzle-burner were considered, viz., 0.7, 1.4 and 2.5. The latter case corresponded to a lifted flame of fuel-rich swirling jet flow, partially premixed with the surrounding air. In all cases the flame front was subjected to deformations due to large-scale vortices, which rolled-up in the inner (around the central recirculation zone) and outer (between the annular jet core and surrounding air) mixing layers.

1. Introduction

Swirl is often used for stabilization of jet-flames. Jet flows with strong swirl are featured by breakdown of the swirling vortex core, formation of the central recirculation zone and presence of spiral/helical vortices [1]. The effect of such flow features as vortex breakdown and precession of the vortex core on the combustion in swirling jets is still not completely understood yet. In particular, impact of the large-scale vortex structures (including precessing vortex core, see [2, 3]) on dynamics and stability of flames in swirling flows is still a debated issue [4]. Thus, analysis of flame-vortex interactions in swirl-stabilized combustors is important for better understanding of unsteady combustion phenomena [5].

The large-scale vortices are known to induce deformations of the flame front, affect the heat release rate and may result in local flame extinction. These features have been recently studied by using the combination of particle image velocimetry and planar-laser induced fluorescence (PLIF) for hydroxyl [6, 7], produced in the flame front and present in the combustion products. Nevertheless, there is a need in a detailed analysis of the reaction zone shape and flame front deformations, including those, which are not induced directly by large-scale flow motions.

The present paper reports on the experimental study of flame stabilisation in a high-swirl jet of methane/air mixture for different equivalence ratios. Three cases of combustion regimes are studied, namely, fuel-lean and fuel-rich premixed flames and fuel-rich partially premixed lifted flame.

Content from this work may be used under the terms of the Creative Commons Attribution 3.0 licence. Any further distribution of this work must maintain attribution to the author(s) and the title of the work, journal citation and DOI.
Published under licence by IOP Publishing Ltd

IOP Conf. Series: Journal of Physics: Conf. Series **980** (2018) 012033 doi:10.1088/1742-6596/980/1/012033

2. Experimental setup

Measurements were carried out for the swirling flames of the methane/air mixtures at atmospheric pressure. The flames were organized in an open combustion rig (see details in [2]) by using a contraction axisymmetric nozzle with the exit diameter $d = 15$ mm. A vane swirl was installed inside the nozzle (more details in [8]) to generate high-swirl flow. The swirl rate based on definition in [1] was 1.0, which is well above the critical value of 0.6 for the vortex breakdown in jet flows. Three cases of the equivalence ratio ϕ of the methane/air mixture outflowing from the nozzle were studied, namely, $\phi = 0.7$, $\phi = 1.4$ and $\phi = 2.5$. The Reynolds number of the air jet without methane was fixed as 5 000 (bulk velocity $U_0 = 5$ m/s).

A photo of the PIV/PLIF experimental setup is shown in Figure 1. To provide PIV measurements, the flow from the nozzle was seeded by 4 μm TiO_2 particles. The surrounding air was seeded by using a fog generator. A system of two CCD PIV cameras (ImperX IGV-B2020) was oriented horizontally as shown in Figure 1. Each camera captured 4 Mpix images. The cameras were equipped with Sigma AF #50 lenses and band-pass optical filters (60% transmittance at 532 nm and with FWHM of 10 nm). The seeding particles were illuminated by the second harmonic of radiation of a double-head pulsed Nd:YAG laser with 200 mJ energy per each pulse. A beam was converted into a laser sheet by using a system of cylindrical and spherical lenses. Duration of each laser pulse was approximately 10 ns. Time separation between two PIV laser pulses was 35 μs.

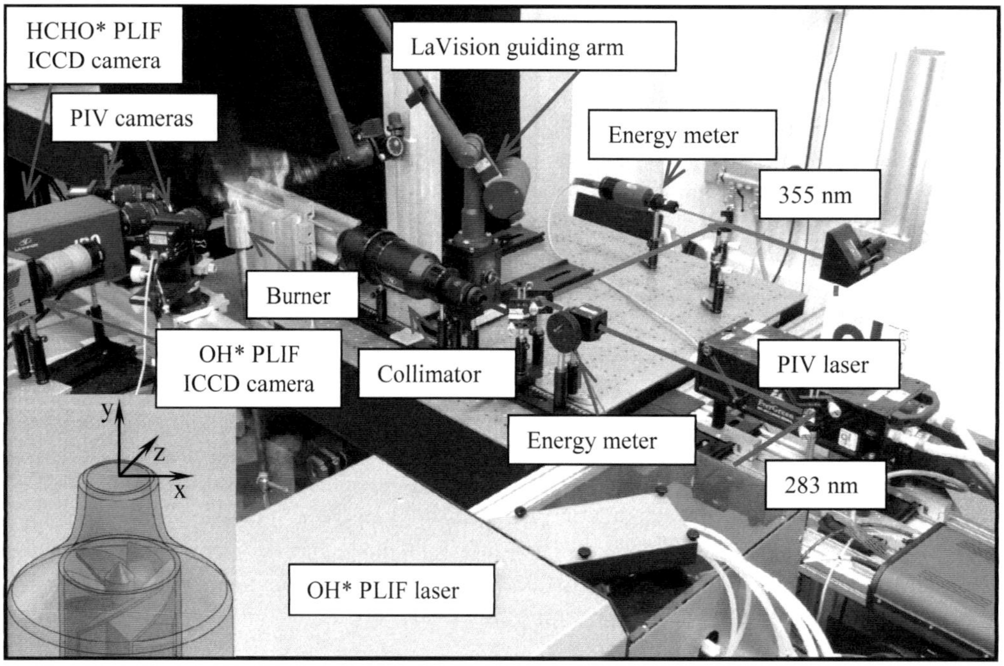

Figure 1. Photo of the PIV/PLIF experimental setup

In the case of the OH PLIF measurements a tunable dye laser (Sirah) was used. The tunable laser was pumped by the second harmonic (532 nm) of radiation of a high-energy pulsed Nd:YAG laser (QuantaRay, 1 J per pulse at 532 nm). The output energy of the dye laser pulses was monitored. The average laser pulse energy in the range of 282-284 nm was approximately 5 mJ. Radiation of the third harmonic (355nm) of a Nd:YAG laser (Quantel Brilliant B with 45 mJ energy per pulse) was used for excitation of HCHO fluorescence. The A–X transition was excited. In both cases, the RMS of the laser pulse energy variation was below 5%. Laser beams for OH* and HCHO excitation were combined by using a dichroic mirror. After a collimator optics, the height of PLIF light sheet was 50 mm. The laser sheet thickness was below 0.8 mm in the measurement region. The sheet illuminated the axial (vertical) plane of the reacting flow.

The fluorescence of OH* was collected by a registration system that consisted of a UV-sensitive image intensifier (LaVision IRO) and sCMOS camera (LaVision, Imager sCMOS, 16 bit images with resolution of 2560×2160 pixels) and was equipped with a UV-lens (100 mm, f# = 2.8). Photocathode (S20 multialkali) of the IRO provided quantum efficiency of about 25% for wavelengths in the range of interest (300-320 nm). The fluorescence of HCHO was collected by a 16-bit ICCD camera (Princeton instruments PI-MAX-4 with GEN II photocathode with quantum efficiency 25% in UV spectral region), equipped with a Sigma AF #50 lens. Appropriate band-pass optical filters were mounted on the lenses to detect OH* and HCHO fluorescence. In both cases the exposure time for each PLIF image was 200 ns.

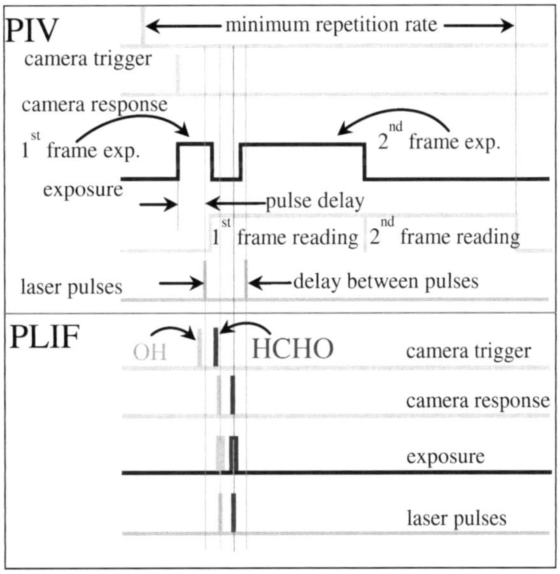

Figure 2. Scheme of temporal synchronization between PIV, OH PLIF and HCHO PLIF systems

For calibration of the dye laser wavelength, excitation spectrum of OH* fluorescence was measured. The excitation spectrum of OH* was compared with the spectrum, calculated by using the LifBase software [9]. The raw PLIF images contained different types of systematic and random error. The systematic errors were caused by spatially non-uniform laser sheet intensity, non-uniform spatial sensitivity of the photocathode and CCD, background signal and dark current. The sequence employed for correcting systematic errors is as follows: background subtraction, laser sheet correction, white sheet correction and laser shot-to-shot fluctuation correction.

To make sure that PIV and PLIF measurement planes coincided, prior to each run a photo-sensitive paper was placed into the measurement volume and exposed to a single shot of each laser. This test was performed to ensure that laser sheet planes were well aligned. Synchronization between PIV, OH and HCHO PLIF systems (see scheme in Fig. 2) was provided by combination of an 8-channel synchronizer (BNC, model 575), LaVision processing timing unit and by an in-house programmable synchronizing processor. The entire measuring system was operated at a repetition rate of three acquisitions per two seconds. The OH* and HCHO images were registered between the first and second pulses of the PIV laser, with a delay after the first pulse of 10 μs and 20 μs, respectively. The time delay between laser pulses of the PIV and PLIF lasers and short PLIF exposure provided a negligible cross-talk between different systems.

IOP Conf. Series: Journal of Physics: Conf. Series **980** (2018) 012033 doi:10.1088/1742-6596/980/1/012033

3. Results

Based on 2D stereo PIV and 2D PLIF measurements, flame front stabilization in the high-swirl flow is analyzed for three considered combustion regimes. The focus is placed on the flow pattern, large-scale flow features (e.g., recirculation zone, flow precession) and shape of the reaction zones. Intensity of HCHO fluorescence provides information about spatial structure of preheat zones, whereas OH* fluorescence is in the flame front and hot combustion products. Pixel-by-pixel multiplication of OH* and HCHO fluorescence signals allows estimating the heat release zone of the flame front. Figure 3 shows flame photographs and examples of the stereo PIV, OH PLIF and HCHO PLIF snapshots which are captured simultaneously. Figure 4 shows an average velocity fields and cross-sections for a high-swirl flow with combustion.

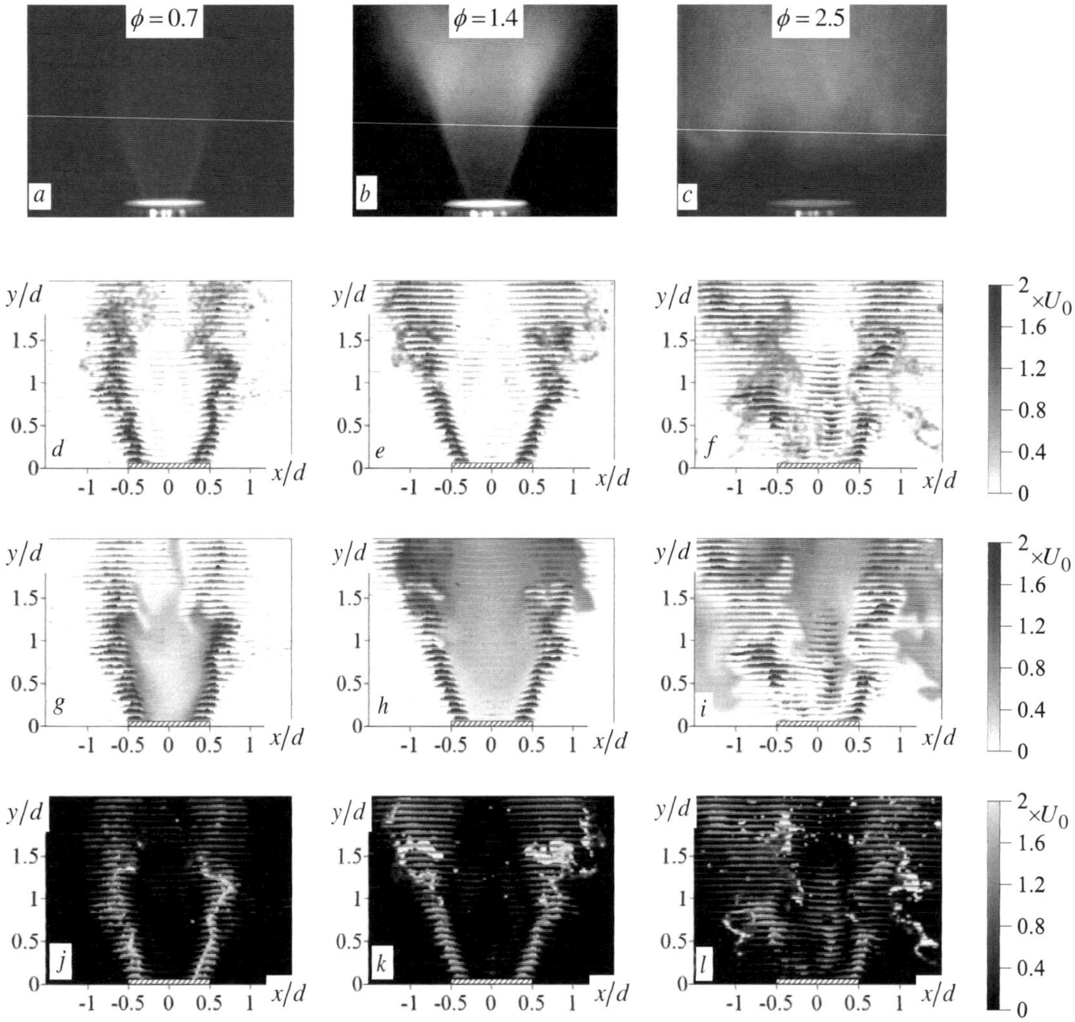

Figure 3. Photographs of three considered combustion regimes (a-c) and examples of the simultaneously instantaneous captured snapshots. Velocity fields and distribution of OH* (d,e,f), velocity fields and distribution of HCHO (g,h,i), velocity fields and region of maximum heat release(j,k,l) for the axial plane of methane/air swirling flames for (a,d,g,j) $\phi = 0.7$, (b,e,h,k) $\phi = 1.4$, (c,f,i,l) $\phi = 2.5$

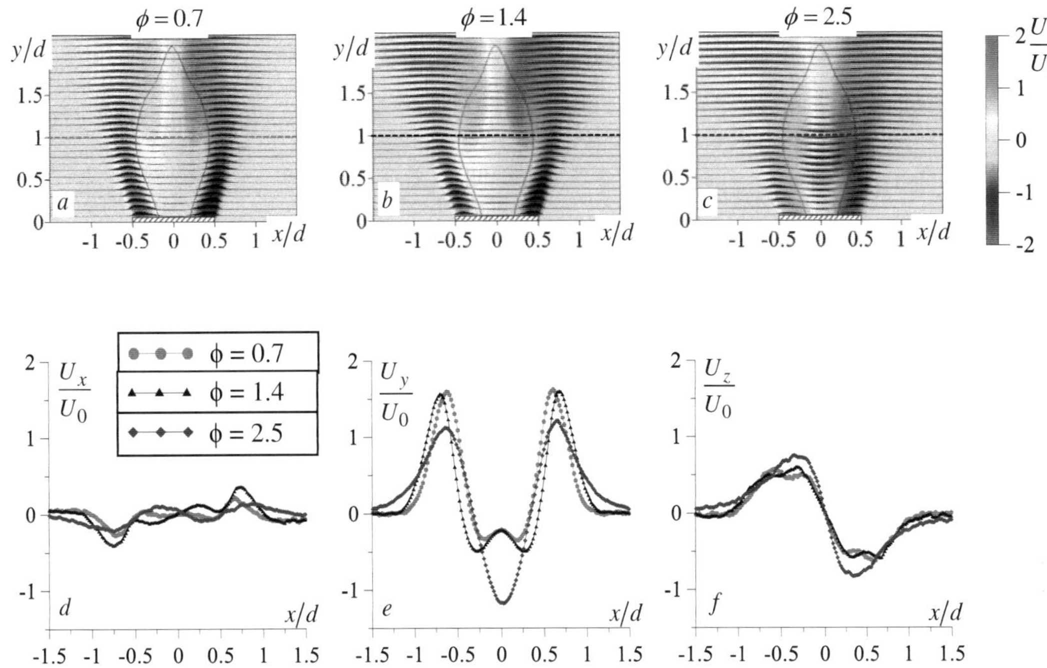

Figure 4. Average velocity fields for different equivalence ratios (a,b,c) and profiles (d,e,f) for the axial plane of the methane/air swirling. For the velocity field, the color shows the normal-to-plane velocity component and the solid line shows the region when the axial velocity component is negative. The dotted lines correspond to cross-section ($y/d = 1$)

An analysis of experimental data shows that for the premixed combustion of fuel-lean ($\phi = 0.7$) and fuel-rich ($\phi = 1.4$) methane/air swirling jets, the flame front has a shape of inverted cone, penetrating inside the nozzle. According to the PIV data, these combustion regimes correspond to an annular swirling jet surrounding the central recirculation zone with weakly turbulent slow reverse flow. The OH PLIF data shows that the hot combustion products concentrate inside the recirculation zone. Pixel-by-pixel multiplication of OH* and HCHO fluorescence signals shows that the methane combustion took place in the inner mixing layer between the central recirculation zone and the annular swirling jet. In addition, in the case of combustion of fuel-rich ($\phi = 1.4$) mixture, the maximum of heat release zone presents downstream the recirculation zone, during unburnt hydrocarbons mixing and reacting with the surrounding air.

Combustion of the fuel-rich mixture with $\phi = 2.5$, viz., when the fuel concentration is above the flammability limit for the homogenous mixture, occurs after the jet mixing with surrounding air. Thus, the flame burns at a certain distance downstream the nozzle exit. The PIV data shows, that the annular swirling jet surrounds the central recirculation zone with unsteady reverse flow. The OH PLIF data shows that there is no OH* in the swirling jet, outflowing from the nozzle and entraining the surrounding air, near the nozzle exit. Hot combustion products with OH* are concentrated inside the recirculation zone and are observed in the outer mixing layer. Simultaneously measurements and multiplication of OH* and HCHO fluorescence signals show that the hydrocarbon combustion occurs both in the inner and outer mixing layers. Fuel preheat takes place during hot gases mixing inside the recirculation zone with the jet outflowing from the nozzle and it burns in the low-speed regions of the mixing layer after mixing with the surrounding air.

4. Conclusions

The present work reports on the combined PLIF/PIV measurements for investigation of premixed and partially premixed combustion of methane in a swirling jet with vortex breakdown and central recirculation zone. The stereo PIV measurements were carried out simultaneously with OH PLIF and HCHO PLIF. Analysis of the reacting zones shapes in the high-swirl flow was carried out for three different conditions, viz., fuel-lean and fuel-rich flames with the flame shape like an inverted cone and lifted flame fuel-rich jet mixing with the surrounding air. For the premixed combustion of fuel-lean ($\phi = 0.7$) and fuel-rich ($\phi = 1.4$) methane/air swirling jets, the flame front was located in the inner mixing layer between the low-speed recirculation zone and annular swirling jet. For the fuel-rich lifted flame ($\phi = 2.5$) the fuel burned in the outer mixing layer after being preheated by the hot gases, recirculating at the jet axis, and mixed with the surrounding air.

Acknowledgments

The research is supported by Russian Science Foundation (grant № 16-19-10566). The authors are grateful to Professor Kemal Hanjalic and Dmitry Markovich for fruitful discussions and recommendations.

References

[1] Gupta A, Lilley D and Syred N 1984 *Swirl Flows* (Tunbridge Wells, Kent: Abacus Press).
[2] Alekseenko S, Dulin V, Kozorezov Yu and Markovich D 2012 Effect of high-amplitude forcing on turbulent combustion intensity and vortex core precession in a strongly swirling lifted propane/air flame *Combust. Sci. Technol.* **184** 1862-190
[3] Oberleithner K., Paschereit C, Seele R and Wygnanski I 2012 Formation of turbulent vortex breakdown: intermittency, criticality, and global instability *AIAA J.* **50** 1437-1452
[4] Syred N 2006 A review of oscillation mechanisms and the role of the precessing vortex core (PVC) in swirl combustion systems *Prog. Energy Combust. Sci.* **32** 93-161
[5] Lieuwen T 2012 *Unsteady combustor physics* (Cambridge University Press)
[6] Stöhr M, Sadanandan R, Meier W 2011 Phase-resolved characterization of vortex–flame interaction in a turbulent swirl flame *Exp. Fluids.* **51** 1153-1167
[7] Boxx I, Stöhr M, Carter C, Meier W 2010 Temporally resolved planar measurements of transient phenomena in a partially pre-mixed swirl flame in a gas turbine model combustor *Combustion and Flame* **157** 1510-1525
[8] Alekseenko S, Dulin V, Kozorezov Y, Markovich D, Shtork S and Tokarev M 2011 Flow Structure of Swirling Turbulent Propane Flames *Flow Turbulence Combust.* **87** 569-595
[9] Luque J, Crosley D 1999 Lifbase: Database and spectral simulation (version 1.5). *SRI International Report*, MP. 99-009

IOP Conf. Series: Journal of Physics: Conf. Series **980** (2018) 012034 doi:10.1088/1742-6596/980/1/012034

Cavitation conditions in the molten metal under the action of the disk agitator

A Tokhmetova, M Khmeleva, L Minkov, V Dammer

Tomsk State University, 36 Lenin Avenue, Tomsk, 634050 Russian Federation

Email: aiko050294@mail.ru

Abstract. This work investigates the conditions for cavitation in molten metal under the action of a mixing device. An estimate of the rotational speed of the agitator is made, at which the cavitation number becomes less than unity. The dependence of the limiting amplitude of the longitudinal oscillations on the oscillation frequency, above which cavitation takes place, has been obtained on the basis of a numerical simulation of the turbulent flow of an incompressible fluid in a crucible, caused by the action of vibrating agitator. It is shown that the bubbles of saturated metal vapors are mainly formed on the leeward side of the agitator elements, and the external discs of agitator are subject to the greatest cavitation effect.

1. Introduction

By now, various technologies for additional alloying and modifying industrial aluminum alloys have been developed, which have reached a certain level, and further attempts to improve them do not allow obtaining the required result [1]. An alternative method (technology) of impact on molten metal is a specially developed mixing device [2], which due to the complex action of vibration, mechanical stirring and introduction of nanoparticles of modifiers (oxides, borides, etc.) into the molten metal enables improving the quality of the metal. The device consists of a disk agitator (figure 1 a) and a source of rotation. The disk agitator consists of three perforated disks and sixteen dowels. The sizes of the holes in the disks are fitted so that when the device operates in a crucible at a melt temperature of 720-780° C, the molten metal can pass through the holes. It is known that when vibrating a mixing device immersed in a molten metal, cavitation [3] may occur, which, on the one hand, has a positive effect on the process of introducing microparticles into the molten metal [2], but on the other hand it can destroy the mixing device.

The purpose of this study is to evaluate the operating conditions of the mixing device [2], in which hydrodynamic cavitation occurs in molten aluminum under the action of a disk agitator.

2. Description of the problem

In figure 1 b the agitator is shown after 5 minutes of operation in an industrial crucible with a capacity of 100 liters with molten aluminum at a temperature of 750° C and a speed of 2500 rpm. As can be seen from the figure, all the dowels are absent. The edges of the disks are "eaten". Considering that the melting point of titanium, from which the agitator is made, is about 1670° C, the temperature influence of molten aluminum can't cause the destruction of the agitator. In addition, the structure of the surface of the disk agitator is typical for a surface that has been subjected to cavitation.

Content from this work may be used under the terms of the Creative Commons Attribution 3.0 licence. Any further distribution of this work must maintain attribution to the author(s) and the title of the work, journal citation and DOI.
Published under licence by IOP Publishing Ltd

Figure 1. Disk agitator.
a – Before the action of cavitation;
b – After the action of cavitation.

Cavitation occurs when fluid flows around poorly streamlined bodies, when the static pressure in the liquid is less than the partial pressure of saturated vapors and dissolved gases. The conditions for the appearance of hydrodynamic cavitation correspond to the cavitation criterion determined by the expression [1, 3]:

$$\kappa = \left[P_1 - P_{SV}(T)\right]\left(0.5\rho V^2\right)^{-1},$$

where ρ is the density of the liquid metal, V is the velocity of the liquid metal, P_1 is the static pressure; and P_{sv} is the pressure of the saturated vapor of the metal at a temperature T, which obeys the law: [4]:

$$P_{SV}(T) = P_1 \cdot \exp\left[LR_u^{-1}\left(T_{boil}^{-1} - T^{-1}\right)\right]$$

It is known that hydrodynamic cavitation occurs when the value of the cavitation criterion is less than unity.

3. Influence of rotational speed

Let us take the heat of evaporation of aluminum $L = 2.943 \cdot 10^5$ J/(mol), and the boiling point of aluminum under normal conditions T_{boil} = 2723 K [4]. Here the universal gas constant is R_u = 8.31 J/(mol·K), and atmospheric pressure P_1 = 101325 Pa. Assume that the temperature of molten aluminum is 750° C (T = 1023 K), the density of liquid aluminum is 2700 kg/m^3, and the outer radius of the agitator disk is R. At a rotational speed of the stirrer per minute – n, the angular velocity is equal to $2\pi n/60$, and the circumferential speed of the edge of disk is $2\pi nR/60$. Then the number of revolutions at which cavitation occurs is determined by the inequality:

$$n > n_{cr}(T,R) = 60(2\pi R)^{-1}\left[2\rho^{-1}\left(P_1 - P_{SV}(T)\right)\right]^{-0.5}.$$

Figure 2 shows the dependence of the critical speed on the temperature of molten aluminum. The region lying above the curve is the region where cavitation occurs, and below is the absence of cavitation. For a disk radius of 40 mm, a speed of rotation above 2100 rpm leads to cavitation at any temperature values, and for a disk radius of 75 mm, this is a speed above 1100 rpm.

Figure 3 shows the dependence of the critical speed on the radius of the agitator disk. The region lying above the curve corresponds to the region of cavitation origin, and the one below – to the absence of cavitation. When the temperature of aluminum changes from the melting point from 660° C to 2000° C, the position of the curve for the dependence of the critical speed of the stirrer on the radius of the disk varies slightly, and at high temperatures of 2200° C and above, the shift of this curve becomes noticeable. It can be seen from the graph that for a disk radius of less than 26 mm – 27 mm cavitation will not occur when the temperature varies from 660° C to 2000° C.

4. Influence of vibration

It is known that the areas of reduced pressure appear in the molten metal near the surface of the vibrator oscillating at certain frequencies and amplitudes, where the pressure becomes less than the saturated vapor pressure of metal [5]. In these areas vapor bubbles are formed, and at their collapse high pressure develops and destroys the surface of the vibrator, i.e. the phenomenon of cavitation occurs.

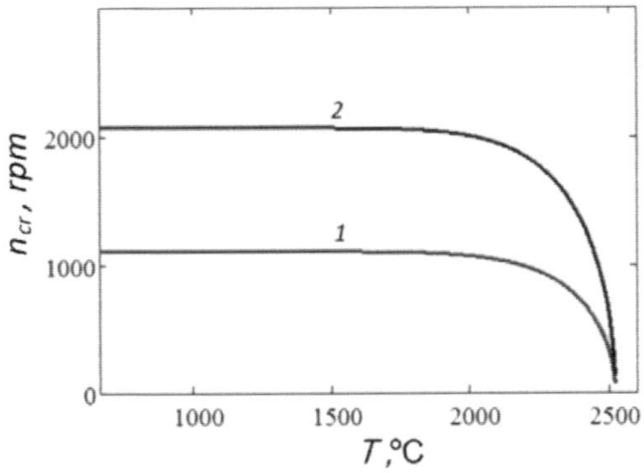

Figure 2. Dependence of critical revolution number on metal temperature. 1 – r = 0.075 m, 2 – 0.04 m

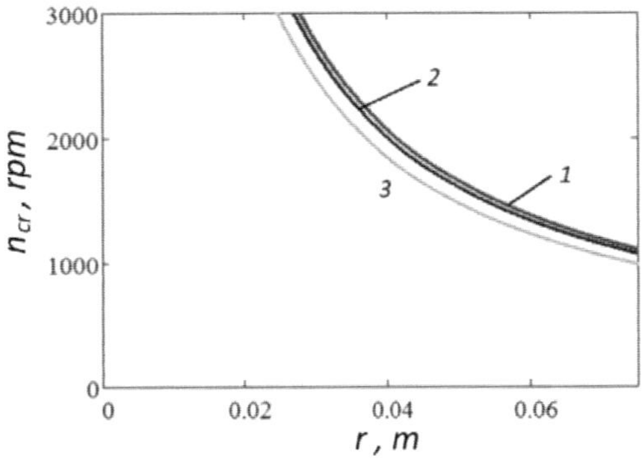

Figure 3. Dependence of critical revolution number on disk radius. 1 – T = 660° C, 2 – 2000° C, 3 – 2200° C

To study the conditions for the appearance of the phenomenon of cavitation one can analyze the flow of molten metal in a cylindrical crucible CABD (figure 4) under the action of vibrator 2 harmonically oscillating along the guide rod 1 by the law $x = A \cdot \cos(\omega t)$. The top bound of the crucible CD is open, and the boundaries of CA, AB and BD are the solid surfaces which are impermeable to molten metal.

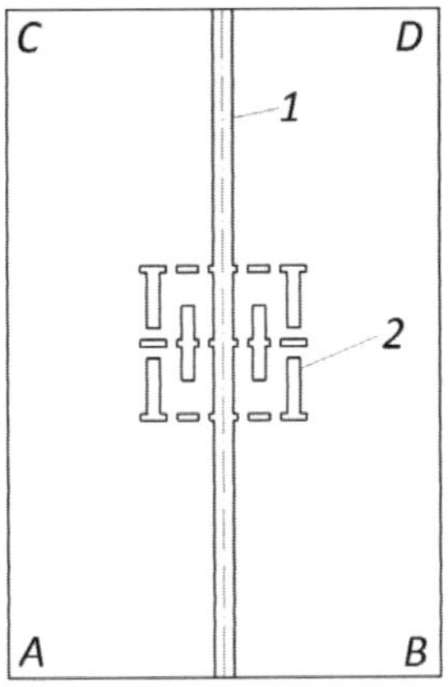

Figure 4. Problem domain

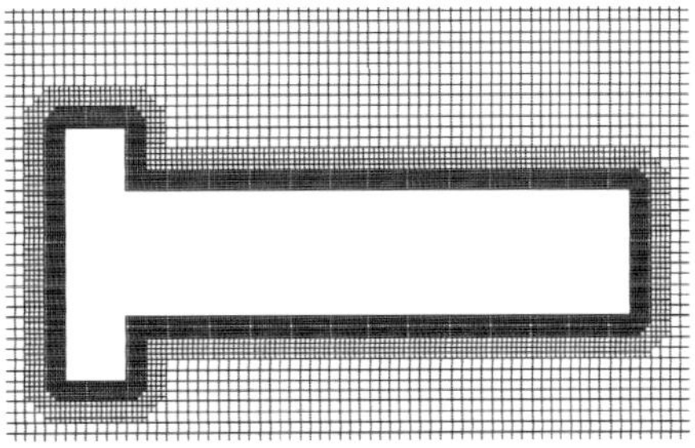

Figure 5. Difference grid element.

The assumptions used to describe the mathematical model are as follows: the molten metal and metal vapor are considered to be an incompressible fluid; the surface tension of a liquid metal, which is significant for bubble radii much less than 1 μm, is neglected; the flow is turbulent, axisymmetric; the vibration is low frequency one; the velocity lag of vapor bubbles in the molten metal is neglected.

A set of governing equations consists of:

– mass conservation equation for the metal vapor

$$\frac{\partial \alpha_v \rho_v}{\partial t} + \nabla \cdot \left(\alpha_v \rho_v \vec{V} \right) = R_e - R_c \tag{1}$$

– mass conservation equation for a mixture of molten metal and metal vapors

$$\frac{\partial \rho}{\partial t} + \nabla \cdot \left(\rho \vec{V} \right) = 0 \tag{2}$$

– momentum equation for the mixture

$$\frac{\partial \rho \vec{V}}{\partial t} + \nabla \cdot \left(\rho \vec{V} \vec{V} \right) = -\nabla p + \nabla \cdot \bar{\bar{\tau}}_{eff} \tag{3}$$

– turbulence kinetic energy equation

$$\frac{\partial \rho k}{\partial t} + \nabla \cdot \left(\rho k \vec{V} \right) = \nabla \cdot \left[\left(\mu + \frac{\mu_{turb}}{\sigma_k} \right) \nabla k \right] + G_k - \rho \varepsilon \tag{4}$$

– turbulence dissipation rate equation

$$\frac{\partial \rho \varepsilon}{\partial t} + \nabla \cdot \left(\rho \varepsilon \vec{V} \right) = \nabla \cdot \left[\left(\mu + \frac{\mu_{turb}}{\sigma_\varepsilon} \right) \nabla \varepsilon \right] + C_{1\varepsilon} \frac{\varepsilon}{k} G_k - C_{2\varepsilon} \rho \frac{\varepsilon^2}{k} , \tag{5}$$

$\rho = \alpha_v \rho_v + (1 - \alpha_v) \rho_l$ is the density of mixture; ρ_l is the density of molten metal; ρ_v is the density of metal vapor; $\bar{\bar{\boldsymbol{\tau}}}_{eff} = \bar{\bar{\boldsymbol{\tau}}} + \bar{\bar{\boldsymbol{\tau}}}_{turb}$ is the effective stress tensor; $\bar{\bar{\boldsymbol{\tau}}}_{turb} = \mu_{turb} \left(\nabla \vec{V} + \nabla \vec{V}^T \right) - \frac{2}{3} \rho k \mathbf{U}$ is the turbulence stress tensor; $\bar{\bar{\boldsymbol{\tau}}} = \mu \left(\nabla \vec{V} + \nabla \vec{V}^T \right)$ is the viscous strain rate tensor; $\mu = \mu_v \alpha_v + \mu_l (1 - \alpha_v)$ is the dynamic viscosity of mixture; $\mu_{turb} = \rho C_\mu k^2 \varepsilon^{-1}$ is the turbulent viscosity; $G_k = 2 \mu_{turb} \mathbf{S} : \mathbf{S}$ is the production of turbulent kinetic energy due to shear flow (gradient of speed); and $\mathbf{S} = 0.5 \left(\nabla \vec{V} + \nabla \vec{V}^T \right)$ is the strain tensor.

The mass transfer process from the vapor phase to the molten metal phase and back is described in accordance with the Zwart - Gerber - Belamri model [6]:

$$R_e = f_{vap} \frac{2 a_{nuc} (1 - \alpha_v) \rho_v}{r_b} \left(\frac{2}{3} \frac{p_{SV} - p}{\rho_l} \right)^{-0.5} , \quad p < p_v$$

$$R_c = f_{cond} \frac{2 a_{nuc} \alpha_v \rho_v}{r_b} \left(\frac{2}{3} \frac{p - p_{SV}}{\rho_l} \right)^{-0.5} , \quad p > p_v$$

p_v is the saturated vapor pressure; $r_b = 10^{-6}$ м is bubbles radius, $\alpha_{nuc} = 5 \cdot 10^{-4}$ is the nucleation site volume fraction; $f_{vap} = 50$ is the evaporation coefficient; and $f_{cond} = 0.01$ is the condensation coefficient.

Calculations in (3) – (4) were carried out for the next parameter values: $C_{1\varepsilon} = 1.44$; $C_{2\varepsilon} = 1.92$; $C_\mu = 0.09$; $\sigma_k = 1.0$; $\sigma_\varepsilon = 1.3$.

No-slip conditions are set on solid surfaces of the crucible and the agitator, the atmosphere static pressure is set at the open boundary.

The solution of the mathematical model is realized with Ansys-Fluent [7]. The finite volume method is used for solving the set of governing equations (1) – (5). A set of governing equations (1)-(5) is calculated by implicit method in which the flow finite-difference scheme of the first order approximation is used for convective terms of the equations. The pressure field is calculated with the standard algorithm. Coupling the pressure and velocity is carried out with the method SIMPLE [8]. The values of turbulence variables at the solid surfaces are determined in accordance with the model of SWF (Standard Wall Function). To simulate the vibrator motion we used the Dynamic mesh approach with the mesh rebuilding by layers (Layering). The time step depends on the oscillation frequency of the vibrator and varies from 10^{-2} s to 10^{-6} s.

The calculations were performed on a finite difference grid consisting of 81848 quadrangular cells, which were refined near the surface of the vibrator (figure 5).

5. Results and discussion

Calculations were carried out for the following parameter values: the density of molten metal $\rho_l = 2700$ kg/m^3 and the vapour density of metal $\rho_v = 0.87$ kg/m^3; the dynamic viscosity of liquid metal $\mu_l = 10^{-3}$ Pa·s and the dynamic viscosity of a metal vapor $\mu_v = 1.8 \cdot 10^{-5}$ Pa·s; and $p_{SV} = 1$ Pa. The sizes of the crucible are as follows: the height CA is 0.3 m, and the diameter AB is 0.2 m.

The limit dependence of the oscillation amplitude of the vibrator A_{lim} on the oscillation frequency $f = \omega/2\pi$ at which cavitation is possible is shown in figure 6. The cavitation occurs if the vibrator oscillates with a frequency f and amplitude $A > A_{\mathrm{lim}}(f)$ and otherwise does not arise.

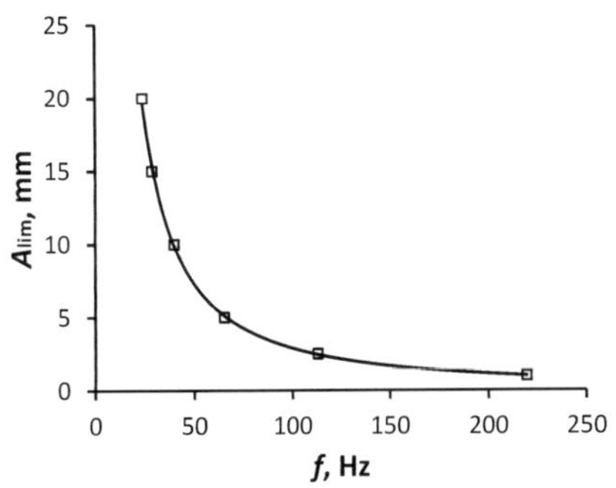

Figure 6. Limit dependence of the amplitude of oscillation on the frequency.

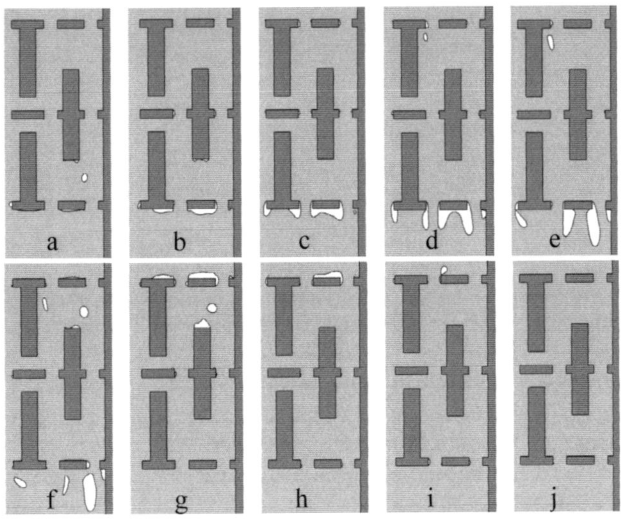

Figure 7. Bubble formation of vapor metal.
a) – t/T =1.0, b) – 1.1, c) – 1.2, d) – 1.3, e) – 1.4, f) – 1.5, g) – 1.6, h) – 1.7, i) – 1.8, j) – 1.9.
light grey – molten metal, dark grey – structural elements of the vibrator, white color – metal vapor.

The results of numerical simulation are denoted by the symbols and approximated by a curve $A = 1371 \cdot f^{-4/3}$ (solid line) using the least squares method. According to the theoretical estimates of Ignat'iev *et al* [5] the limiting amplitude of vibrations is inversely proportional to the vibration frequency. The difference in the results may be explained by the fact that Ignat'iev does not take into account the viscosity of the molten metal when estimating the filling time of the cavity, which is of great importance in this case.

The patterns of the formation of metal vapor bubbles near the vibrator elements oscillating along the vertical axis with a frequency of 100 Hz and amplitude of 7 mm after the first oscillation period in the time interval $1 \leq t/T < 2$ are shown in figure 7. The limiting value of the oscillating amplitude of the vibrator for such a frequency is ≈3 mm, at which cavitation occurs.

Metal vapor bubbles growing with time are formed at the periphery of the holes of the lower disk when the vibrator moves upwards (figure 7 b-e) and break out of the disk in the maximum up position of the vibrator, figure 7 f. The bubble sizes are comparable with the cross-sectional sizes of the vibrator elements. In addition, bubbles are formed on the elements of the middle and upper disks, the sizes of which are much smaller than the ones of the vibrator elements. They break out of the disk before the vibrator takes the maximum up position (figure 7 c-f). When the vibrator moves downward, the formation of bubbles occurs mainly on the elements of the upper disk (figure 7 g-j); and at the same time one can see small bubbles that are formed on the elements of the lower and middle disks.

6. Conclusions

The hydrodynamic cavitation is shown to be the cause of destruction of the 80 mm agitator's disks rotating at a speed of 2500 rpm. Increasing the diameter of the disk agitator allows one to reduce the number of revolutions of the source of rotation at which cavitation occurs.

To create a cavitation effect in molten aluminium with a temperature of more than 660° C at a speed of about 1000 rpm, an agitator with a disc radius of more than 75 mm should be used.

It is shown that at a temperature of molten aluminium in the range from 660° C to 2200° C and a rotational speed of the agitator below 3000 rpm, a cavitation effect does not occur if the radius of the agitator disk is less than 26 mm.

The critical revolution number of the agitator remains constant when the molten metal temperature rises to 70% of the boiling point, after which a sharp drop in this value is observed.

According to the patterns of formation and development of bubbles obtained from the numerical simulation, the greatest cavitation impact on the surface of the vibrator should be expected at the edges of the upper and lower disks that is confirmed by the results of the experiment (figure 1 b).

The limiting amplitude of the axial vibration of the agitator, at which cavitation occurs, is obtained to be inversely proportional to the vibration frequency raised to the four-thirds power.

Acknowledgments

This work was carried out with financial support from the Ministry of Education and Science of the Russian Federation (State assignment No. 9.9625.2017/8.9).

References

[1] Efimov V A and El'darkhanov A S 2004 *Technologies of modern metallurgy* (Moscow: Novye Technologii)

[2] Vorozhtsov A B, Arkhipov V A, Shrager E R, Dammer V Ch, Vorozhtsov S A, Khmeleva M G Ustroistvo dlya smeshivaniya zhidkostey i poroshkov s zhidkostyu [Device for mixing liquids and powders with liquid]. Zayavka na patent. The patent application № 2016130836 RF / Zayavleno. Is claimed 26.07.2016

[3] Franc J-P and Michel J-M 1982 *Fundamentals of Cavitation* (New York: Kluwer)

[4] Nikolskiy B P 1982 *Chemical Handbook* vol 1 (Moscow-Leningrad: Chemistry) pp 682–93.

[5] Ignat'ev I E, Dolmatov A V, Ignat'eva E V, Istomin S A, Pastukhov E A 2012 *Russian Metallurgy* **2** 97–101.

[6] Zwart P J, Gerber A G, Belamri T A 2004 *Proc. Fifth Int. Conf. on Multiphase Flow* (Yokohama, Japan: Tsukuba).

[7] ANSYS FLUENT. Tutorial Guide: Release 14.0. 2011. ANSYS Inc.

[8] Patankar S 1980 *Numerical heat transfer and fluid flow* (Washington: Hemisphere Publ. Corp.)

Mathematical modeling of swirled flows in industrial applications

A A Dekterev[1,2], A A Gavrilov[1, 2] and A V Sentyabov[1, 2]

[1]Institute of Thermophysics SB RAS, Novosibirsk, Russia
[2]Siberian Federal University, Krasnoyarsk, Russia

Email: dekterev@mail.ru

Abstract. Swirled flows are widely used in technological devices. Swirling flows are characterized by a wide range of flow regimes. 3D mathematical modeling of flows is widely used in research and design. For correct mathematical modeling of such a flow, it is necessary to use turbulence models, which take into account important features of the flow. Based on the experience of computational modeling of a wide class of problems with swirling flows, recommendations on the use of turbulence models for calculating the applied problems are proposed.

Swirled flows are widely used in technological devices such as pumps, hydraulic turbines, cyclone units, separators, burners and furnace devices, chemical reactors, etc. At the same time, the flow regimes essentially depend on the way the swirl is organized, the intensity of the swirl, the fluid properties and the process conditions (single and multiphase flows, heat exchange, chemical reaction, mass forces, etc.). Swirling flows are characterized by a wide range of flow regimes such as a quasi-solid rotation with a maximum tangential velocity at the periphery of the rotation region; vortex flows with central recirculation zones; a rotation with one or several concentrated vortices; a flow with bubble or spiral vortex breakdown; a reconnection of the vortex filaments and the formation of vortex rings etc.

In recent years, 3D mathematical modeling of flows is widely used in research and design. However, the lack of validity of the models can lead to unreliable numerical results and errors in the devices design. For correct mathematical modeling of such a flow, it is necessary to use turbulence models, which take into account important features of the flow. In this case, widely used RANS models of eddy viscosity (k-ε, k-ω SST, k-ε-ζ-f) usually do not allow obtaining correct results. Methods of Direct Numerical Simulation (DNS) and Large Eddy Simulation (LES) with a resolution of the near wall layer are still rarely used to solve the applied problems due to enormous computational costs.

The aim of the paper is to generalize the experience of the modeling of the turbulent quasi-stationary and transient swirling flows. In-house CFD code SigmaFlow based on the finite volume method and SIMPLE-like procedure for solving the Navier-Stokes equations used for numerical simulations.

Simulation of laminar swirling flows in a cylindrical container with a rotating lid (figure 1) shows that, using the second-order discretization schemes and a sufficient mesh resolution, calculations can reproduce such a structure observed experimentally in the swirling flow [1, 2] as multiple bubble vortex breakdown [3].

Content from this work may be used under the terms of the Creative Commons Attribution 3.0 licence. Any further distribution of this work must maintain attribution to the author(s) and the title of the work, journal citation and DOI.
Published under licence by IOP Publishing Ltd

For turbulent flows with a weak swirl, a comparison of computational results with the experimental data for a swirling flow in a pipe (figure 2) [4] and a flow in a model tangential combustion chamber (figure 3) shows that it is quite acceptable to use eddy viscosity models.

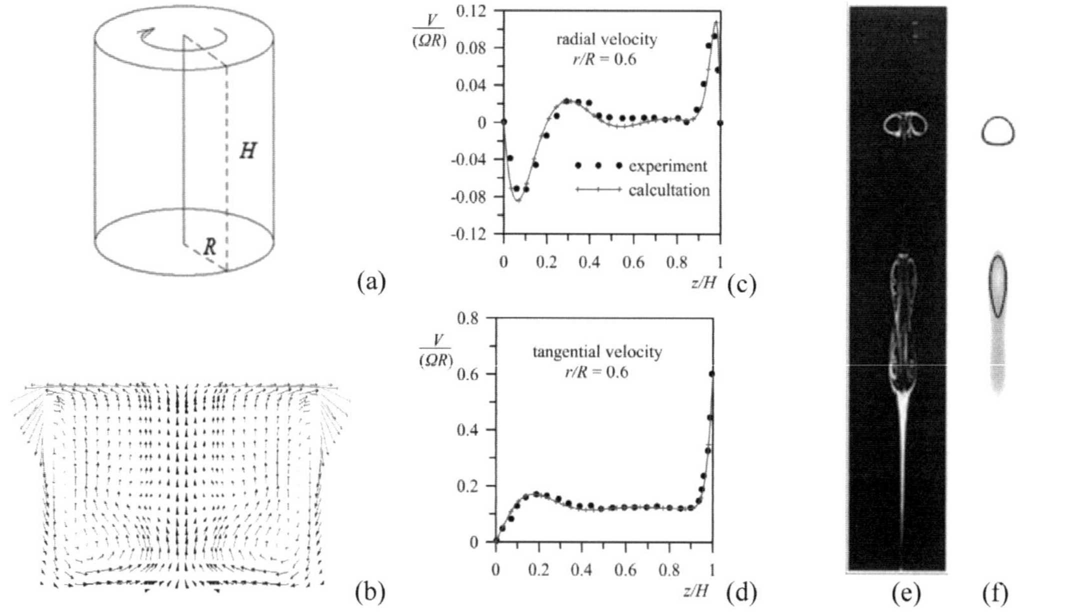

Figure 1. Flow in a container with the rotating lid: (a) the geometry; (b) the flow in the central cross-section; (c, d) velocity components along the vertical line (H/R = 1.0, Re = 1800, numerical grid contains 0.8 million cells); (e, f) H/R = 3.25, Re = 2752, the grid contains10 million cells ((e) is the experiment, and (f) is the zero isoline of the vertical velocity component in the calculation).

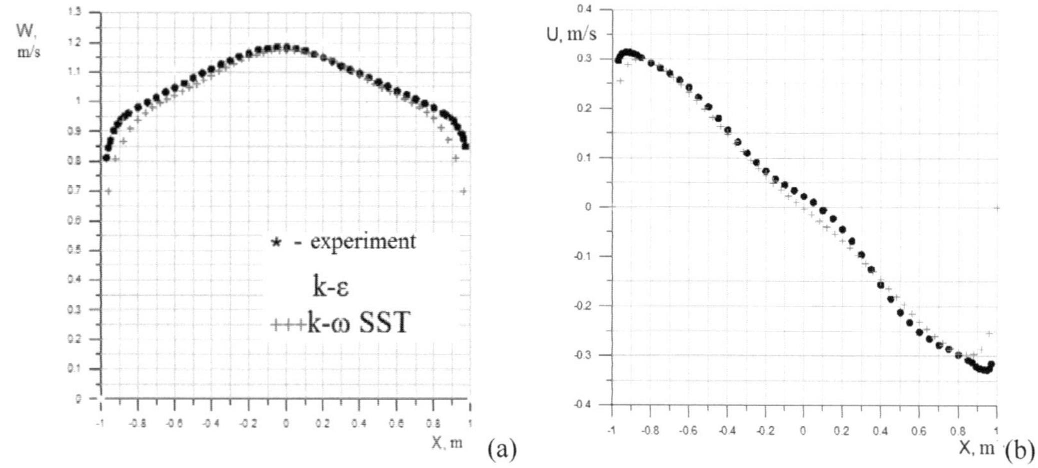

Figure 2. Swirling flow in the pipe (S = 0.17, Re = 3.103): (a) axial velocity; (b) tangential velocity.

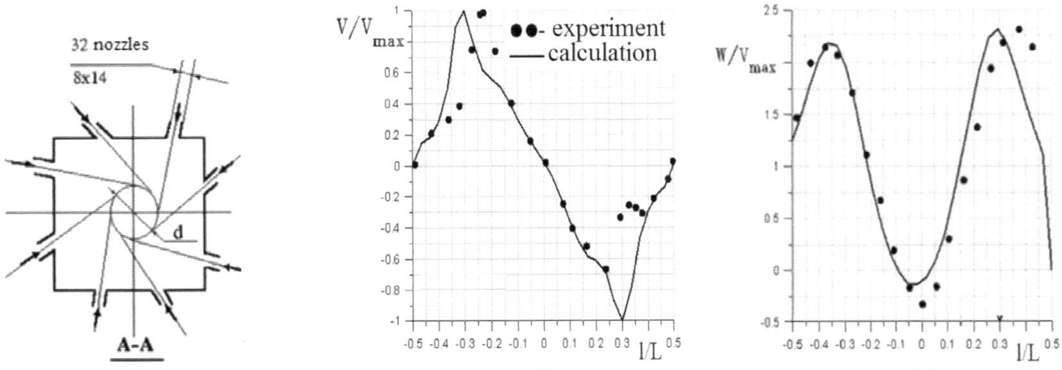

(a) (b) (c)

Figure 3. The flow in the model tangential combustion chamber of the IT SB RAS: (a) the chamber scheme, (b) tangential velocity, (c) axial velocity.

When there are concentrated vortices in the flow, the eddy viscosity models lead to large errors (figure 4) due to certain limitations of the Boussinesq approximation associated with misrepresentation of the turbulent anisotropy. For this class of flows, Reynolds stress models, based on the differential equations for the Reynolds stresses (RSM), are the most optimal turbulent models.

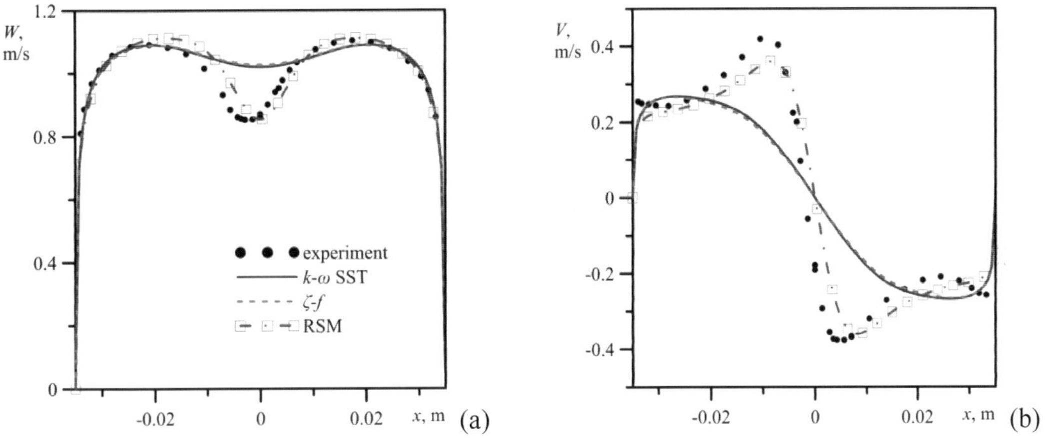

Figure 4. Swirling flow in a pipe with a concentrated vortex (S = 0.17, Re = $5{\cdot}10^4$): (a) axial velocity; (b) tangential velocity.

In some cases, an increase in the swirl number leads to the formation of unsteady precessing vortices in the swirling flow. For such flows, the URANS models with isotropic eddy viscosity do not describe the flow even qualitatively, as it can be seen from the results of the calculation for the flows in the model vortex burner (figure 5) and for the flow downstream the runner of the hydraulic turbine (figure 6, 7). URANS RSM models with acceptable computational costs give correct results comparable in accuracy with eddy-resolving LES and hybrid LES/RANS models [5, 6].

IOP Conf. Series: Journal of Physics: Conf. Series **980** (2018) 012035 doi:10.1088/1742-6596/980/1/012035

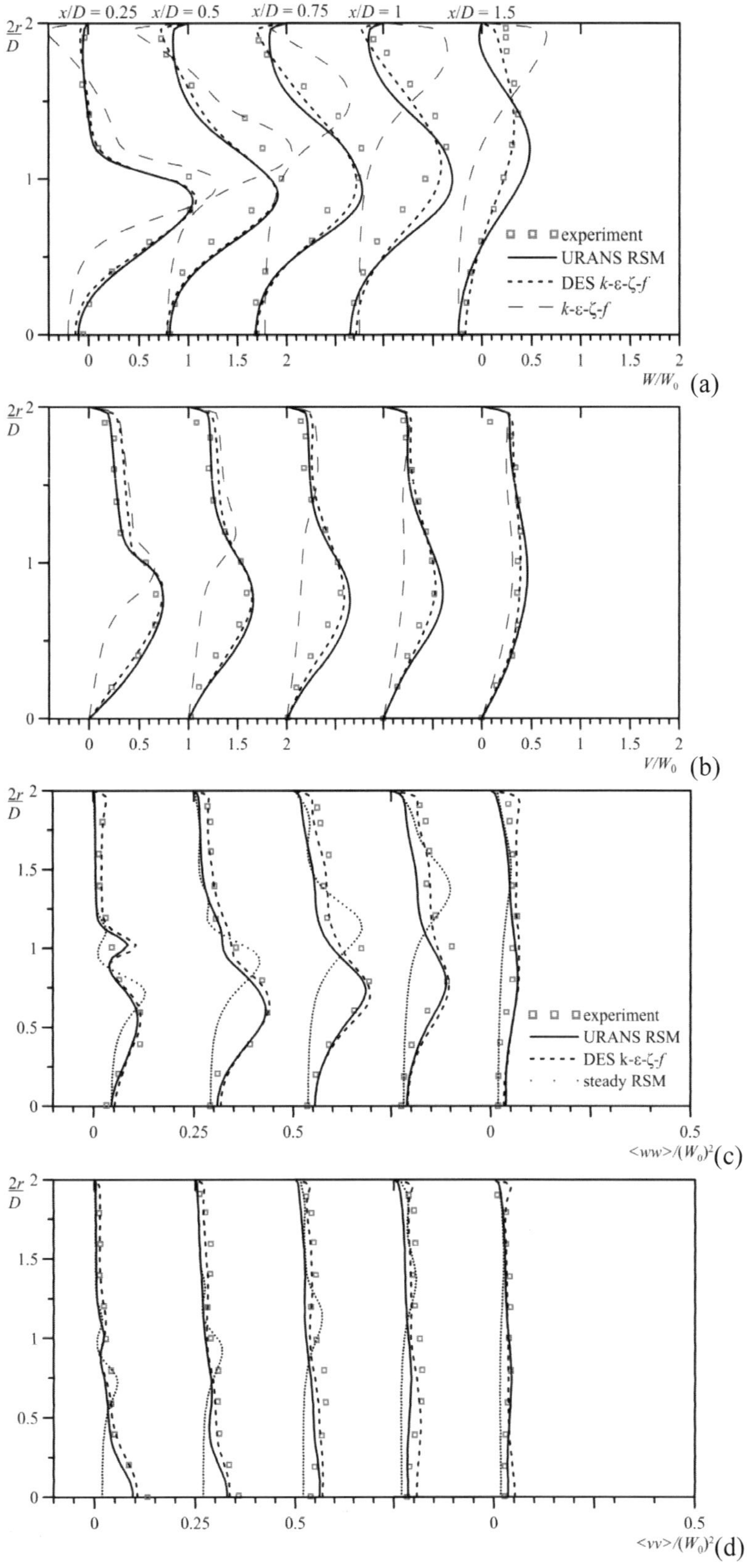

Figure 5. Swirling flow in a vortex burner: (a) axial velocity, (b) tangential velocity, (c) pulsations of the axial velocity, (d) pulsations of the tangential velocity.

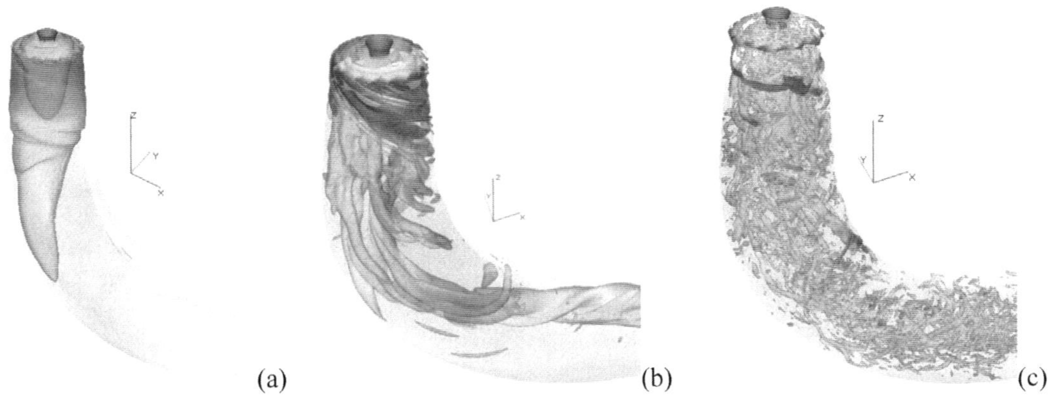

Figure 6. Vortex structures behind the runner of a hydraulic turbine. Turbulence models: (a) EVM (ζ–f), *b*) RSM, (c) hybrid LES/RANS.

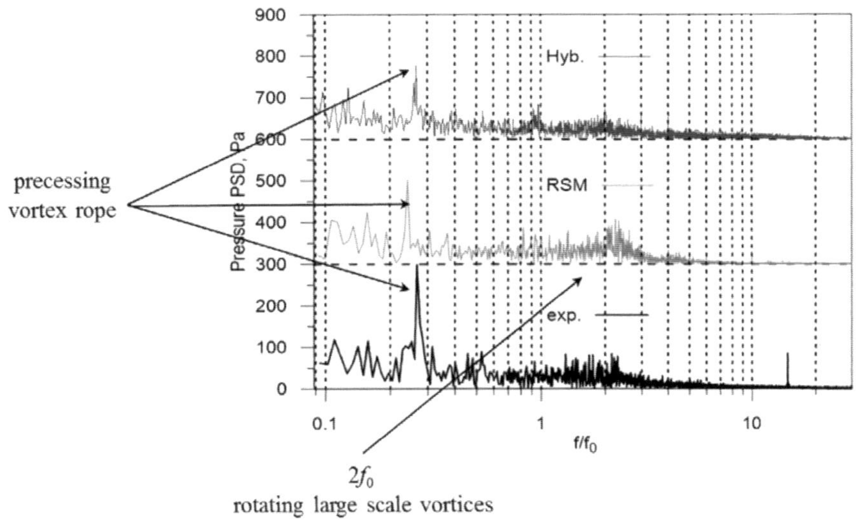

Figure 7. Power spectrum density of pressure signals at the cone wall.

It is interesting to use the so-called URANS models of the second generation, in particular, the model of partially averaged Navier-Stokes equations (PANS), proposed in [7]. Within the approach, the width of the averaging filter controlled by a given ratio of modelled turbulence energy to the total kinetic turbulence energy f_k. Depending on the value of f_k, seamless transition from RANS at $f_k = 1$ to DNS at $f_k = 0$ is realized. The physical resolution of the flow structures is independent of the mesh spacing. In case of sufficient mesh resolution, the PANS can resolve the most important energy-containing vortex structures (figure 8). However, the PANS method needs both additional development and comprehensive testing, especially for near-wall flows.

Preliminary calculations by means of URANS RSM should be used to determine the flow nature and to justify the selection of the turbulence model, such as RANS, URANS or LES, which will be used for the massive simulations in the engineering design process. An eddy-resolving turbulence model, that can accurately resolve all the energy-containing scales of the flow, is preferable if there are sufficient computational resources.

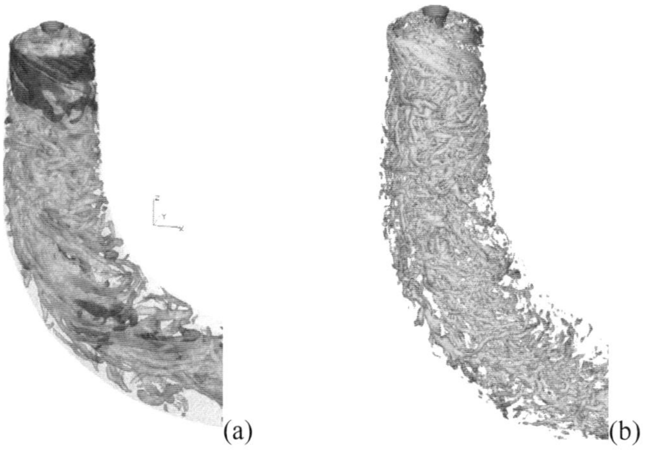

Figure 8. Vortex structures downstream the runner of the hydraulic turbine visualized by the λ_2 criterion. Turbulence models: (a) PANS RSM, f_k = 0.6, (b) hybrid LES/RANS.

Conclusion

Based on the experience of computational modeling of a wide class of problems with swirling flows, recommendations on the use of turbulence models for calculating of applied problems has been proposed.

Acknowledgments

The work was financially supported by the Ministry of education and science of the Russian Federation, grant agreement No 14.607.21.0150 (RFMEFI60716X0150) and by the Russian Foundation for Basic Research, the Government of the Krasnoyarsk Territory, the Krasnoyarsk Territory Fund for Support of Scientific and Technical Activity within the framework of the scientific project No. 16-48-242085.

References

[1] Michelsen J A 1986 Modeling Incompressible Rotating Fluid Flow. *Ph.D. thesis* Technical University of Denmark

[2] Escudier M P 1984 Observation of the flow produced in a cylindrical container by a rotating endwall *Exp. in Fluids* **2(4)** 189–196

[3] Sentyabov A, Gavrilov A, Krivov M, Dekterev A and Pritula M 2017 Development of a high-performance code for hydrodynamic calculations using graphics processor units *Communications in Computer and Information Science* (In: Sokolinsky L., Zymbler M. (eds.) Parallel Computational Technologies. PCT 2017, Springer, Cham) **753** 288-300

[4] Steenbergen W 1995 Turbulent pipe flow with swirl. Eindhoven University of Technology *PhD Thesis*

[5] Gavrilov A A, Dekterev A A and Sentyabov A V 2015 Modeling of Swirling Flows with Coherent Structures Using the Unsteady Reynolds Stress Transport Model *Fluid Dynamics* **50**(4) 471 – 482

[6] Gavrilov A A, Sentyabov A V, Dekterev A A and Hanjalic K 2017 Vortical structures and pressure pulsations in draft tube of a Francis-99 turbine at part load: RANS and hybrid RANS/LES analysis *International Journal of Heat and Fluid Flow* **63** 158-171

[7] Girimaji S S 2006 Partially-averaged Navier–Stokes model for turbulence: a Reynolds-averaged Navier–Stokes to direct numerical simulation bridging method *Journal of Applied Mechanics* **73** 413-421.

IOP Conf. Series: Journal of Physics: Conf. Series **980** (2018) 012036 doi:10.1088/1742-6596/980/1/012036

Possibilities of application of the swirling flows in cooling systems of laser mirrors

Yu Shanin[1] and A Chernykh[1]

[1] FSUE SRI SIA «LUCH», Podolsk, Moscow region, Russia, 145103
E-mail: syi@luch.podolsk.ru

Abstract. The paper presents analytical investigations into advanced cooling systems of the laser mirrors with heat exchange intensification by methods of ordered vortex impact on a coolant flow structure. Advantages and effectiveness of the proposed cooling systems have been estimated to reduction displacement of an optical mirror surface due to a flexure.

1. Introduction

Cooled laser mirrors consist of a thin substrate, a cooling system and a thick base. The task of the laser mirror is to reflect as much as possible the incident radiation and minimize the distortions caused by the heating of the mirror. Thermal expansion and bending movement of the optical surface depend on the temperature level and its distribution over the thickness of the mirror. The task of the cooling system is to reduce the overtemperature level and to protect the mirror base from the penetration of heat into it. The cooling system is characterized by two integral parameters: the reduced heat-transfer factor α_{pr} taken to a heat exchange surface of a substrate and "thermal insulation coefficient" of base K_{ti} – the relation of excess temperatures of heat exchange surfaces of a base and a substrate. For improvement of mirror operability it is necessary to maximize α_{pr} and to minimize K_{ti}, but these requirements are contradictory. The value of α_{pr} is influenced by both the surface heat-transfer coefficient and the degree of finning of the surface of the cooling system. For increase in α_{pr} it is necessary to intensify the heat transfer so that its growth would exceed hydroresistance growth. One of the effective intensification methods is impact on structure of the heat carrier flow by well-ordered vortex formation.

In [1] it is shown that for channel cooling systems of laser mirrors (at Biot's numbers of $Bi=\alpha_0 d_g/\lambda<1$, where d_g - the hydraulic diameter of the channel) the value of the reduced heat transfer coefficient α_{pr} and the temperature distribution over the mirror thickness are practically independent of the relations coefficients of heat transfer on finned α_0 and unfinned α surfaces of the cooling system (within $0.7<\alpha/\alpha_0<1.5$), but is determined only by the average coefficient of heat transfer. In [2] the results of an experimental study of the hydraulic resistance and heat transfer are presented in the trial application of some methods of vortex intensification of heat transfer in cooling systems of laser mirrors.

In this work perspective cooling systems with heat transfer intensification were analyzed, and the advantages and efficiency of the proposed cooling systems are estimated. Estimated calculations were performed using programs written in mathematical editor Mathcad.

Content from this work may be used under the terms of the Creative Commons Attribution 3.0 licence. Any further distribution of this work must maintain attribution to the author(s) and the title of the work, journal citation and DOI.
Published under licence by IOP Publishing Ltd

2. Results of estimated calculations of the proposed cooling systems

The cooling systems taken for the calculated analysis are schematically given in figure 1. In comparison with the models of cooling systems [2] tested earlier it is offered:

a) make smaller the structure of the cooling system following the path of reducing the hydraulic diameter and the increase of the heat exchange surface (compactness of the system);

b) to use the processes of vortex formation intensifying heat transfer in the energetically favorable region of Reynolds numbers (as a rule, this is the transient flow and the initial development of the turbulent flow);

c) to take for the comparative variant of the porosity-optimized cooling system, which ensures maximum α_{pr} and "to subject" its intensification of heat transfer.

For the calculations of the hydraulic resistance and heat transfer both literature data and our results [2] were taken.

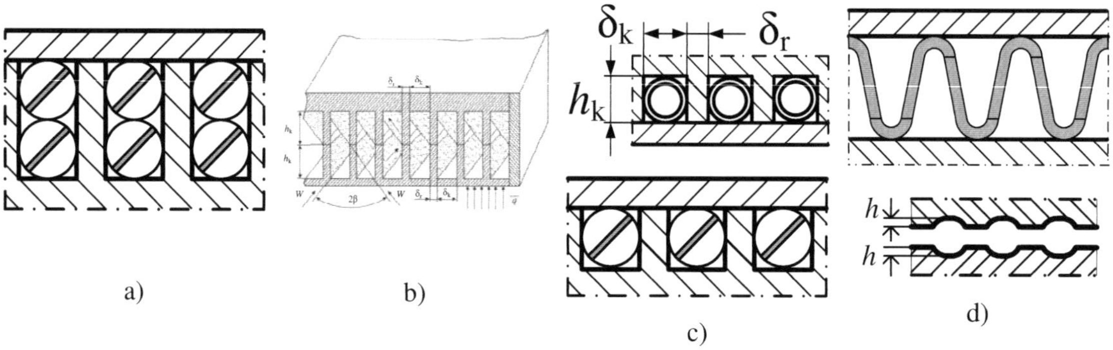

Figure 1. Schemes of the studied cooling systems: a) the slit channel with spiral tapes; b) the crossing (coplanar) channels [3] - 2β - angle of the relative crossing of channels; h_r=h_k - height of the rib, channel; a=δ_k - width of the channel; δ_r - thickness of the rib; w – heat carrier speed; q – specific heat flux; c) a spring insert and a spiral tape in a square channel; d) corrugation with holes.

2.1. Slit channels with spiral tapes

In the case of technological impossibility of obtaining narrow high channels with thin fins, it is proposed to increase the surface heat transfer of the finned surface by installation to the rectangular canal of several spiral tapes on height of the channel. Calculations were made for two cases of slotted channels made in copper (λ= 385 W·m^{-1}·K^{-1}):

1) in the channel with a width and height of 1.5×3.0 mm^2 and a rib thickness of 0.5 mm are inserted at each other two spiral tapes from a copper strip 0.25 mm thick with two different steps - 12.5 and 6.25 mm, which are placed along the entire length of the channel;

2) in the channel 1.5×4.5 mm^2 with rib thickness of 0.5 mm are inserted at each other two spiral tapes identical stated above.

For the two cases considered, the porosity of the initial cooling system was optimized in the Reynolds number range Re=(2-10)·10^3 and was ε=0.75. At the same time, the coefficient of heat transfer intensification (K_{in}=α_{pr}/α_0) was: 1) for the first case K_{in}=4.0 (Re=2·10^3) and K_{in}=2.7 (Re=1·10^4); 2) for the second case K_{in}=4.8 (Re=2·10^3) and K_{in}=2.9 (Re=1·10^4). The insert of tapes leads to double decrease of d_g, but at the same time the tape surface practically does not develop the heat exchange area of fins.

For calculated estimates of hydraulic resistance and heat transfer in the channel with tapes are used: a) our results [2] obtained on a square channel of 2.65×2.65 mm^2 with an inserted spiral tape; b) literary data for round channels [4, 5].

For the first case and a step of the tape t=12.5 mm increase in relation to the smooth channel is received: 1) hydraulic resistance from 2 to 1.6 times (Re=(2-10)·10^3); 2) α_{pr} from 1.4 to 1.1 times; 3)

α_0 from 1.7 to 1.36 times; 4) the energy efficiency calculated from the ratio of the increase in the relative reduced heat transfer to the increase in the relative hydraulic resistance was less than 1 (0.44-0.71); 5) reduction in the thermal insulation coefficient base 1.09-1.16 times. The coefficient of thermal insulation in the channel without tapes varied from 0.68 ($Re=2\cdot10^3$) to 0.36 ($Re=1\cdot10^4$).

For the first case and a step of the tape t=6.25 mm increase in relation to the smooth channel is received: 1) hydraulic resistance from 3.2 to 2.3 times ($Re=(2\text{-}10)\cdot10^3$); 2) α_{pr} from 1.6 to 1.2 times; 3) α_0 from 2.3 to 1.7 times; 4) the energy efficiency was less than 1 (~0.5); 5) reduction in the thermal insulation coefficient base 1.26-1.44 times.

For the second case and a step of the tape t=12.5 mm increase in relation to the smooth channel is received: 1) hydraulic resistance from 2.9 to 1.55 times ($Re=(2\text{-}10)\cdot10^3$); 2) α_{pr} from 1.36 to 1.14 times; 3) α_0 from 1.9 to 1.53 times; 4) the energy efficiency was less than 1 (0.47-0.73); 5) reduction in the thermal insulation coefficient base 1.27-1.52 times. The coefficient of thermal insulation in the channel without tapes varied from 0.53 ($Re=2\cdot10^3$) to 0.213 ($Re=1\cdot10^4$).

For the second case and a step of the tape t=6.25 mm increase in relation to the smooth channel is received: 1) hydraulic resistance from 3.2 to 2.3 times ($Re=(2\text{-}10)\cdot10^3$); 2) α_{pr} from 1.54 to 1.24 times; 3) α_0 from 2.6 to 1.9 times; 4) the energy efficiency was less than 1 (0.48-0.54); 5) reduction in the thermal insulation coefficient base 1.62-2.13 times.

The calculated increase in the surface heat transfer coefficient was 1.5-2.6 times. In this case, the maximum relative increase of α_{pr} with decreasing Reynolds number decreased from 1.7 to 1.14.

2.2. Cooling system with coplanar (crossing) channels

An attempt was made to construct an optimal cooling system with a perpendicular arrangement of crossed (coplanar) square channels, a kind of a two-storeyed cooling system. As a basis channels of square section 1.5×1.5 mm^2 were assumed made in a copper blank with 2.1 mm pitch (thickness of a rib δ_r=0.6 mm, porosity ε=0.75, d_g=1.5 mm). As one of the limiting cases of crossing of channels (intersection angle $2\beta=0°$) the system with the doubled height of channels (the size of channel 1.5×3.0 mm^2, d_g =2.0 mm) is considered. Channels in this case are "stacked" on each other, a continuous rib. The case of mutually perpendicular crossed channels ($2\beta=90°$) was also calculated.

In the calculations of a single square channel and a "double" channel, formulas were used for the hydraulic resistance and heat transfer in smooth channels for laminar and turbulent flow of water at room temperature in them:

- for a laminar flow regime (Re<2300)

$$\xi=A/\mathrm{Re},\ (A=64 \text{ for } h_k/\delta_k=1;\ A=62 \text{ for } h_k/\delta_k=2),\ \ \mathrm{Nu}_l=1.4\,\mathrm{Pr}^{0.33}\left(\mathrm{Re}\frac{d_g}{L}\right), \tag{1}$$

- for turbulent flow regime (Re>2300)

$$\xi=0.316/\mathrm{Re}^{0.25}, \qquad \mathrm{Nu}_t=0.021\mathrm{Re}^{0.8}\,\mathrm{Pr}^{0.43}. \tag{2}$$

For crossed channels, generalizations of other researchers using generalized geometric characteristics for hydraulic tracts have been used [3, 6, 7]. In [3] to hydraulic resistance and heat transfer in the turbulent flow regime proposed the following formulas:

$$\xi=\left(\frac{2512}{\mathrm{Re}^{1.32}}\right)+0.137,\ \ \mathrm{Nu}=0.16\mathrm{Re}^{0.68}\,\mathrm{Pr}^{0.4}. \tag{3}$$

In [6,7] as the characteristic dimension at generalization of results the equivalent diameter of $d_e=V_b/F_b$ is accepted (where V_b – the volume occupied by the heat carrier in a path, F_b – the heat exchange surface of the hydraulic path). In relation to our configuration formulas will take a form:

$$V_b = \delta_k(\delta_k + \delta_r)\left[2h_k\left(1 - \frac{\delta_r}{\delta_k + \delta_r}\right)\right],\ F_b = 2\delta_k(\delta_k + \delta_r)\left[1 + \left(\frac{2h_k}{\delta_k + \delta_r}\right) + \frac{2h_k\delta_k}{(\delta_k + \delta_r)^2}\right] \quad (4)$$

and the equivalent diameter is equal d_b =1.44 mm.

For hydraulic resistance and heat transfer in the turbulent flow regime in the [6, 7] was proposed the following formulas:

- for the coefficient of hydraulic resistance:

$$\xi = B/\mathrm{Re}^n, \quad (5)$$

where $n = H/(6S) + 1.8\bar{\beta}^2 - 2\bar{\beta} + 0.55$, $B = 3.65H/S + 150\bar{\beta}^2 - 120\bar{\beta} + 21.15$, $H=h_{k1}+h_{k2}=2h_k$, $S=\delta_r+\delta_k$ for $2\beta=90°$ ($\bar{\beta} = 0.5$).For our calculated case of n=0.25, B=4.13.

- for surface heat transfer:

$$\mathrm{Nu}_e = \left(0.24 - 0.185\frac{H}{S}\right)\mathrm{Re}^{0.16\frac{H}{S}+0.62}, \quad (6)$$

and in our case $\mathrm{Nu}_e=0.07\mathrm{Re}^{0.765}$

Also for calculation of the surface heat transfer the formula for heat exchange of a turbulent flow in short channels [8] was used. As applied to our conditions it was transformed to the form:

$$\mathrm{Nu}_{kk} = 0.22 \cdot 1.57 \cdot \left(\frac{\delta_r}{d_g}\right)^{-0.12} \mathrm{Re}^{0.8}\,\mathrm{Pr}^{0.43}. \quad (7)$$

Earlier we in [2] received the following formula for reduced and surface heat transfer of coplanar channels with sizes $\delta_k \times h_k = 1.5 \times 1.0$ mm^2 which are crossed at an angle $2\beta=120°$, porosity of a cooling system $\varepsilon=0.5$:

$$\alpha_{pre} = 587.8\mathrm{Re}^{0.594},\ \alpha_e = 63\mathrm{Re}^{0.7}. \quad (8)$$

Our data on surface heat transfer were closer to the data obtained in [3, 8].

The results of the calculations on the resistivity showed that the relative excess of the resistance in the developing turbulent flow regime (Re = $(2.3\text{-}10)\cdot 10^3$) is ~5 times for [3], 13 times for [6], and is practically independent of the Reynolds number.

The data on surface heat transfer calculations showed that the relative excess of heat transfer in the regime of developing turbulent flow (Re = $(2.3\text{-}10)\cdot 10^3$) at application the formulas of different authors is 1.5 to 3 times and decreases with increasing Re.

In the future, in order to find the best energy efficiency of the system (within a certain range of Reynolds numbers) different crossing angles, and dimensions of the channels, the porosity and the material can be chosen for calculations. To calculate the thermal and hydraulic characteristics of the systems can be recommended the calculated dependences given in [3, 8].

2.3. Spring insert and spiral tape in a square channel

In this subsection on the basis of ours [2] and literary data [5] calculations and comparisons of the springs and spiral tapes installed in the square channel are carried out. In calculations, the diameter of the spring and its step, as well as the step of the tape and the dimensions of the channel, were varied.

At the beginning, to compare the spiral and spring swirls, a square-section channel was taken with a spring inserted into it [2], for which there were data on the hydro resistance and heat transfer. A spring with an external diameter D_{vn}=3.5 mm was wound from a nichrome wire [λ=13.4 W·m^{-1}·K^{-1}] with a diameter d_{sp}=0.5 mm with step h=1.0-1.1 mm. It is inserted into the cooling system model with square shaped channels $\delta_k=h_k$=3.5 mm chopped cutter with step $t=\delta_k+\delta_r$=4.4 mm in the copper blank, and acted as an intensifying element. In the calculations it was assumed that a spiral tape was inserted into such a square channel with the maximum possible degree of its twist (step H=6.25 mm). This case corresponded to one of the experimental cases considered in [2], with a square channel size of 2.65 mm.

Experimental and calculated data showed that the hydraulic resistance of the channel with a spring is 16-26 times greater than the channel resistance without a spring (in the interesting range of Reynolds numbers Re=(1-10)·10^3, α_{pr} is 2-3 times more, and the energy efficiency is at the level of η=0.14-0.1. The hydraulic resistance of the channel with the tape at the step H= 12.5 mm is 2-1.5 times greater than the resistance of the channel without the tape (in the range of numbers Re=(2.3-10)·10^3), α_{pr} is 1.1-0.82 times more, and the energy efficiency is at the level η=0.56-0.53. The hydraulic resistance of the channel with the tape at the step H= 6.25 mm is 3.1-2.3 times greater than the resistance of the channel without the tape (in the range of numbers Re = (2.3-10)·10^3), α_{pr} is 1.4-1.0 times more, and the energy efficiency is at the level η=0.45-0.43.

A case of a smaller channel with insertion of a tape and a spring was also considered. In an optimized square channel of 2×2 mm^2 (δ_r=0.5 mm, ε=0.8, d_g=2 mm), a spring "was inserted" from the wire d_{pr}=0.2 mm and a step t=1.5 mm. The increase in the surface heat transfer coefficient K_{pr} was estimated from the formula [5] proposed for the turbulent flow regime in round tubes:

$$K_{pr} = 1.85 + 2.5\left(2d_{pr}/d_g\right) - \left[\frac{0.85 + 2.5\left(2d_{pr}/d_g\right)}{2.8 + 12.6\left(2d_{pr}/d_g\right)}\right]\frac{t}{d_g}. \quad (9)$$

For our case K_{pr}=2.16 takes place. At the same time the increase in the reduced heat transfer in comparison with the channel without insert at 3.4-3.1 time was observed.

For a case when the spiral tape 0.3 mm thick "was inserted" into the same channel diameter of D_l=δ_k=2 mm, and step of H=4.7 mm (at this d_g=1.13 mm, and the relative step of the tape $2\pi H/D_l$ corresponds to one of experiments in [2]). For estimates of hydraulic resistance and the surface heat transfer the formulas [5] proposed for turbulent flow regime in circular pipes were used. Calculations at change of Reynolds numbers in the range of Re= (2.3-10)·10^3 established excess of heat transfer:

- channel with a spring above a smooth channel - 2.16 times (see K_{pr}=2.16);
- in the channel with the tape in 6-5.1 times with respect to the smooth channel.

2.4. A cooling system from corrugations with the small holes applied on them

This system "was formed" hypothetically of a corrugated insert on which the small holes of necessary optimum depth and density located in chess packing are previously applied. The corrugation from the high-temperature material had the optimum step in terms of the porosity of the system, providing a maximum of α_{pr} in the region of interesting Reynolds numbers. In calculations are used our experimental [2] and literary data.

To carry out the calculations a corrugation was taken for which there were experimental data [9]: the height of the corrugation h_k=1.73 mm, the average width of the channel δ_k=1.73 mm, the thickness of the corrugation δ_r=0.1 mm, the porosity of the cooling system ε= 0.89.

This corrugation was "coated" with holes of diameter D=1 mm and depth h=0.1 mm (the characteristic ratios H/D=1.73, H=δ_k, h/D=0.2, and f=0.7 is the relative area of a covering of a corrugation the small holes). The calculation was taken our data slit channel obtained in [2]. For the heat exchanged surface coated with the holes (f=0.7) was established an increase in heat transfer (it was normalized to the heat transfer of an identical reference specimen without holes) to 1.5 times (with Re~10^4). This value was increased surface heat transfer for the hypothetical calculated cooling system. At the same time, because of the small contribution of the finning in α_{pr} (which decreases rapidly with increasing Reynolds number), it has increased approximately by the same amount. The coefficient of thermal insulation is reduced by 2-7 times (with Re changing from 2300 to 10^4). This system in the Reynolds number region $4\cdot10^3 \le Re \le 14\cdot10^3$ is energetically favorable, since in it the heat transfer exceeded the hydraulic resistance (by 1.4 times (at $Re=1.4\cdot10^3$) with a drop to 1 (at $Re=14\cdot10^3$)).

Conclusion

The calculations and analytical comparison demonstrate the usefulness of some methods of heat transfer enhancement in the proposed designs of cooling systems.

References

[1] Fedoseev V, Shanin O, Shanin Yu and Afanasyev V 1989 Heat transfer in rectangular channels with heat-conducting walls under one-sided heating *Thermal physics of high temperatures* **27**(6) 1132-38

[2] Shanin Yu and Shanin O 2015 Heat transfer and hydraulic resistance for an ordered vortex-transformation in cooling systems of laser mirrors *Proc. Heat transfer and hydro dynamics in swirling flows (Kazan)* (St.-Petersburg: Svoye izdatel'stvo) 106-07

[3] Pelevin F, Ilyinskaya O and Orlin S 2014 Application of coplanar channels in engineering *Aero-kosmicheskaya tekhnika* **37** 71-85

[4] Ibragimov M, Nomofilov E and Subbotin V 1961 Heat transfer and hydraulic resistance during screw motion of liquid in a pipe *Teploenergetika* **7** 57-60

[5] Migai V 1980 *Increase of efficiency of modern heat exchangers* (Leningrad: Energy) p 143

[6] Sokolov N, Polishchuk V and Andreev K 2013 Hydraulic resistance in channels of rectangular cross section with crossed finning *Vedomosti S.-PSPU. Energy. Electrical engineering* **2**(171) 85-94

[7] Sokolov N, Polishchuk V and Andreev K 2013 Heat transfer in channels of rectangular cross section with crossed finning *Vedomosti S.-PSPU. Energy. Electrical engineering* **3**(178) 17-27

[8] Sukomel A, Velichko V and Abrosimov Yu 1979 *Heat transfer and friction in turbulent gas flow in short channels* (Moscow: Energia) p 216

[9] Shanin Yu and Shanin O 2013 Heat transfer and hydraulic resistance of cooling systems for laser mirrors from corrugations *Engineering and Physics Journal* **86**(4) 774-84

IOP Conf. Series: Journal of Physics: Conf. Series **980** (2018) 012037 doi:10.1088/1742-6596/980/1/012037

The Effect of Flow Swirling on the Safety and Reliability of Nuclear Power Installations of New Generation

O V Mitrofanova[1], O A Ivlev[2] and D S Urtenov[2]

[1]National Research Nuclear University MEPhI, Moscow 115409, 31 Kashirskoe sh., Russia
[2]National Research Centre "Kurchatov Institute", Moscow 123182, 1 Akademika Kurchatova Sq., Russia

Email: omitr@yandex.ru

Abstract. Hydrodynamics and heat exchange in the elements of thermal hydraulic tracts of ship nuclear reactors of the new generation were numerically simulated in this work. Parts of the coolant circuit in the collector and piping systems with geometries that may lead to generation of stable large-scale vortexes, causing a wide range of acoustic oscillations of the coolant, were selected as modeling objects. The purpose of the research is to develop principles of physical and mathematical modeling for scientific substantiation of optimal layout solutions that ensure enhanced operational life of icebreaker's nuclear power installations of new generation with reactors of integral type.

1. Introduction

The study of processes leading to generation of sustainable vortex structures in the channels of complex geometry, is connected with the need to exclude resonant effects from the elements of thermal hydraulic circuits of new generation in marine nuclear power units with nuclear reactor of the integral type. This is expedient in case of coincidence of frequencies of acoustic fluctuations of the coolant with natural frequencies of the elements of the nuclear reactor internals. This may cause a risk of vibration processes and regimes of thermal cycling, leading to reduction in strength characteristics of heat exchange surfaces, the interloop leakages and other negative consequences.

It is known that the main circulation pumps and non-stationary hydrodynamic processes in the coolant (acoustic waves, vortex formation, turbulence, cavitation, etc.) are the main sources of exciting oscillations in primary circuits of nuclear power installations. The vibrational properties of nuclear power installation as a hydro-mechanical system are not be reduced to simple superposition of frequency characteristics of its elements (the reactor, steam generator, pumps, pressurizer, pipelines, etc.). The possibility of arising of acoustic resonances in the mono-block design may be connected with nonlinear superposition of frequency spectra of the major fluctuations and their overtones, due to the influence of complicated turbulent internal and external flows in the tube bundles of the steam generator and the influence of the other sources of disturbances, in particular, centrifugal pumps of the primary circuit and the pipeline system of pressurizer.

The considered here several test problems correspond to the cases of flows that can occur due to loop supply of the coolant, start-up and transient operation of the reactor, the design features, resulting in the creation of torque, or the inconsistent work of the pumps.

Content from this work may be used under the terms of the Creative Commons Attribution 3.0 licence. Any further distribution of this work must maintain attribution to the author(s) and the title of the work, journal citation and DOI.
Published under licence by IOP Publishing Ltd

IOP Conf. Series: Journal of Physics: Conf. Series **980** (2018) 012037 doi:10.1088/1742-6596/980/1/012037

2. Simulating vortex formations in the pressure header of a reactor unit of integral type

One of the first meaningful attempts to mathematically model complex vortex flows with large-scale circulation flow in the collectors of nuclear reactors was made by A. N. Patrashev [1]. The analysis of the current state of the problem of numerical modeling of complex turbulent flows held in the monograph [2] showed that the use of turbulence models of different classes to calculate the swirling flows is still a separate issue and should be based on a deeper understanding of the physics of swirl flows.

The appearance of spontaneous flow swirling which is not provided for nominal flow regime in the manifold system can be attributed to the effect of secondary flows. By definition, the secondary flow is a time-averaged flow developing in the plane perpendicular to the primary flow direction.

Calculations of velocity fields and pressure areas of the inlet coolant manifold and the transition from the downward annular section of the manifold to the pressure chamber in front of the entrance to the core were performed for a simplified model of the manifold using ANSYS CFX codes. The configuration of the three-dimensional computational domain is shown in figure 1 a. Four inlet pipes of the coolant supply loop are shown in the upper part of the figure.

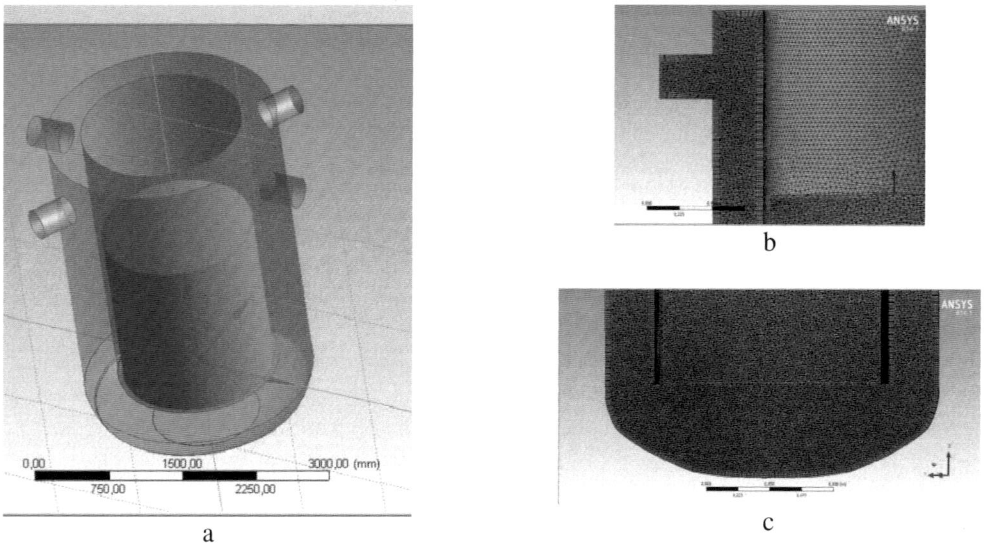

Figure 1. Scheme of a simplified model of the manifold: a) 3D geometry of the computational model of the investigated part of the collector system; b, c) - the structure of the computational mesh: b) – in the area of the inlet pipe, c) – in the bottom part of pressure.

Non-transparent allocated volume shows the space occupied by the core. The size of cells in near-wall areas was chosen based on the specified Reynolds number, the geometrical data and thermal properties of the water coolant for the nodes of cells closest to the walls to fall in the region of the logarithmic law of flow velocity distribution and for the use of standard near-wall functions. For this calculation we have chosen the mesh that consists of 1.3 million nodes and 7.1 million elements. The number of elements and the size of cells of the computational mesh are changed depending on the complexity of geometry of various parts of the manifold. To improve the accuracy of calculations near the streamlined surfaces thickening of the computational grid was created (figure 1 b,c).

Different turbulence models were used for mathematical description. It enables the comparative analysis of results and choice of the most appropriate model on the basis of comparison with the experiments or, in their absence, with the alternative analysis. In particular, the monograph [2] contains theoretical description of the conditions favorable for the development of secondary vorticity and the transition to helical motion with generation of large-scale vortex structures.

There were 3 series of computational experiments: 1 - without swirl flow in the input branch pipes; 2 - with flow swirling in the inlet pipes when the influence of centrifugal pumps was considered; 3 – with irregular coolant flow rate in the loop inlets. For the 3rd series of calculations, it was assumed that the difference in coolant flow, provided by the pumps in the neighboring nozzles situated at 60^0 or 120^0 from one another along the arc perimeter, was changed by 12.5%, while the total flow rate remained unchanged.

The calculations were carried out using the following turbulence models: Eddy Viscosity Transport Equation, SSG Reynolds Stress and Shear Stress Transport, as well as models of inviscid fluid No Viscosity.

According to the selected analysis algorithm and formulation of boundary and initial conditions, the process of vortex formation was considered only in the region of the downward part and in the pressure chamber of the collector system. The space of the core was represented as a porous body. The distribution of hydraulic resistance met the required predetermined pressure differential in the core. It was accepted that the transverse motion of the coolant in the core is negligible.

Computational experiments have shown that the results of calculations on different turbulence models, although being significantly different, indicate the same qualitative result: the selected scheme of coolant supplying leads to arising the large-scale vortices.

At that the frequency of rotation in these vortices and the amplitude of oscillations generated by them depend on the chosen geometry and on the coolant flow rate. On the basis of the comparative analysis of applicability of turbulence models it may be concluded that the most adequate are the results obtained with the model of Shear Stress Transport.

The illustrations presented in figures 2, 3 show that the development of large-scale vortex motion under the downward coolant motion causes the flow swirling in all area of the pressure chamber, which entails a reduction in pressure and, consequently, a decrease in the flow rate of coolant in the central part of the pressure chamber directly under the core of nuclear reactor.

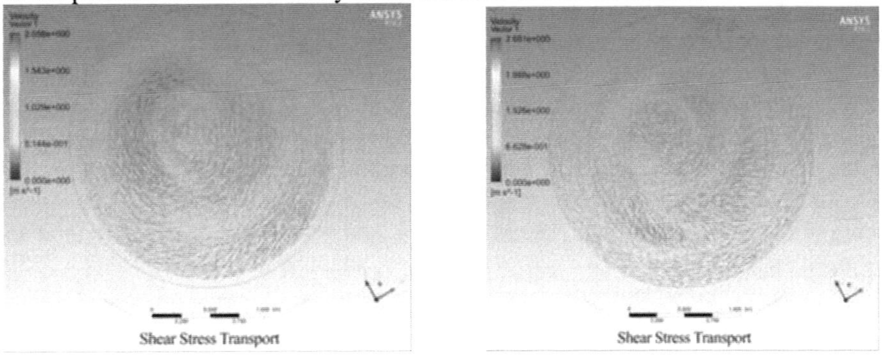

a) b)

Figure 2. The velocity distribution of the coolant with account of the swirl flow in the input branch pipes: a) in cross section at a distance of 0.2 m under the core, b) – at the entrance to the core.

The results of calculations made with account of the radial vorticity introduced by the influence of centrifugal pumps (figure 2) indicate the formation of large-scale vortex with an offset relative to the center vertical axis. The presence of such vortex in the pressure header in front of the entrance to the core can lead to reduced coolant flow rate in the most energy-stressed central part of the core.

The results of calculations presented in figure 3 show that under irregular distribution of the coolant flow rate in the loop inlets, there is also the large-scale vortex motion. When supplying a flow in the pair loop inlets from one side of the reactor shell perimeter more of 12.5% compared to the other pair loop inlets at a constant total flow rate a powerful vortex with tangential velocity, comparable with average flow velocity, is formed in the cross section of pressure chamber under the core. Formation of large-scale vortex leads to the decrease of coolant flow rate in the center of the reactor core.

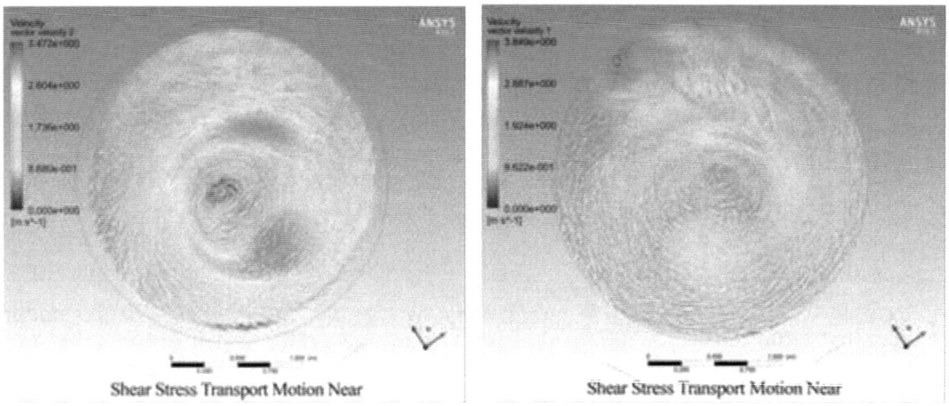

a) b)

Figure 3. The coolant velocity distribution for the case of asymmetric supplying of coolant flow rate in the loop inlets: a) in cross section at a distance of 0.2 m under the core, b) – at the entrance to the core.

3. Generation of large-scale vortices in the pipe system of the pressurizer

The nuclear power installation design includes a large number of auxiliary systems and mechanisms, ensuring reliable operation of the reactor in different modes. To main sources of vibration may be related rotating elements of pumps exposed to dynamic loads as well as generation of large-scale vortices, leading to the coolant flow swirling. The experience of previous years has showed that one of the most vulnerable parts of the construction of ship nuclear unit is the pipe system of the pressurizer.

In this work the generation of low frequency acoustic waves in the pipeline system of the pressurizer was considered by the example of the turbulent flow modeling in the pipeline section of complex geometry for the cases of increasing and decreasing power of the reactor unit.

Two series of computational experiments corresponding to these cases were conventionally called "hot" and "cold" tasks. In the "hot" task, we considered an instant increase of nuclear reactor power, corresponding to a pressure jump $\Delta P = 0.3$ MPa at nominal pressure $P_0 = 16.0$ MPa. Under the action of pressure difference hot water from a nuclear reactor at a temperature of 300^0 C moved from the first to the last cross section along the pipeline, filled in initial time with water at a temperature of 100^0 C.

In the "cold" task, the reverse initial condition was used. Under the action of negative pressure difference $\Delta P = - 3$ MPa cold water motion started at a temperature of 100^0 C from the last to the first cross section. The water temperature in the pipeline dropped from 300^0 C to 100^0 C while filling the entire pipeline with cold water from the pressurizer.

The results of calculations have revealed that in both cases, under the conditions of "hot" and "cold" tasks a large-scale flow swirling is formed in the entire cross section of the channel. Herewith, the angular flow velocity is comparable with the natural frequency of pump operation.

The calculations have showed that the field distribution of vorticity in the study of unsteady process of the coolant flow field topologically coincides with the field of local flow temperature. Therefore, the identification of conditions for the appearance of stable large-scale vortices allows avoiding the regimes of thermal cycling and enhances the reliability and lifetime of pipeline systems.

4. Conclusion

Computational experiments have showed that generation of large-scale vortices in pressure header is connected with selection of the scheme of coolant supply in the pressure chamber. The frequency of rotation in these vortices and the amplitude of the acoustical oscillations generated by them are dependent on the geometry of channels and on the distribution of the coolant flow rate.

Computational experiments have also revealed that the vortex flow structure in the channels of piping systems significantly depends on the geometry of the pipeline and the sequence of bends of different orientations.

The presence of areas with three-dimensional curvature of the channels leads to the generation of large-scale vortex motion and low frequency acoustic oscillations in the output section of the pipeline.

Large-scale spiral-vortex flow structure remains for over twenty calibers after passing the bends and has an effect on the flow pattern, which is formed in the bends downstream.

The formation of vortex movement of coolant flow in significant degree influences the propagation of the temperature field in the pipe wall. The study of this question is necessary to increase the strength and reliability of pipeline systems and the pressurizers of the ship nuclear power installations.

This work was supported by RFBR grant 16-08-00687 and the program of increasing the competitive ability of National Research Nuclear University MEPhI (agreement with the Ministry of Education and Science of the Russian Federation of August 27, 2013, project no. 02.a03.21.0005).

References

1. Patrashev A N 1980 Turbulent flows at the fluid with large-scale vortices *Mechanics of turbulent flows* pp 166-187.
2. Mitrofanova O V 2010 *Hydrodynamics and heat transfer of swirl flows in channels of nuclear power installations* (Moscow: Fizmatlit) p 288.

IOP Conf. Series: Journal of Physics: Conf. Series **980** (2018) 012038 doi:10.1088/1742-6596/980/1/012038

Experimental modeling of swirl flows in power plants

S I Shtork, I V Litvinov, E S Gesheva, M A Tsoy, S G Skripkin

S.S. Kutateladze Institute of Thermophysics SB RAS,
630090, Russia, Novosibirsk, 1 Lavrentyev Ave.

shtork@itp.nsc.ru

Abstract. The article presents an overview of the methods and approaches to experimental modeling of various thermal and hydropower units - furnaces of pulverized coal boilers and flow-through elements of hydro turbines. The presented modeling approaches based on a combination of experimentation and rapid prototyping of working parts may be useful in optimizing energy equipment to improve safety and efficiency of industrial energy systems.

1. Introduction

Since experiments in full-scale power plants are extremely expensive and time-consuming, experimental modeling using reduced laboratory models is widely used to develop energy unit designs [1,2]. Strict fulfillment of the similarity condition by all criteria is practically impossible. Therefore, the most commonly used is the approximate modeling only on certain parameters [3]. For example, in the case of combustion chambers, isothermal modeling on geometrically similar reduced models based only on one similarity criterion, the Reynolds number, enables the relatively low-cost study of the combustor aerodynamics, which, in turn, cardinally determines the combustion process [4]. Although the results of isothermal modeling are somewhat limited when it comes to full-scale firing conditions, they still allow determining the direction of the design optimization and, thus, reducing the equipment development costs, by diminishing the number of design options, studied on large-scale fire stands and real size facilities [5].

Owing to rapid progress in the computer technique, the numerical modeling is now increasingly becoming the main tool for the power plant development and design [5]. The role of physical modeling is still very important, while the main objective of the experiment is to provide empirical information for the verification of numerical codes [6]. With respect to experimental modeling, the reliability and detailed character of the data obtained on the model become crucial. To this end, it is important to precisely specify the boundary conditions, which implies using a simplified, generalized geometry of the working area. To obtain reliable data, it is also important to use non-contact, non-perturbing measurement techniques [7]. Another relevant objective of the experiment, in addition to verifying the computer simulation, is to reveal the physical mechanisms of the processes and to obtain fundamentally new physical effects that may be further used in practice. To accomplish this task the laboratory model should allow for variation within a wide range of geometry and operating modes.

In this context, the article presents some overview of methods and approaches of experimental modeling of various thermal and hydropower devices - combustion chambers of pulverized coal boilers and flow elements of hydro turbines. The focus is on the study of large-scale concentrated vortices, determining the global structure of the flow and critically influencing the technological processes in industrial equipment.

Content from this work may be used under the terms of the Creative Commons Attribution 3.0 licence. Any further distribution of this work must maintain attribution to the author(s) and the title of the work, journal citation and DOI.
Published under licence by IOP Publishing Ltd

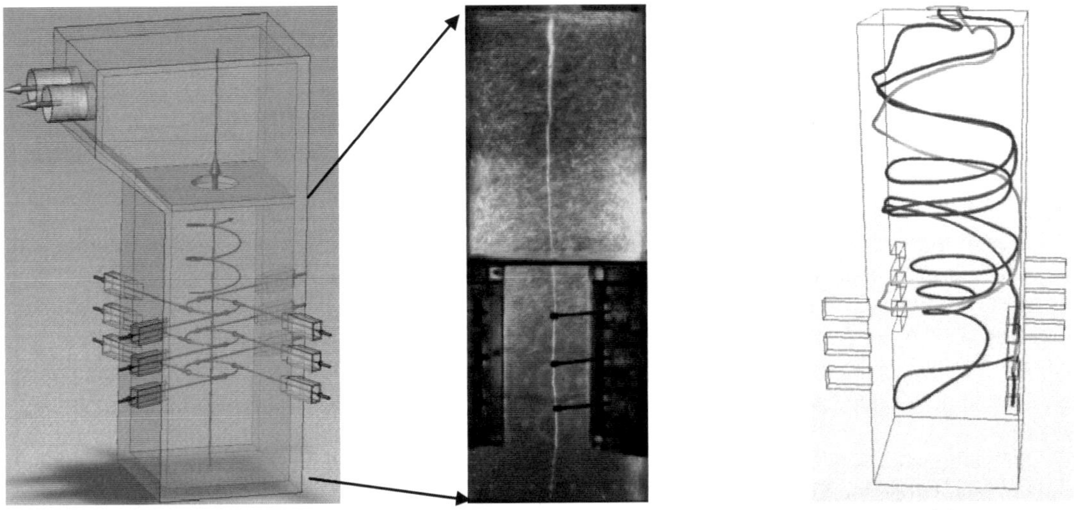

(a) (b) (c)

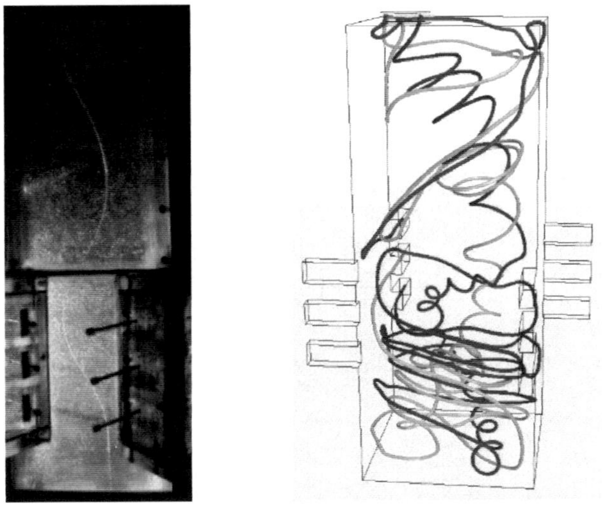

(d) (e)

Figure 1. Scheme of hydraulic model of vortex combustor (a) and flow visualization for test section with central (b) and shifted outlet orifice (d). Trajectories of fluid particles at generation of stationary vortex structures: c – rectilinear vortex, e – spiral vortex.

2. Experimental techniques

2.1. Flow structure diagnostics

A specific feature of the used modeling technique is the replacement of working media: gas by liquid, as it was done in the modeling of furnaces for pulverized coal boilers [4-6], and water by air in flow modeling of hydro turbine equipment [7,8]. In the first case, the use of liquid as a working medium simplified the visual studies, which provide important qualitative information on the flow structure. Small air bubbles were used as liquid flow markers (figure 1), which were illuminated by a light source (a powerful LED spotlight for analyzing the general flow pattern or a narrow light "sheet" formed by sweeping on a cylindrical lens of the laser beam from a solid-state laser to visualize the flow structure in a specific flow cross-section).

To obtain quantitative information about the flow field a two-component laser-Doppler anemometer LAD-06i was used [9,10]. To measure pressure pulsations in the experiments with air the acoustic sensors based on precision measuring microphones were used [7]. In hydrodynamic models, sensitive piezoelectric pressure sensors were employed [6]. To measure the instantaneous velocity fields, the PIV system was applied [6,11].

IOP Conf. Series: Journal of Physics: Conf. Series **980** (2018) 012038 doi:10.1088/1742-6596/980/1/012038

2.2. Test rigs

2.2.1. Hydraulic model of the vortex combustor. The model test section (figure 1a) is a square-section channel with dimensions of $188\times188\times625 mm^3$. This geometry corresponds to the design of the tangential furnace E-500, used to burn a pulverized-coal fuel [6]. The square shape of the cross section in the case of large tangential furnaces is conditioned by the fact that the flat side walls are easier to manufacture than the cylindrical ones. It has been shown experimentally that the use of a square shape does not result in vortex destruction or distortion [11]. It is also important that the flat side walls of the model do not give optical aberrations facilitating the use of non-contact optical methods for investigating the flow. Therefore, this unit is widely used not only to simulate the flows in tangential furnaces but also to study the general physical effects, governing the dynamics of vortex flows. To control the structure of the vortex flow the boundary conditions at the ends of the chamber were changed. In particular, a diaphragm with an axisymmetric or offset hole was installed at the chamber outlet [6].

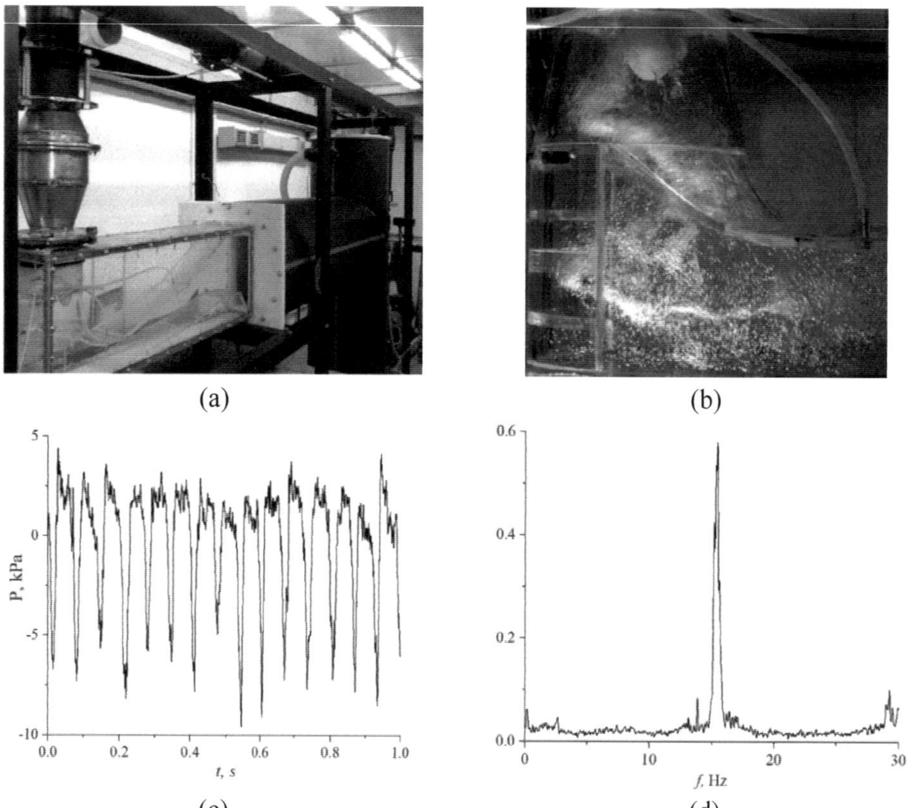

Figure 2. View of the hydraulic stand with model TURBINE-99 (a) and visualization of precessing vortex rope (b). Pressure pulsations recorded with piezoelectric sensor: (c) - time series; (d) FFT spectrum (vertical axis is in arbitrary units).

2.2.2. Hydraulic models of hydroturbine draft tubes. Figure 2a shows the hydraulic stand with the model of a draft tube with the geometry TURBINE-99, widely used for testing computer codes [12]. The internal geometry of the model is formed by relatively thin Plexiglas plates. The whole construction is inserted into a liquid-filled rectangular casing, made of thicker Plexiglas, taking up fluid pressure inside the working area [13]. Such a design with plane external walls, similar to the model of vortex furnace, provides minimal optical distortion. To simulate the regimes with a large

IOP Conf. Series: Journal of Physics: Conf. Series **980** (2018) 012038 doi:10.1088/1742-6596/980/1/012038

residual swirl at the input to the working section, a stationary blade swirler with the design swirl parameter S=1.1 is installed [14]. Such regimes characterized by the development of strong flow nonstationarity behind the impeller (in the conical part of the draft tube) are realized in the operation under non-optimal conditions of underload or overload of the hydroturbine [7].

Figure 3a presents the scheme of the working section, simulating only the inlet conical part of the draft tube. This enables the detailed study of the processes of vortex formation behind the impeller, not complicated by the influence of the draft tube elbow. To ensure optical access the working area is made of Plexiglas and has plane external walls. A stationary blade swirler is used for swirl generation similar to the previous case. For better matching the real conditions of velocity and pressure distributions at the draft tube inlet, a rotating element is added to the swirler design. At the first stage of research, a freely rotating runner was used. This, in principle, can provide distributions of axial and tangential velocity components in the model to be fairly close to the flow distribution under the working wheels of full-scale hydroturbines [15]. However, similar to the previous case, this combination of swirlers ensures simulation of conditions only for a single operating regime of the hydroturbine. In the presented study, we use a stationary swirler with inclination angle of the blades, corresponding to the design swirl parameter S=1.1.

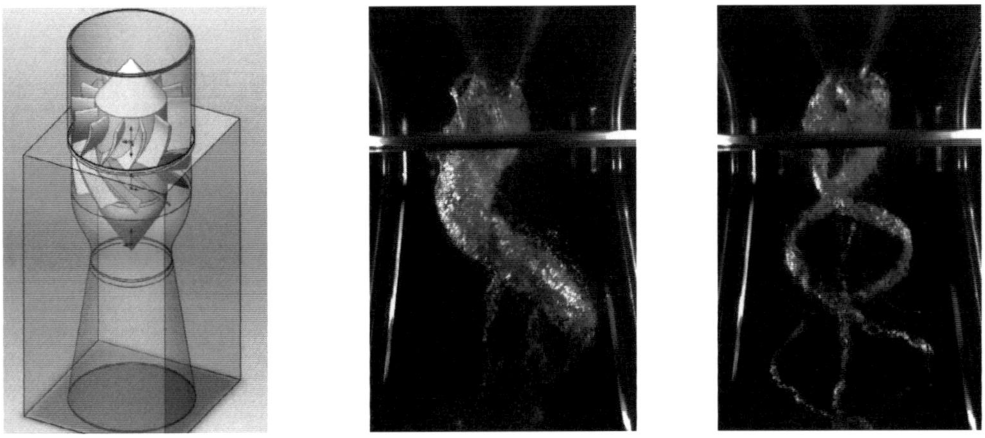

(a) (b) (c)

Figure 3. Sketch of axisymmetric model of the draft tube (a) and visualization of single spiral (b) and double spiral (c) vortex ropes.

2.2.3. Air models of hydroturbine draft tubes. The replacement of liquid media by air has been widely used for the simulation of hydraulic machines from the previous century up to present [7,8,17]. The advantages of an experiment with air are due to much less difficulties in manufacturing experimental sections (avoiding problems of joint sealing and test section strength; providing easier optical access). In our case, the use of air enables 3-d printing based on available plastic materials for rapid and accurate reproduction of complex geometry of the channel elements.

The scheme of the working section is shown in figure 4a. In this unit, similar to the hydraulic model, the velocity distribution close to the distribution behind a real hydroturbine is reached by means of a system of two blade swirlers [16,17]. One of the swirlers is fixed (stator) and functions as a spiral chamber and a guiding vane of the hydroturbine (figure 4b). The air flow with a volume flow rate Q (m^3/h) is uniformly flowing onto it. The second swirler (runner) being an analog of the hydroturbine impeller rotates. Unlike the above design of the hydraulic model, the runner in the air model is rotated at a frequency n (rpm) with the aid of an external servo drive. Thus, the unit with the air model allows simulating various operating regimes of the hydroturbine based on specifying the parameters Q, n and the geometry of the swirler arrays.

3. Sample results

3.1. Stationary spiral vortices in the combustor model.

The initial geometry of the tangential chamber with no diaphragm is characterized by the development of large-scale flow instability. It is expressed in the vortex core shift relative to the channel center and its high-amplitude precession motion around the channel center [6]. Diaphragming the chamber allows stabilizing the flow by fixing the vortex core near the channel axis [11] and obtaining a generally axisymmetric pattern of the flow (figure 1b). The flow below the diaphragm is, in general, insensitive to the conditions behind the diaphragm, in particular to the asymmetrical side exit from the working channel. However it drastically changes with the introduction of asymmetry at the ends of the vortex [6]. For example, when the outlet hole of the diaphragm is displaced the axis of the vortex takes the form of a stationary (immobile) single helix (figure 1d).

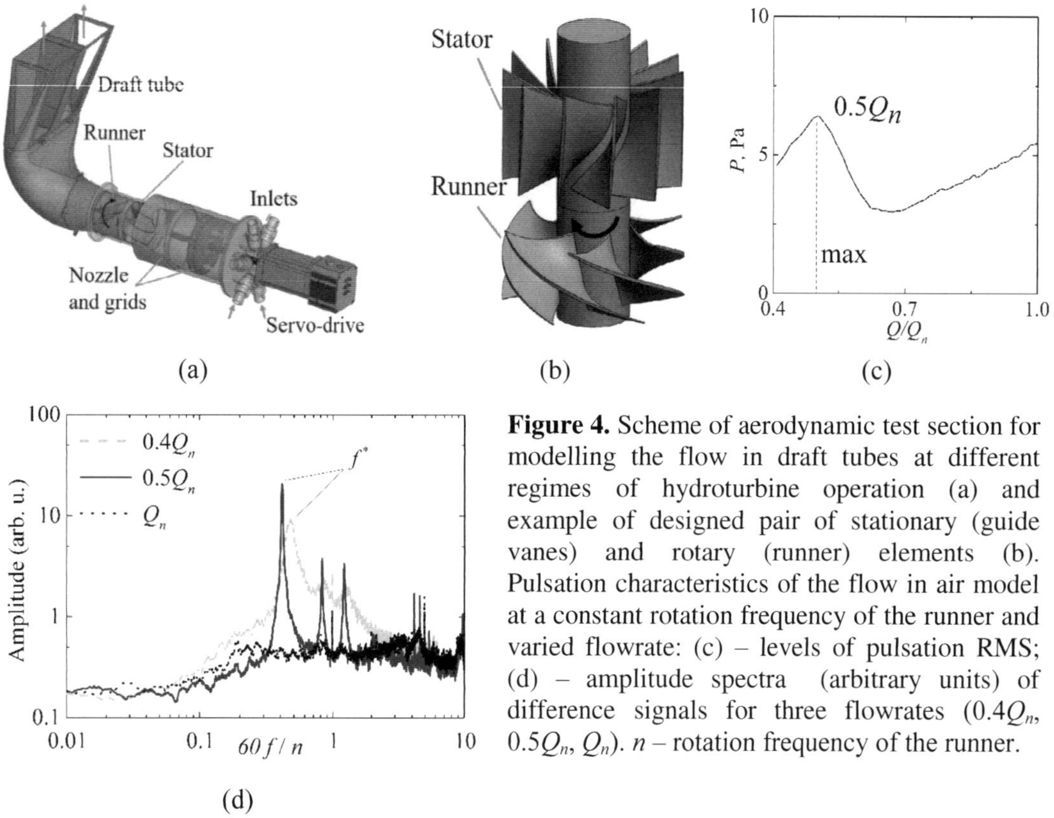

Figure 4. Scheme of aerodynamic test section for modelling the flow in draft tubes at different regimes of hydroturbine operation (a) and example of designed pair of stationary (guide vanes) and rotary (runner) elements (b). Pulsation characteristics of the flow in air model at a constant rotation frequency of the runner and varied flowrate: (c) – levels of pulsation RMS; (d) – amplitude spectra (arbitrary units) of difference signals for three flowrates ($0.4Q_n$, $0.5Q_n$, Q_n). n – rotation frequency of the runner.

The results obtained on the tangential furnace model for the conditions of forming the vortex structures with spiral axes may be interesting from a practical point of view to demonstrate possibilities of control over the combustion chamber aerodynamics. In the case of the formation of spiral vortex structures the fluid particles move along spiral trajectories around the spiraled rotation axis. Therefore, it is a case of elongation of the trajectory (in comparison with the rectilinear vortex) and accordingly of an increasing residence time. This is important for a more complete burn-out of fuel and improved combustion efficiency. Modes with a single and especially a double spiral vortex are characterized by a greater involvement of the internal chamber volume in the intense vortex motion that may assumingly intensify the mixing processes and improve the heat exchange in the combustion chamber. These assumptions were verified by calculating the length of trajectories and the residence time using numerical LES modeling [18]. The comparison of the calculation results shows (figure 1c, e) that the trajectory lengths for regimes with spiral vortices are noticeably larger than for a rectilinear vortex. For example, in the case of a single-helical vortex, the average residence time increases almost 2 times.

3.2. Precessing spiral vortices in hydraulic models of draft tubes.

Under conditions of strong swirling an instability develops in the form of a precessing vortex rope (figure 2b), which generates powerful pressure pulsations (figure 2c). This strong unsteadiness is expressed in the pulsation spectrum by the presence of a dominant peak (figure 2d). The frequency of precession is linearly dependent on the volumetric flow rate and the amplitude of the pulsations at the precession frequency varies quadratically with the flow rate. In a dimensionless form, the dependence of the precession frequency demonstrates a self-similarity with respect to the Reynolds number [15].

Studies show that under conditions of strong swirling the dominant mode is a single-helical vortex rope [14]. However, in some regimes there is an abrupt transition from single-spiral (figure 3b) to a double-helical mode (figure 3c). This kind of instability has a special practical and scientific interest in view of the random changes in the characteristics of the vortex rope: vorticity field, period and frequency of precession [15]. This behavior of the precessing rope in draft tubes is poorly predicted and can cause additional unexpected vibrations of hydraulic equipment.

3.3. Unsteady regimes in air model of a draft tube.

The experiments with the air model included the determination of regimes with an increased level of flow pulsations, caused by the precessing vortex rope. Such regimes arise, in particular, at moving away from the nominal flow rate Q_n into the region of lower flowrates, corresponding to partial loads of the hydroturbine. To determine the modes with maximum pressure pulsations, the flow rate was varied from $0.4Q_n$ to Q_n with a step of $0.05Q_n$ at a constant n. Based on the pressure pulsation signals recorded with the precision B&K microphone on the cone wall, the level of RMS fluctuations of the flow was determined as a function of the ratio Q/Q_n (figure 4c). The data shows the formation of a local extremum (labeled as "max" on the graph). This extremum arises due to generation of a precessing vortex rope in the draft tube cone. To prove this two acoustic sensors (Behringer microphones) were mounted flush with the wall opposite each other in the mean horizontal section of the cone [7,16,17]. Figure 4d shows the spectra of the difference signals of the pressure pulsations, obtained from two sensors for the modes $0.4Q_n$, $0.5Q_n$ and Q_n. On the difference spectra, there is the dominating frequency $f^*(Q)$, which differs from the frequency n. Starting at a flow rate $Q=0.4Q_n$ a peak with a frequency f^* is formed, then at a flow rate of $0.5Q_n$ the peak amplitude f^* reaches its maximum. When approaching the nominal mode (Q_n) the rope frequency f* completely disappears from the spectrum.

Thus, the studies have resulted in determining the regimes with maximum pressure pulsations, which appeared in the region of low (relative to the optimal mode) flowrates. Based on the analysis of pressure pulsation signals, the development of the precessing vortex rope in the cone of the draft tube is demonstrated. The modeling approaches presented in this paper and based on a combination of experiment and rapid prototyping of working parts may be useful for optimizing the geometry of the flowing parts of a hydroturbine with the view of increasing the dynamic reliability and efficiency of a hydropower plant operation.

Conclusions

The presented sample results demonstrate the ability of the developed approaches in experimental modelling of power equipment to reveal new physical phenomena, which may drastically influence the technological processes. Among them, the formation of large-scale stationary spiral vortices noticeably increasing the residence time has been studied. This is important for a more complete burn-out of fuel and improved combustion efficiency in thermal power plants. Regarding the hydropower engineering, the new features of emergence of spiral precessing vortices at off-design operation regimes of hydroturbines, such as sudden switching between single and double helical modes, have been described. Most dangerous regimes associated with maximum pressure pulsations have been determined using a novel laboratory model, allowing for simulation of different conditions of a hydroturbine work.

Acknowledgments
The work was partially supported by the Russian Foundation for Basic Research (Grant No. 15-08-05220).

References

[1] Kutateladze S S, Lyakhovskii D N, Permyakov V A 1966 *Modeling of Thermal Power Equipment* (Energiya, Moscow) [in Russian]
[2] Migai V K 1987 *Modeling of Heat Transfer Power Equipment* (Energoatomizdat, Moscow) [in Russian]
[3] Lyakhter V M, Prudovskii A M 1984 *Hydraulic Modeling* (Energoatomizdat, Moscow) [in Russian]
[4] Markovich D M, Shtork S I, Semenov V I, Vasechkin V N, Bil'skii A V, Serant F A 2004 *Thermal Eng.* **51**(1) 58-65.
[5] Alekseenko S V, Burdukov A P, Dekterev A A, Markovich D M, Shtork S I 2011 *Therm. Eng.* **58**(9) 779-785.
[6] Gesheva E S, Litvinov I V, Shtork S I, Alekseenko S V 2014 *Therm. Eng.* **61**(9) 649–657.
[7] Shtork S I, Skripkin S G, Tsoy M A, Litvinov I V 2016 In: *Advances in Energy Engineering Simulations. K Hanjalic, DM Markovich, DPh Sikovsky Eds.: ISBN: 978-5-4437-0572-9* (Publishing Center of Novosibirsk State University, Novosibirsk) 147-175 [in Russian]
[8] Povkh I L 1974 *Aerodynamic Experiment in Machinery Construction* (Mashinostroenie, Leningrad) [in Russian]
[9] Shtork S I, Yusupov R R, Alekseenko S V 2017 *Bulletin of the Tomsk Polytechnic University. Geo Assets Engineering* **328**(7) 6–18. [in Russian]
[10] Cala C E, Fernandes E C, Heitor M V, Shtork S I 2006 *Exp. Fluids.* **40** (2) 267-276.
[11] Gesheva E S, Shtork S I, Alekseenko S V 2014 *Siberian Journal of Physics* **9**(3) 39-48. [in Russian]
[12] Cervantes M J, Andersson U, Lövgren H M 2010 *IOP Conf. Ser.: Earth Environ. Sci.* **12** 012014.
[13] Alekseenko S V, Dekterev A A, Litvinov I V, Minakov A V, Pylev I M, Shandro A I, Shtork S I 2011 *Journal of Siberian Federal University. Engineering & Technologies* **4**(5) 489-503. [in Russian]
[14] Skripkin S G, Kuibin P A, Shtork S I 2015 *Technical Physics Letters* **41**(7) 638-640. [in Russian]
[15] Skripkin S G, Tsoy M A, Shtork S I 2015 *Siberian Journal of Physics* **10**(2) 73–82. [in Russian]
[16] Kuibin P A, Litvinov I V, Sonin V I, Ustimenko A S, Shtork S I 2016 *Siberian Journal of Physics* **11**(1) 56-65. [in Russian]
[17] Litvinov I V, Mitryakov A S, Shtork S I 2016 *Power Technology and Engineering* **50**(1) 1-5
[18] Gesheva E S, Shtork S I, Alekseenko S V 2017 *IOP Conf. Series: Journal of Physics* **899** 022007

Investigation of inner aerodynamics of the four-vortex furnace model

I S Anufriev, E Yu Shadrin* and O V Sharypov

Kutateladze Institute of Thermophysics SB RAS, 630090 Novosibirsk, Russia

*e-mail: evgen_zavita@mail.ru

Abstract. The internal aerodynamics of a perspective vortex furnace chamber of a pulverized coal boiler with a diagonal arrangement of burners is studied using the non-contact optical method of flow diagnostics. The results of laser Doppler anemometry, characterizing the complex spatial structure of a swirling flow in an isothermal laboratory model of the furnace device, are presented. The velocity distribution in the vortex chamber volume is obtained, and the flow structure in the form of four conjugate closed vortices with curved axes is visualized.

The solution of the topical thermal power engineering problem of expansion of the fuel and raw materials base through the use of low-grade (non-projected) coals involves the development of new types of combustion devices that meet modern requirements for energy efficiency and environmental safety. One of the promising technologies in this area is the combustion of fuel in the vortex flow. Achievement of high indicators for flaring of fuel in the vortex flow is mainly provided by the optimal organization of the flow structure in the combustion chamber. The justification of the choice of constructive and regime parameters requires complex modern scientific methods of physical and mathematical modeling of transport processes in bounded multiphase turbulent swirling reacting flows.

The purpose of this paper is to obtain data on the flow structure in a perspective vortex combustion device for burning pulverized coal with solid ash removal [1]. The investigated design with the original aerodynamic scheme (Fig. 1), applied on a number of boilers of TPP in the Krasnoyarsk region, ensures stable combustion of high-ash coals. At the same time, in practice a number of shortcomings, related to the slagging of heat exchange surfaces, have been revealed. Increasing the effectiveness of this technology requires detailed scientific studies to optimize the design and operation modes of the combustion device [2] based on modern methods.

An experimental study of the inner aerodynamics of this vortex furnace has been carried out on a laboratory isothermal model (Fig. 2) made of 10 mm thick Plexiglas at a scale of 1:25 (internal dimensions of 290x880x730 mm). On the side walls in three tiers there are two diagonally directed nozzles at an angle of 6° (dimensions of 28x50 mm), their axes are directed to the center of the furnace. The front and rear nozzles are also installed in three tiers on the front and back walls (at the same height as the side nozzles) and directed towards the side walls at an angle of 20 °. The size of the front nozzle is 23x66 mm, the rear one is 11x64 mm.

Content from this work may be used under the terms of the Creative Commons Attribution 3.0 licence. Any further distribution of this work must maintain attribution to the author(s) and the title of the work, journal citation and DOI.
Published under licence by IOP Publishing Ltd

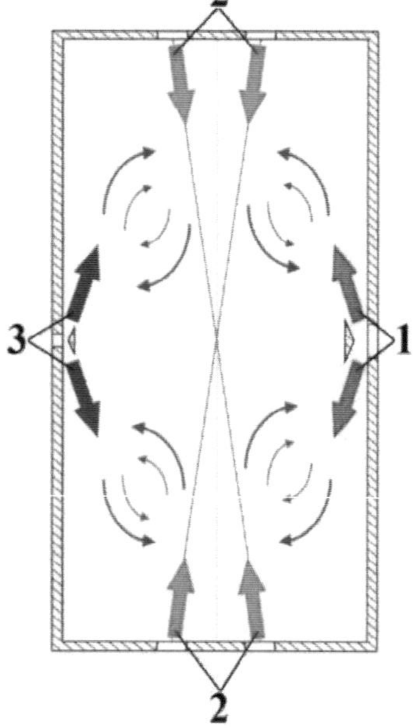

Figure 1. The scheme of flow in the furnace:
1 - front nozzles; 2 - side nozzles; 3 - rear nozzles (tertiary air).

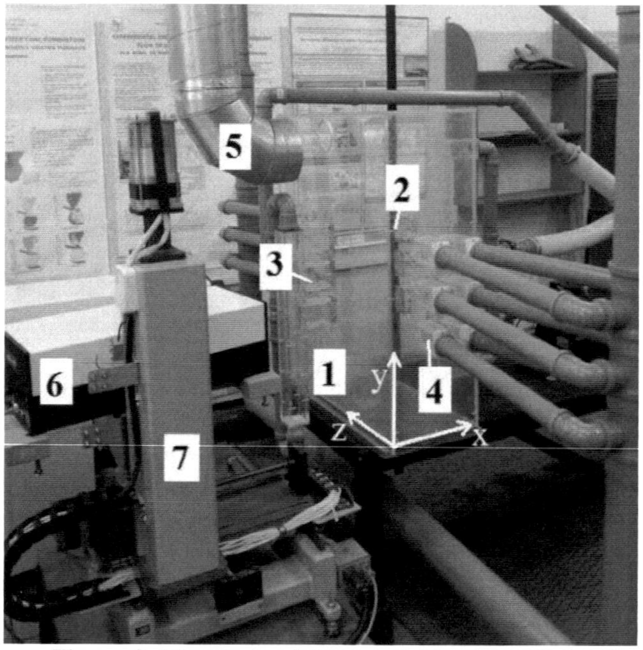

Figure 2. The photo of the experimental stand: 1 - four-vortex furnace model; 2 - front nozzles; 3 - rear nozzles; 4 - side nozzles; 5 - ventilation; 6 - measuring system LAD-056; 7 - coordinate-moving device.

The studies were carried out on an experimental stand [3], the main elements of which are (Fig. 2): an automated complex for regulating the supply of compressed air; isothermal laboratory model of a four-vortex furnace; and controlling instruments. The stand is connected to the compressed air supply and ventilation system; and it is equipped with a device for seeding the flow by tracers - microdroplets of glycerin-based liquid (Martin Magnum 1800 fog generator).

To visualize the structure of the flow, a three-component laser Doppler anemometer LAD-056 produced by IT SB RAS was used. The method is based on the measurement of displacements of particles suspended in a flow (tracers). Crossing the interference field, the particles generate an optical signal whose frequency is directly proportional to their velocity.

Measurements of the averaged velocity are carried out in a number of sections of the model at grid nodes with a spatial step of 10 mm. At each node, for at least 20 seconds, at least 1000 measurements are taken for each of the three velocity components (up to 400 samples per second). The regimes with different values of the flow velocity at the nozzle outlet are investigated. The characteristic value of the Reynolds number calculated for the length of the vortex chamber is Re> 105, i.e. the results are obtained in the Re range, which ensures the self-similarity regime of the flow [4], and is applicable to analysis of the flow structure in real-scale furnaces.

Fig. 3 shows the characteristic velocity field in the horizontal section of the model, demonstrating the flow structure with four closed vertical vortices with mutually opposite directions of rotation (the regime at which the velocity at the exit from the side and central nozzles was V0 = 5 m/s is shown). In the vertical section (Fig. 4), one can see that the flow is characterized by ascending jets.

IOP Conf. Series: Journal of Physics: Conf. Series **980** (2018) 012039 doi:10.1088/1742-6596/980/1/012039

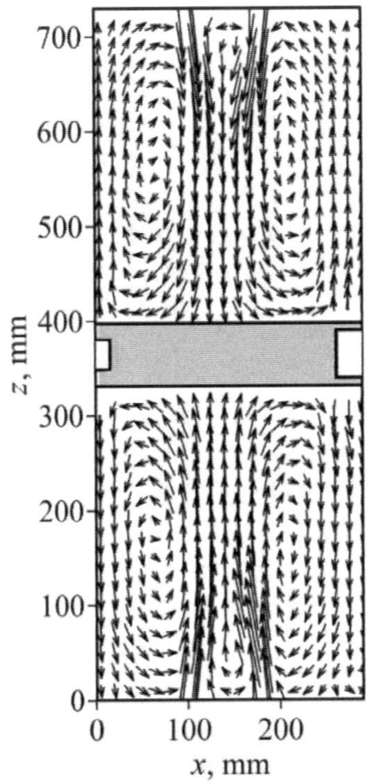

Figure 3. The field of average velocity in the horizontal section at the level of the lower tier of nozzles (top view).

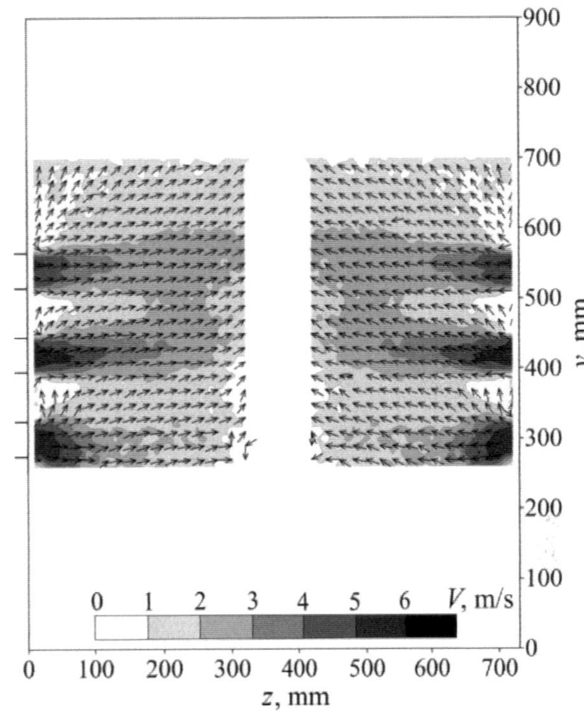

Figure 4. The field of average velocity in a vertical section passing through the centers of the side nozzles (side view).

The value of the z component of the velocity Vz changes its sign twice along the horizontal line x = 150 mm (Fig. 5), which corresponds to the location of the conjugate vortices with vertical axes (in the figures there are no velocity measurements in areas not available for optical measurements). The position of the vortex centers (Vz = 0) depends on the vertical coordinate y, which indicates the curvature of the rotation axes of the vortices. It is known that the screw-like vortex structures observed in the models of power apparatuses turn out to be nonstationary under certain conditions - it is a precession of vortices [5, 6]. In practice, this means the presence of low-frequency pulsations of pressure, which adversely affect the operation of the equipment.

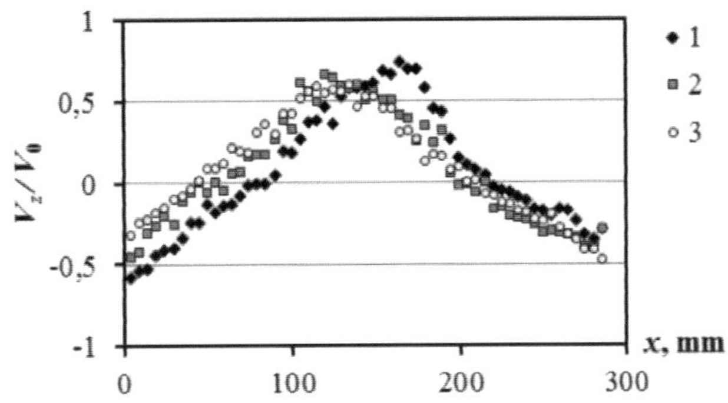

Figure 5. The distribution of the z-component of the velocity: 1 - the lower tier of the nozzles; 2 - the middle tier; 3 - the upper tier.

The results obtained on the basis of high-precision non-contact measurements allowed analyzing the spatial structure of the flow in a vortex furnace, to visualize the characteristic scheme of the averaged flow with four conjugate screw vortices. A modern methodological base for further experimental studies of the flow stability and the effect of regime parameters on the local characteristics of heat and mass transfer has been created with the aim of improving the technical and economic performance of the combustion device. The presented data are applicable for verification of mathematical models used for full-scale numerical calculations of furnace processes [7] for scientific substantiation of effective ways to control the flow structure in order to reduce toxic emissions and prevent slagging of heat exchange surfaces.

The study was carried out with financial support of the Russian Science Foundation (project No. 14-29-00093).

References

[1] Sryvkov S V, Protsailo M Ya, Dekterev A A, Kozlov S G, Pronin M S, Kovalevsky A M and Popov V P 1995 *RF Patent No. 2032853*
[2] Brikman I A 2013 *Thermophysical basis of energy technologies* (Tomsk: TPU Publ.) p 240
[3] Anufriev I S, Strizhak P A, Chernetskiy M Yu, Shadrin E Yu and Sharypov O V 2015 *Tech. Phys. Lett* **41** 727
[4] Kutateladze S S 1986 *Similarity analysis and physical models* (Novosibirsk: Nauka) (in Russian).
[5] Alekseenko S V, Kuibin P A, Okulov V L and Shtork S I 1999 *J. FluidMech.* **382** 195
[6] Gesheva E S, Litvinov I V, Shtork S I and Alekseenko S V 2014 *Thermal Engineering* **61** 649
[7] Chernetskiy M Yu, Kuznetsov V A, Dekterev A A, Abaimov N A and Ryzhkov A F 2016 *Thermophys. Aeromech.* **23** 591

High effective heterogeneous plasma vortex reactor for production of heat energy and hydrogen

N K Belov[1,2], I P Zavershinskii[1], A I Klimov[1,2], N E Molevich[1,3], D P Porfiriev[1,3] and B N Tolkunov[1,2]

[1] Samara National Research University, Samara, 443086, Russia

[2] Joint Institute for High Temperatures RAS, 125412, Moscow, Russia

[3] Lebedev Physical Institute, Samara, 443011, Russia

E-mail: ipzav63@mail.ru

Abstract. This work is a continuation of our previous studies [1-10] of physical parameters and properties of a long-lived heterogeneous plasmoid (plasma formation with erosive nanoclusters) created by combined discharge in a high-speed swirl flow. Here interaction of metal nanoclusters with hydrogen atoms is studied in a plasma vortex reactor (PVR) with argon-water steam mixture. Metal nanoclusters were created by nickel cathode's erosion at combined discharge on. Dissociated hydrogen atoms and ions were obtained in water steam by electric discharge. These hydrogen atoms and ions interacted with metal nanoclusters, which resulted in the creation of a stable plasmoid in a swirl gas flow. This plasmoid has been found to create intensive soft X-ray radiation. Plasma parameters of this plasmoid were measured by optical spectroscopy method. It has been obtained that there is a high non-equilibrium plasmoid: $T_e > T_V >> T_R$. The measured coefficient of energy performance of this plasmoid is about COP = 2÷10. This extra power release in plasmoid is supposed to be connected with internal excited electrons. The obtained experimental results have proved our suggestion.

1. Introduction

Swirl flow control and vortex control near a flying vehicle realized by non-equilibrium plasma formation is a key task in plasma aerodynamics today. Experimental results obtained during this study can be used in aviation and plasma-assisted combustion [1-11]. Physical properties of a longitudinal plasmoid created by the capacity-coupled high-frequency discharge in the high-speed swirl flow were studied in our works [1-6] in detail. Measuring power budget in this plasmoid in swirl flow is very important for aluminium-hydrogen power generation [6] and extra power production [9]. One of the most promising technologies for hydrogen production is based on PVR [6]. This work is a continuation of our previous studies [1-10].

2. Energy release inside heterogeneous plasmoid in swirl gas flow

PVR was designed and elaborated on the basis of 20 years of experimental background [1-10]. Heterogeneous metal cluster plasmoid was created by combined discharge (HF+DC) in different gas flows. High value of COP = 2-10 was obtained namely by PVR [5]. Some of the experimental results obtained using PVR are considered in this work.

Scheme of this PVR is shown in figure 1. The experimental setup was described in the previous work [5]. Plasma-vortex reactor (PVR) is made of a quartz tube with a diameter of 60 mm. A swirl gas flow is created both by a tangential gas injector and an axial gas injector in this reactor. The swirl flow

Content from this work may be used under the terms of the Creative Commons Attribution 3.0 licence. Any further distribution of this work must maintain attribution to the author(s) and the title of the work, journal citation and DOI.
Published under licence by IOP Publishing Ltd

in PVR is excited and heated by the combined discharge (HF+DC). A mass gas flow rate is measured in this experiment. The tested gases used in this set up are as follows: helium, nitrogen, argon, water steam and their mixtures. This mixture is used to obtain dissociated water steam and hydrogen atom and ions in the argon plasma and to study their interaction with metal clusters. The typical parameters of pulsed repetitive HF generator used in this work are as follows: maximal output pulsed voltage is ~ 60kV, a pulsed HF power is ~ 1÷10 kW, a HF frequency is F_{HF}~ 0.3-10 MHz, a pulse repetitive frequency is $F_{P1}=10 \div 10^4$ Hz, and pulse duration is T_p=10μs÷100ms. Typical parameters of high-current pulse repetitive discharge are as follows: a pulse current is up to 100 Amp, a pulse duration is 1-100 μs, and a pulse repetitive frequency is $F_{P2}=10 \div 10^4$ Hz. Typical parameters of the DC discharge are as follows: a current of ~1-2 Amp, and a mean power of 1-2 kW. Experimental conditions are as follows: gas mixture argon (or helium): water steam, mass gas flow rate < 10 g/sec, gas flow Mach number M < 0.2, static pressure P_{st} < 2 Bar, mean electric power input < 1 kW, mean thermal output power < 10 kW, and electrode mass erosion rate < 1 mg/sec.

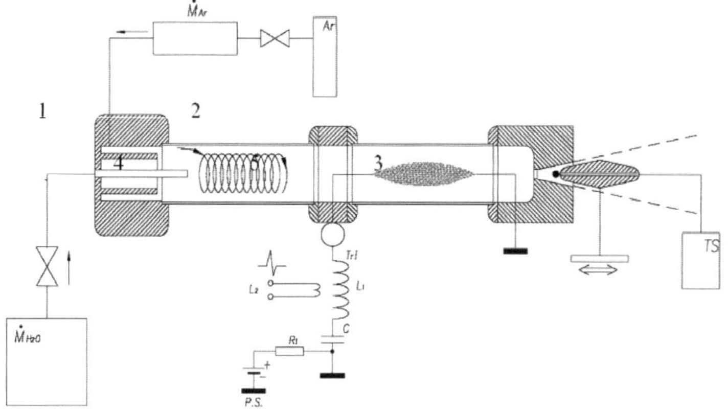

Figure 1a. Schematic of experimental set-up. 1- swirl generator, 2- argon high-pressure tank, 3-nozzle-cathode, 4- quartz glass tube, 5- anode, 6 (TS) - valve with thermocouple, M_{H2O}- water steam generator, M_{ar}- mass argon flow meter

Heterogeneous metal cluster plasmoid (HP) created by combined discharge in swirl flow is shown in figure 2. This plasmoid is revealed to consist of many metal charged micro-droplets and metal nanocluster particles (cathode electrode erosion). The typical size of these micro-droplets is about 10-100 μm. The typical size of metal nano-particles is about 10-100 nm. It is obtained that these nano-clusters absorb hydrogen atoms and ions very effectively.

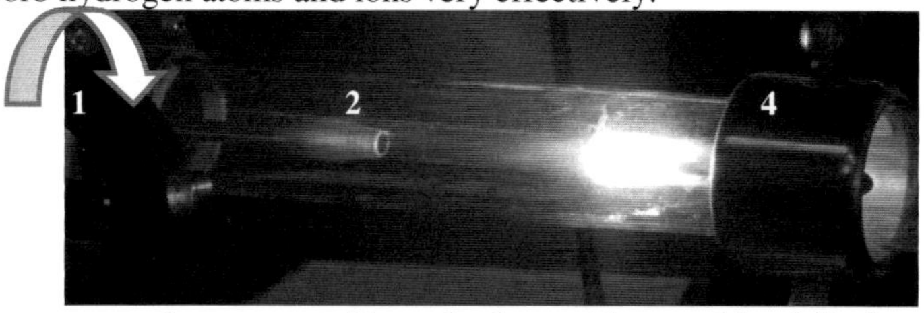

Figure 1b. Heterogeneous plasma created by pulsed repetitive combined discharge in swirl flow. Gas mixture Ar : H_20 = 10:1. Axial velocity V_x is closed tangential velocity V_t: V_x~ V_t ~ 30m/s, P_{st} ~1.5 Bar. 1-swirl generator, 2- water steam injector, 3- erosive metal clusters, 4- cathode.

Analysis of the optical spectra proves this conclusion. Measured gas temperature of this heterogeneous plasmoid is about T_g= 2000÷5000K. It is important to note that high value of COP=2÷10 was obtained in this PVR namely. The value COP = 0.5÷0.7< 1 was obtained in a straight gas flow (non-swirl flow). So, swirl flow plays an important role in obtaining high COP in PVR.

Measurement of power budget in the PVR. Calorimetric method is used to measure power budget in the PVR. The measured parameters are as follows: thermal power N_t of gas flow behind PVR's nozzle, electric power input in plasma N_e, and mechanical power of swirl flow N_f. It is well-known that there are many technical difficulties of power budget measurement in non-homogeneous swirl flows. They result from reverse swirl flow creation in PVR. So, it is important to measure the output gas flow parameters by several independent diagnostic methods.

It is important to measure power losses in this reactor (calibration regime) at the first step of any calorimetric experiment. Traditional electric heater with the same geometry and the same power as plasma formation ones is used in this experiment. The typical temperature of this electric heater is about 900C. It is close to gas temperature in argon flow at plasma on. An overall view of the experimental setup with this electric heater is shown in figure 2. The heater is arranged between two electrodes. The measured electric parameters in this experiment were: - electric current I, electric voltage U and average electric power N_e of the heater. The measured gas flow parameters were: - flow mass rate M_g and average output gas flow temperature T_g.

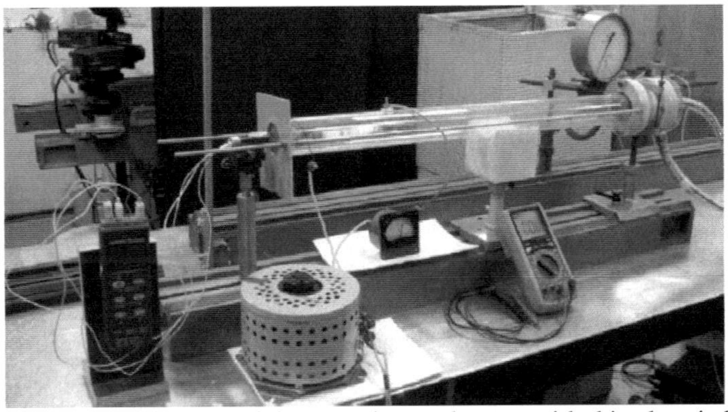

Figure 2. An overall view of the experimental setup with this electric heater.

The calorimetric results obtained in calibration regime are shown in table 1. One can see that there are large power losses in this vortex reactor of about 20-60% of electric heater power. This heat loss results from reverse flow creation in this reactor at power on and heat transfer from hot reverse gas flow to cold metal walls of this PVR.

Table 1. Calibration regime with electric heater
Swirl airflow. Initial airflow temperature- T_{g1}=23,8 C

M_g, G/s	δP_{st}, Bar	T_{g3},C	I, Amp	U, V	N_e, W	N_T, W	**K_0,%**
10.5	*0,2*	*92*	*4*	*119*	*475*	*182*	*40*
15.5	*0.4*	*104*	*4*	*117*	*467*	*373*	*80*

where T_{g3} is the final airflow temperature behind the nozzle, δP_{st} is the excess static pressure in the vortex reactor, N_e is the electric power input to plasma, N_T is the heat power of hot gas flow behind nozzle, and $K_o = N_T / N_e$

Power budget in the PVR. The typical experimental results on heat power output in the PVR at plasma on obtained by calorimetric method are shown in table.2. One can see that power output release exceeds unity: **COP=1.3÷ 1.8** (taking into account heat losses in this reactor). It is revealed that there is the value of **COP=0.8÷0.9< 1** in non-swirl airflow at the same experimental conditions. Note that maximal value of COP measured in our calorimetric experiments with PVR was about 10-12.

Table 2. Heat power of output swirl flow measured by calorimetric method

M_g, G/s	T_{g1} ,C	T_{g3}, C	δP_{st}, Bar	N_e, kW	N_T, kW	**COP**
12.5	21	80	0.4	2.63	2.97	**1.33**
20.4	21	95	0.6	3.17	4.13	**1.54**
11.4	18	100	0.4	1.56	2.87	**1.84**

where M_g, G/s is the mass gas flow rate, T_{g1} is the initial airflow temperature, T_{g2} is the final output gas flow temperature behind the nozzle, δP_{st} is the excess static pressure in vortex reactor, N_e is the electric power input to plasma, N_T is the total heat power of hot gas flow behind the nozzle, and COP= N_T/N_e.

3. Analysis of the optical spectra obtained in vortex heterogeneous plasmoid

It has been revealed that there is chemical element transmutation in cluster plasmoid. This transmutation may be connected with LENR or chemical reaction with excited internal electrons in the PVR [10-14]. Optical spectra, ion mass spectra, and soft X-spectra prove this conclusion.

The typical optical spectra form heterogeneous plasmoid obtained in the experimental setup are shown in figure 2. The tested gas mixture used in this experiment was helium+ water steam. This gas mixture is very interesting in terms of obtaining lithium **Li** and zinc **Zn** in the PVR. In fact the following transmutation (synthesis) reactions are possible according to [13,14]:

$$Ni^{28}{}_{60,\,64} + He^{2}{}_{4} = Zn^{30}{}_{64,\,68},\ He^{2}{}_{3,4} + H^{1}{}_{1,2} = Li^{3}{}_{6,7} \qquad (1)$$

So, registration and study of the optical lines of **Li** and **Zn** are very important in our experiment.
The typical optical spectra obtained in the setup with PVR are shown in figures 3, 4. One can see that there are the optical lines of *Zn, Ni, Ca, Ca II, Mg, He, Na, H, O, K* and molecular band *OH* in the spectrum in figure 3. Note that the optical lines of **Zn** in heterogeneous plasmoid can be explained by equations (1). This chemical element (**Zn**) is absent in initial *Ni* electrodes (99.99%) and reactor's wall material (quartz). Note that this element (**Zn**) is not detected/recorded in heterogeneous plasma in the gas mixture flow: argon+ water steam. Remind that it is intensive optical lines of cuprum ions *Cu II* that were registered in gas mixture: argon+ water steam (but not zincous lines), [10]. This element (cuprum) is the final chemical element in the following synthesis reaction [see Rossi A.& Co]: $Ni^{28}{}_{64} + H^{1}{}_{1} = Cu^{29}{}_{65}$. So, possible transmutation (first reaction (1)) is realized in this experiment and it explains the appearance of zincous optical lines in the spectrum, figure 3.

Processing of this optical spectrum gives the value of gas temperature T_g~T_R~ 3000K (where T_R is the rotation temperature of OH-radical) and the value of electron temperature T_e~7000-11000K. So, there is non-equilibrium plasmoid created in the setup with PVR. Typical optical spectra obtained in the heterogeneous plasmoid (helium-water steam mixture flow) at high concentration of H_2O molecules (up to 50%) are shown in figure 4. One can see that there are optical lines of *Mg I, Na I, H I,* **Li I**, *K I* and continuous spectrum connected with the large number of metal nano-clusters created by cathode's erosion in these figures. It is interesting to note that the optical **Li I** line appears in this spectrum. Note that the optical line *Li I* is recorded in rich helium-water steam mixture flow only. This line is absent in argon-water steam mixture flow. So, second reaction (2) is realized in this experiment and it explains the appearance of zincous optical lines in the spectrum, figure 3. Processing of this optical spectrum gives the value of black-body temperature T_b~T_g~ 5000K and the value of electron temperature T_e~7000K. So, there is a non-equilibrium plasmoid also created in this experiment.

4. Analysis of soft X-ray spectra obtained in vortex heterogeneous plasmoid

It was mentioned in our previous work [10] that an extra heat release in the PVR can be connected with LENR (low energy nuclear reactions) in the heterogeneous cluster plasmoid. In order to study this question in detail it is necessary to measure the X-radiation from this plasmoid [10,13,14].

The calculated wave range of this X-radiation is within 100÷10000 eV. A spectrometer X-123 is used to measure the X-radiation from the heterogeneous plasmoid in the swirl flow. The typical X-spectra are shown in the figure 5. It was revealed that the intensive X-radiation (figure 5) is created namely by the heterogeneous plasmoid. This radiation is absent in the diffuse plasma. Remember that the heterogeneous plasmoid is created by the combined discharge in the following gas mixture: Ar : water steam ~ 1 : 1. Dissociated hydrogen atoms are created by the cathode plasma in this regime (2, figure 1). Recorded optical spectra proved this conclusion. There are some maxima in the recorded X-spectra. The first main maximum is located near λ_1~ 1.3 keV. The second maximum is located at λ_2~ 4.6 keV. The X-spectra are also recorded behind the reactor nozzle. Note that an electric discharge is absent in this region but there the excited hydride metal clusters are drifted away by swirl flow. So, these excited and charged cluster particles are responsible for X-radiation generation. An intensity of X-radiation is controlled by the polarity of the erosive (or evaporated) electrode. The maximal

IOP Conf. Series: Journal of Physics: Conf. Series **980** (2018) 012040 doi:10.1088/1742-6596/980/1/012040

intensity of X-radiation is measured near the cathode electrode. A positive hydrogen ion flux (protons) bombarded its surface in this regime.

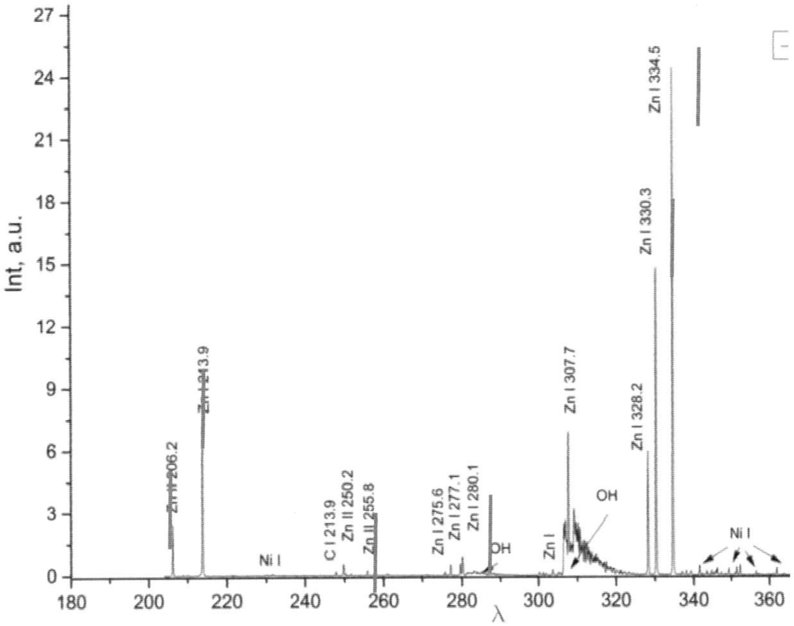

Figure 3. The typical optical spectrum of heterogeneous plasma in the test gas mixture flow: helium+water vapour, (small concentration of water steam, about 1%).

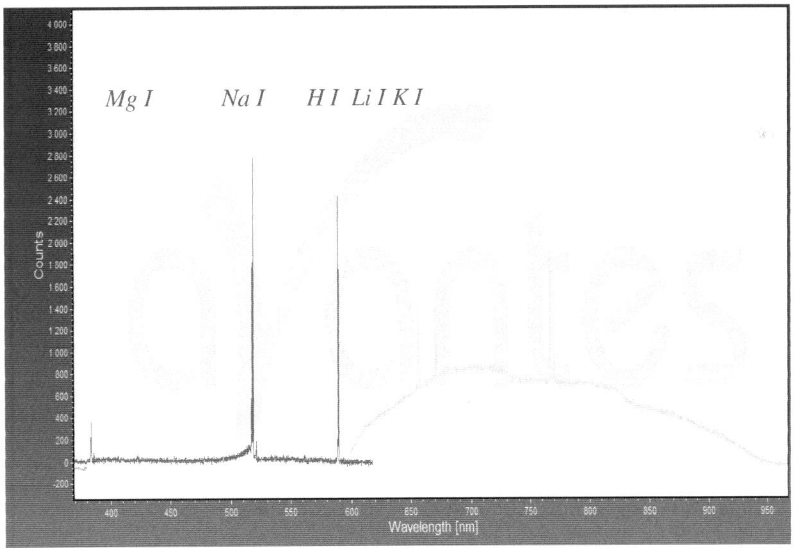

Figure 4. Typical spectrum obtained in heterogeneous helium-water steam plasma. Concentrations He : H_2O= 1:1.

The maximal extra energy release (maximal COP) is measured in the PVR at the maximal X-radiation intensity. The X-radiation is absent in the swirl plasmoid without the water steam injection at the same electric discharge and the argon flow parameters.

Summarizing all obtained experimental results one may suppose that the soft X-radiation results from metal hydride cluster creation near the erosive electrode. So, the theoretical model considered in the work [13,14] may be used to clarify the physical mechanism of extra energy release in our experiment. According to this model bi-nuclear atoms may be created by proton-heavy metal atom interaction with high extra energy release (several 100 eV/ atom).

Conclusion

Parameters of heterogeneous non-equilibrium plasma were measured in the experimental set up with PVR. An electronic temperature estimated by processing the optical spectra is about T_e~7000-10000 K. Plank's temperature estimated by the continuous cluster spectrum is about T_b~ 2000-5000 K. So, it has been revealed that there is non-equilibrium heterogeneous plasma in the swirl flow: $T_e >> T_b$. Supposedly, plasma-chemical kinetics in cluster plasmoid may be connected with extra energy release in the PVR [10, 13, 14]. It has been found that the heterogeneous non-equilibrium H_2O-Ar plasmoid ctreates soft X- radiation with quantum energy E~1-10 keV. High threshold chemical reactions with internal excited electrons in metal clusters are found to play an important role in extra energy release in the PVR or LENR.

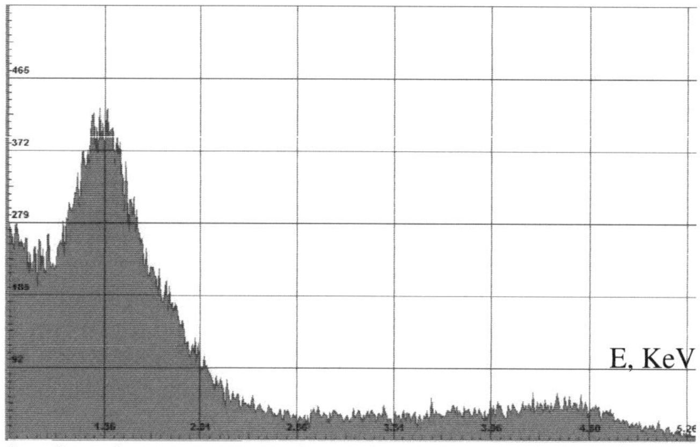

Figure 5. X-radiation from heterogeneous plasma in PVR. Combined discharge (DC+HF); mean power – 500W; hot electrode-cathode.

Acknowledgments

The study was supported in part by the Ministry of Education and Science of Russia by state assignment to educational and research institutions under Project 3.1158.2017/4.6.

References

[1] Klimov A 2013 *Vortex Plasmoids Created by High-Frequency Discharges, Atmosphere and Ionosphere: Dynamics, Processes, Monitoring* (Berlin: Springer)
[2] Klimov A 2006 *Proc. 14th Conf. on Cold Nuclear Transmutation and Ball Lightning* (Dagomys, Sept. 19-26) p 246
[3] Klimov A, Moralev I et.al. 2008 *Proc. 46th AIAA Conf.* p 11
[4] Klimov A, Moralev I et.al. 2009 *Proc. 47th AIAA Conf.* p 1
[5] Klimov A, Bityurin V, Grigorenko A et.al. 2009 *Proc. 47th AIAA Conf.* p 12
[6] Klimov A, Bitiurin V, Tolkunov B, Chinnov V, Godin S, Efimov A, Kutuzov D and Polyakov 2011 *AIAA Paper* 2011-3285
[7] Gorbunova A, Klimov A, Molevich N, Moralev I, Porfiriev D, Sugak S and Zavershinskii I 2016 Precessing vortex core in a swirling wake with heat release *International Journal of Heat and Fluid Flow* **59** 100-108
[8] Klimov A, Bityurin V 2009 Study of Plasma-Assisted Combustion of Hydrocarbon Fuel in High-Speed Airflow, Combustion Laws *AIAA Paper* 2009-411
[9] Klimov A, Grigorenko A. Efimov A et.al 2011 *J. Condensed Matter Nucl. Sci.* vol 19 no 1
[10] Klimov A., Evstigneev N., Moralev I., et.al. 2013 *AIAA Paper* 2013-1046
[11] Van Wie D, Garscadden A, Gunguly B, at.el., 2005 *J. Phys. D: Appl. Phys.* vol 38 p 33-57
[12] Widom A, Larsen L 2006 *Eur. Phys. J. C – Particles and Fields* vol 46 issue 1 pp 107-111
[13] Chalyi V P, Gurevich V L and Pogorelsky M Y 2009 *Techn. Phys.* vol 54 pp 159-164

IOP Conf. Series: Journal of Physics: Conf. Series **980** (2018) 012041 doi:10.1088/1742-6596/980/1/012041

Numerical research of the swirling supersonic gas flows in the self-vacuuming vortex tube

V T Volov[1], A S Lyaskin[2]

[1]Department of Natural sciences, Samara State Transport University, Samara, Russia
[2]Department of Computational Fluid Dynamics, CADFEM-CIS, Samara, Russia

Email: vtvolov@mail.ru

Abstract. This article presents the results of simulation for a special type of vortex tubes – self-vacuuming vortex tube (SVVT), for which extreme values of temperature separation and vacuum are realized. The main results of this study are the flow structure in the SVVT and energy loss estimations on oblique shock waves, gas friction, instant expansion and organization of vortex bundles in SVVT.

Keywords: self-vacuuming vortex tube, supersonic swirling gas flows, oblique shock wave.

Nomenclature

t – time, x_i – Cartesian coordinates, u_i – Reynolds-averaged components of instantaneous velocity vector in the Cartesian coordinate system, u_i' – pulsation components of instantaneous velocity vector in the Cartesian coordinate system, ρ – density, p – pressure, h_t – total enthalpy, T – temperature, λ – coefficient of thermal conductivity, C_p – specific heat capacity at constant pressure, μ – dynamic viscosity, $\overline{\rho u_i' u_j'}$ – components of turbulent stress tensor, k – specific turbulent kinetic energy, ω – specific rate of dissipation of k, P^* - total (stagnation) pressure, T^* - total (stagnation) temperature.

1. Introduction

Vortex flows have been of significant interest since the mid-20th century because of their occurrence in industrial applications such as furnaces, gas turbines and collectors (Gupta et al. [1]). In addition, vortex (or high degree of swirling) can also produce a hot and cold stream separation via Ranque-Hilsch vortex tube [2, 3]. Nash [4] and Dobratz [5] provided extensive reviews of vortex tube utilizations and enhancements. Intense experimental and numerical studies of Ranque-Hilsch tubes have begun since 1931 and continue even nowadays [6, 7, 8]. The most famous scientific school studying the Ranque-Hilsch effect in Russia was led by Professor A.P. Merkulov [9, 10, 11]. In addition to traditional applications of vortex tubes, this school proposed a number of new utilizations in aviation, chemical industry, power industry, agriculture, etc. Special merit of this scientific school is the creation of vortex electro-discharged devices: CO_2-lasers and plasmatrons [12, 13]. A significant number of theories have been proposed to explain the vortex effect of «temperature separation» since its initial observations by Ranque [2]. Despite all proposed theories, none has been able to fully explain the one. This fact is partially compensated by the numerical models of flow in vortex tubes. Numerical studies of swirling gas flows in the vortex tube were presented by following researches [6,

Content from this work may be used under the terms of the Creative Commons Attribution 3.0 licence. Any further distribution of this work must maintain attribution to the author(s) and the title of the work, journal citation and DOI.
Published under licence by IOP Publishing Ltd

7, 8]. Simulations were made using turbulent models ranging from simple RANS (as standard k-**ε**) [6] to large eddy simulation (LES) [7]. At the same time the least developed type of vortex tubes in theoretical and experimental aspects is the self-vacuuming vortex tube (SVVT), where extreme values of temperature separation (ΔT_{cold}=152*K*, gas - air [1]) and vacuum ($\pi^*=P^*/P_{axis}$=40, where P^*- total gas pressure at the inlet of SVVT, P_{axis}- gas pressure near the axis of SVVT) are realized. In contrast to classic vortex tubes [2, 3] SVVT have neither hot nor cold flow exits. Instead, SVVT is used for its internal body cooling. There are some engineering models for calculation of flows and thermodynamic performances of SVVT [13, 14]. Though they allowed creating a new class of electro-discharged devices (CO_2-lasers and plasmatrons [12]), still there are no theoretical studies of the flow behaviour in SVVT, based on modern computational fluid dynamics (CFD) simulation techniques. The present paper is devoted to the CFD study of strongly swirling gas flows in SVVT based on the RSM turbulence model.

2. Schematics of the problem geometry

Schematics of the problem geometry is shown in figure 1. The basic dimensions of SVVT are the following: tube's radius and height are R_t = 5 *mm*, and H_t = 11.5 *mm*, respectively. Central body radius, fillet radius, diffusor radius and diffusor height are R_{cb} = 0.5 *mm*, R_f = 1.5 *mm*, R_d = 25 *mm*, H_d = 1 mm, respectively. The width of inlet is a = 2.1 *mm,* and its height is b = 1.4 *mm*.

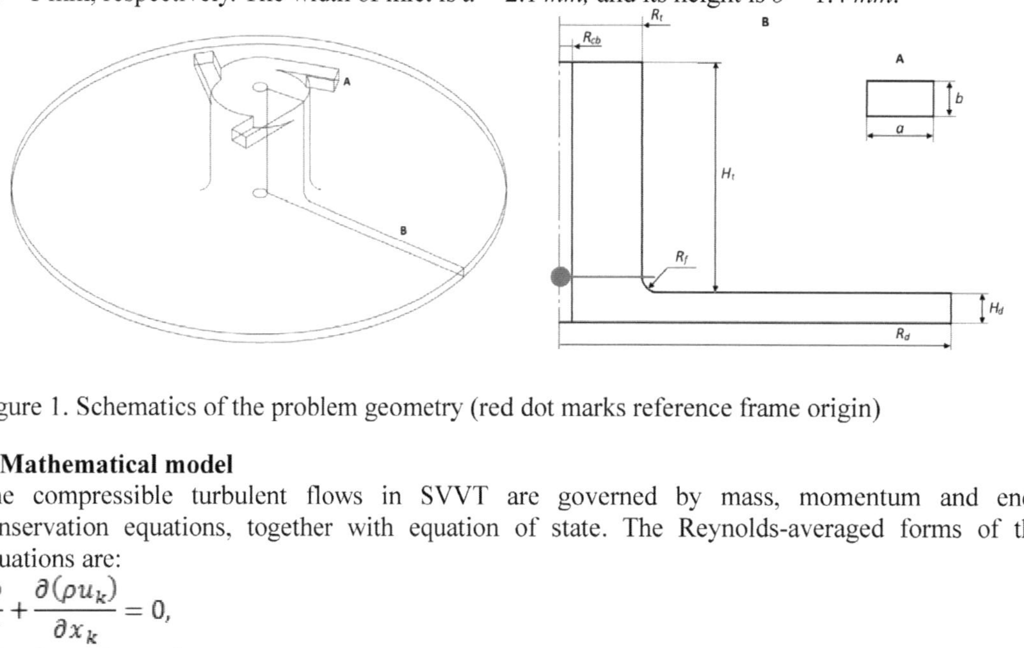

Figure 1. Schematics of the problem geometry (red dot marks reference frame origin)

3. Mathematical model

The compressible turbulent flows in SVVT are governed by mass, momentum and energy conservation equations, together with equation of state. The Reynolds-averaged forms of these equations are:

$$\frac{\partial\rho}{\partial t}+\frac{\partial(\rho u_k)}{\partial x_k}=0,$$

$$\frac{\partial(\rho u_i)}{\partial t}+\frac{\partial(\rho u_i u_k)}{\partial x_k}=-\frac{\partial p}{\partial x_i}+\frac{\partial}{\partial x_k}\left(\tau_{ki}^{eff}\right),$$

$$\frac{\partial(\rho h_t)}{\partial t}-\frac{\partial p}{\partial t}+\frac{\partial(\rho u_k h_t)}{\partial x_k}=\frac{\partial}{\partial x_k}\left[\lambda\left(\frac{\partial T}{\partial x_k}\right)\right]+\frac{\partial}{\partial x_k}\left(u_i\tau_{ki}^{eff}\right),$$

$$\rho=\frac{Wp}{R_0 T},$$

where $h_t=h+\frac{1}{2}u_k u_k$ is the total enthalpy, $h=C_pT$ is the static enthalpy, $\tau_{ij}^{eff}=\tau_{ij}-\rho\overline{u_i'u_j'}$ is the effective stress tensor, $\tau_{ij}=\mu\left(\frac{\partial u_i}{\partial x_j}+\frac{\partial u_j}{\partial x_i}\right)-\frac{2}{3}\frac{\partial u_i}{\partial x_j}\delta_{ij}$ is the viscous stress tensor, W is the molar mass, and $R_0=8.314472\ m^2\cdot kg/(s^2\cdot K\cdot mol)$ is the universal gas constant.

In this study, the flow was considered fully turbulent. For the description of turbulence, Omega Reynolds Stress model was used [14]. It comprises six transport equations for turbulent stresses as well as an additional transport equation for ω, which is similar to the equation of *k*-ω model.

4. Numerical model

For the numerical solution of governing equations, ANSYS CFX commercial CFD software was used. It utilizes a finite volume method. The simulation was carried out with the following boundary conditions: specified values of total (stagnation) pressure and total (stagnation) temperature at inlets, with flow direction at inlets normal to the boundaries; specified values of pressure and temperature at the exit. These values were interpreted as values of static parameters at those parts of boundary where the flow leaves the simulation domain and as values of total parameters at those parts of boundary where the flow enters the simulation domain in the case of the reversed flow. All the walls of SVVT were considered adiabatic. The inlet parameters varied: P^* from 0.15 *MPa* to 0.7 *MPa*, and T^* from 500*K* to 5000*K*. The gas was helium, with $W = 4.0$ *g/mol*, $C_p = 5193.0$ *J/(kg·K)*. Transport properties λ and μ were considered as functions of temperature according to [15].

The calculation was carried out in a pseudo-transient formulation with fixed pseudo-time step equal to 10^{-7} s. Convergence criteria, besides the standard ones based on average normalised residuals, were the global imbalances of mass and energy (the threshold value 0.1%) and constancy of mass flow rate.

The initial computational mesh featured $4.3 \cdot 10^6$ elements ($1.5 \cdot 10^6$ nodes) with the following adaptation, based on pressure gradient, up to $11.3 \cdot 10^6$ elements ($3.0 \cdot 10^6$ nodes).

5. Results and discussion

Simulations were carried out for different cases with different values of the inlet pressure and temperature. Figures 2-7 show the results for T^*=500*K* and P^*=0.3 *MPa* (which was the main case).

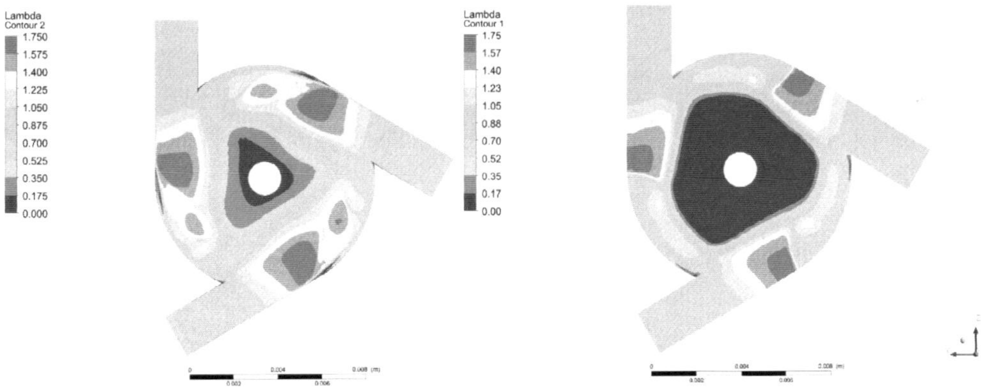

a) Viscous flow a)

nviscid flow

Figure 2. Comparison of viscous and inviscid simulation results (velocity coefficient in the inlet plane)

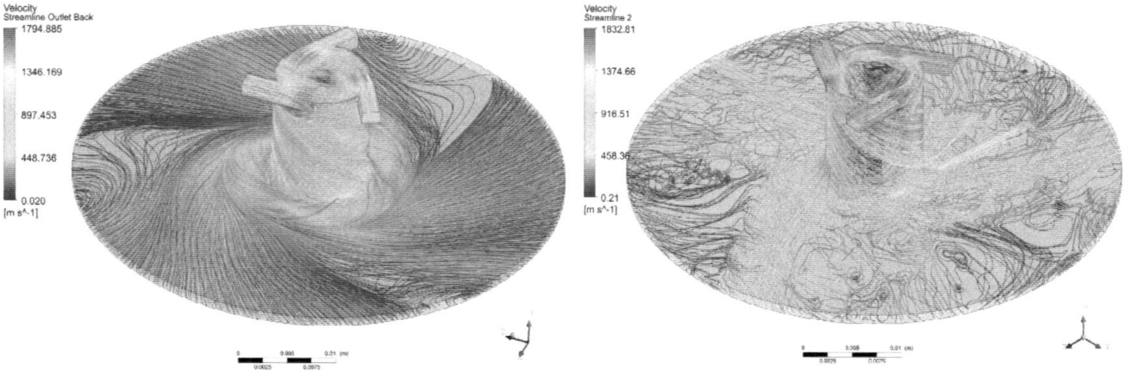

Viscous flow Inviscid flow

Figure 3. Comparison of viscous and inviscid simulation results (streamlines colored with local velocity)

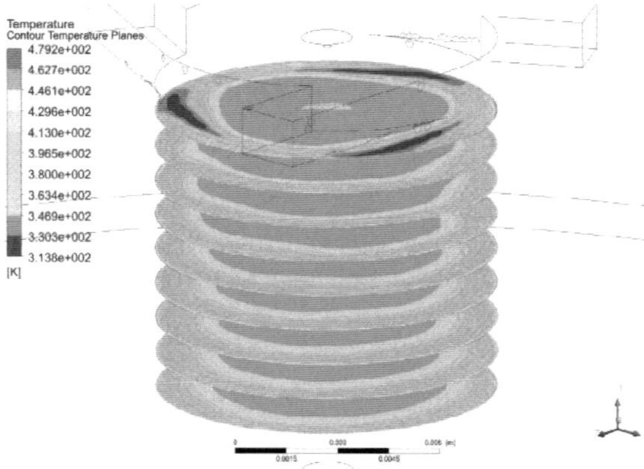

Figure 4. Temperature

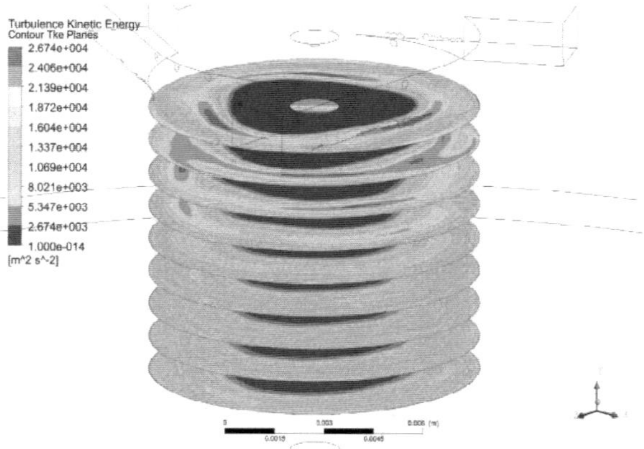

Figure 5. Specific turbulent kinetic energy

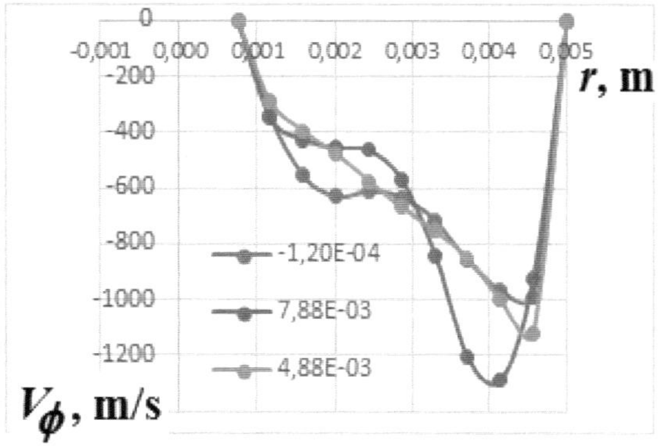

Figure 6. Distribution of circumferentially averaged azimuthal velocity vs radius in three horizontal sections (section position listed in the legend)

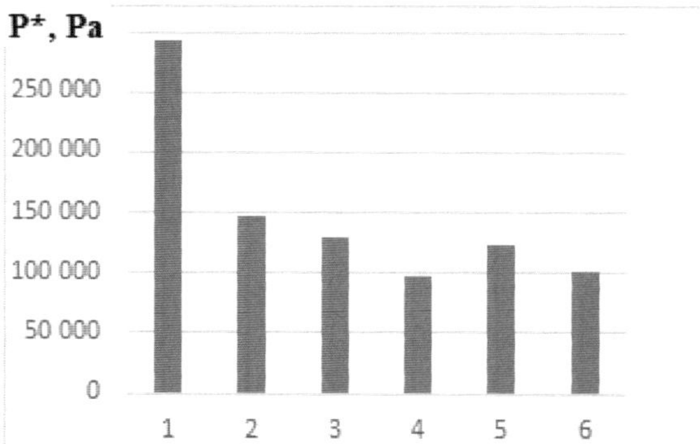

Figure 7. Total pressure in the SVVT: 1 – inlet, 2 – horizontal section at the lower edge of the nozzle, 3 – horizontal section 1 mm down from the lower edge of nozzle, 4 – lowermost horizontal section of the main tube, 5 – annular section at the diffuser inlet, 6 - exit

For the main case, a comparative analysis of viscous and inviscid (i.e. $\mu = 0$) flows was made (figures 2 and 3). Figure 2 shows that in the nozzle section of SVVT for both viscous and inviscid gas the flow entering the main tube from the nozzle accelerates to supersonic and then decelerates to subsonic velocity. For the viscous flow, the deceleration occurs in a system of three oblique shocks, while for the inviscid flow it occurs in a single normal shock. This leads to a significant decrease of the total pressure losses in viscous flow comparing to inviscid one. However, the sum of losses in case of viscous flow is slightly larger: the resulting mass flow rate was about $1.5 \cdot 10^{-4}$ *kg/s* for the viscous flow and about $1.6 \cdot 10^{-4}$ *kg/s* for the inviscid. Figure 3 shows the difference in the general flow pattern. The "jets" produced by nozzles in the viscous flow case exhibit internal secondary swirl due to interaction, which is not observed for the inviscid flow. There is now a toroidal vortex at the inlet into the diffusor in case of inviscid flow. In both cases, there are zones of reversed flow in the diffusor, but for the inviscid flow these zones are much more extensive.

Note that the flow in SVVT has a rather complex character: tangential inlets form the "jets" in which the gas first is accelerated to supersonic velocities and then decelerated in the system of oblique shocks. Besides, the interaction between the "jets" causes secondary swirling. In the end, the large peripheral vortex in the main part of SVVT consists of three (by the number of inlets) vortex cores. In spite of the fillet in the area of transition from the main part of tube to the diffusor, flow separation occurs, resulting in the toroidal vortex with double rotation. All these secondary vortex structures are significant sources of power losses, together with shocks. Unfortunately, these losses are difficult to assess explicitly.

Detailed study of simulated flow fields have showed some unexpected results: local increase of total pressure and total temperature (up to 5%) were observed. It may be considered as an error of the specific computational model. However, in [16] it has been suggested that these local disturbances can take place not due to errors of numerical simulation but because of the action of friction forces. The local increase of total thermodynamic parameters can be explained by interaction between swirled "jets" (vortex cores), because the observed increase of total parameters along the streamline belonging to one vortex core corresponds to the decrease of these parameters along the streamlines belonging to the other vortex cores. Note that in spite of somewhat questionable local variation, integrally all conservation laws are satisfied.

6. Conclusion

For the first time flow pattern in SVVT was studied with state-of-the-art CFD techniques. It has been found that this pattern is rather complex: it features both subsonic and supersonic regions, systems of oblique shocks, primary and secondary vortices with strong interaction. An attempt was made to estimate the sources of power losses. It has been found that only a small amount of losses could be attributed to friction and turbulence (from 2% to 10%, depending on inlet conditions, with losses due to turbulence making less than 1% for all cases). It is assumed that most of the losses occur due to shocks and secondary swirl generation.

7. References

[1] Gupta A K, Lilley D G and Syred N 1984 *Swirl Flows* (Turnbridge Wells: Abacus Press) p 475
[2] Ranque G J 1933 Experiences sur la détente giratoire avec simultanées d'air chaud et d'un enchappement d'air froid *J. Phys. Radium* **4**(7) 112–114
[3] Hilsch R 1947 The use of expansion of gases in a centrifugal field as a cooling process *Rev. Sci. Instrum* **18**(2) 108–113
[4] Nash J M 1972 The Ranque–Hilsch vortex tube and its application to spacecraft environmental control systems *Dev. Theor. Appl. Mech.* **6** 35–90
[5] Dobratz B M 1964 *Vortex Tubes: a Bibliography* (Livermore: University of California) p 68
[6] Eiamsa-ard S and Promvonge P 2007 Numerical investigation of the thermal separation in a Ranque–Hilsch vortex tube. *Int. J. of Heat and Mass Transf.* **50** 821–832
[7] Farouk T and Farouk B 2007 Large eddy simulations of the flow field and temperature *Int. J. of Heat and Mass Transf.* **50** 4724–35
[8] Gao C M, Bosschaart K J, Zeegers J C H and de Weyl A T A M 2005 Experimental separation on a simple Ranque-Hilsch vortex tube *Cryogenics* **45** 173–187

[9] Merkulov A P 1969 *Vortex Effect and its Applications in Engineering* (Moscow: Mashinostrojenie) p 187
[10] Merkulov A P 1964 Hypothesis of the interaction of vortices *Energetic* **3** 32–37
[11] Merkulov A P and Kolishev N D 1969 Investigation of the vortex tube with diffusor. *Proc. of Kuibyshev aviation institute* **37** 35–51
[12] Volov V T 1983 Method of the vortex diffusor device *Inzhenerno-physicheskiy zhurnal* **1** 35–42
[13] Volov V T 2006 *Thermodynamic and heat and mass transfer of the strongly twitted gas flows in the energy devices and apparatuses* (Samara: Samara scientific center of RAS) p 316
[14] Wilcox D C 2006 *Turbulence Modeling for CFD* (La Canada: DCW Industries, Inc.) p 522
[15] Green D W and Perry R H 2008 *Perry's Chemical Engineers' Handbook* vol 2 ed Poling B E, Thomson G H et al. (New-York: McGraw-Hill Companies, Inc.) p 517
[16] Yan X, Li J, Song L and Feng Z 2009 Investigations on the Discharge and Total Temperature Increase Characteristics of the Labyrinth Seals with Honeycomb and Smooth Lands *J. Turbomach.* **131**(4) 041009-1–8

Modeling of atomization and distribution of drop-liquid fuel in unsteady swirling flows in a combustion chamber and free space

A A Sviridenkov[1] P D Toktaliev[1] V V Tretyakov[1]

[1]Central Institute of Aviation Motors named after P.I. Baranov, Moscow 111116, Aviamotornaya 2, Russia

E-mail: tretjak@ciam.ru

Abstract. Numerical and experimental research of atomization and propagation of drop-liquid phase in swirling flow behind the frontal device of combustion chamber was performed. Numerical procedure was based on steady and unsteady Reynolds equations solution. It's shown that better agreement with experimental data could be obtained with unsteady approach. Fractional time step method was implemented to solve Reynolds equations. Models of primary and secondary breakup of liquid fuel jet in swirling flows are formulated and tested. Typical mean sizes of fuel droplets for base operational regime of swirling device and combustion chamber were calculated. Comparison of main features of internal swirling flow in combustion chamber with unbounded swirling flow was made.

1. Introduction

Operating process in combustion chamber has multidisciplinary nature, it includes gas dynamic, phase transition, reaction kinetics and others subjects. Correct numerical modeling of combustion process should reproduce physical features of the flow and should be based on verified models in each of these fields of study. Among them, atomization and propagation of drop-liquid fuel in turbulent flow, typically swirling flow.

The aim of this work is to study by simulation and experiment unsteady swirling flows in the combustion chambers of gas turbine engines and the effect of nonstationarity of the flow on the atomization and distribution of fuel in the primary volume of the combustion chamber. The object of these studies is the swirling frontal device of the model combustion chamber with a gas-dynamic stabilizer mounted at the inlet [1]. The scheme of device is shown at figure 1. Main air enters into primary volume of combustion chamber through blade swirler by annular gap with height h_0=20 mm, additional air is injected by transverse jets at the end of cylindrical stabilizer, fuel is supplied in section between swirler and additional air holes.

Content from this work may be used under the terms of the Creative Commons Attribution 3.0 licence. Any further distribution of this work must maintain attribution to the author(s) and the title of the work, journal citation and DOI.
Published under licence by IOP Publishing Ltd

IOP Conf. Series: Journal of Physics: Conf. Series **980** (2018) 012042 doi:10.1088/1742-6596/980/1/012042

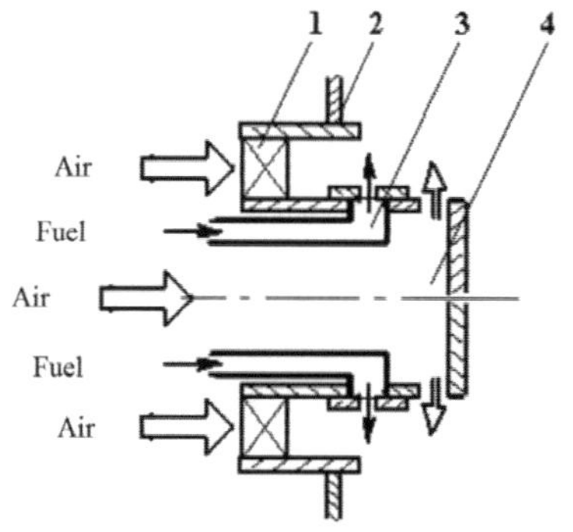

Figure 1. Scheme of a gas-dynamic stabilizer with transverse fuel supply. Numbers: 1 - axial swirler, 2 - frontal device, 3 - channel for feeding liquid fuel, 4 - transverse air jets channels.

2. Numerical simulation

The method for calculating velocity fields is associated with modeling the unsteady nature of airflow and fuel droplets. First, the air velocity fields are calculated from the steady-state Reynolds equations solution with imposed k-ε turbulence model at effective Reynolds numbers of the air flow Re_{eff}> 300 (based on h - annular gap of main channel). In this range, the flow under consideration, as can be shown both from calculations and experiments, is unstable, and repeated patterns of velocity distributions occur at different stages of the iterative calculation process. Random samples of these structures are accepted as instantaneous fields of gas flow velocity, and the series of these instantaneous fields is a model of unsteady flow. This procedure is similar to described in [2]. At the second stage, instantaneous concentration fields corresponding to the instantaneous velocity fields are calculated. And, finally, in the third stage, the instantaneous concentration fields are averaged, resulting in concentration fields that are accepted as averaged. The averaging of the instantaneous velocity and concentration fields is realized without weights. Detailed comparison of the calculated and experimental profiles of velocities can be found in [1] and it shows acceptable agreement. An example of the calculated axial velocity field in the longitudinal section of the combustion chamber with gas dynamic stabilizer obtained using the described approach for regime with swirl intensity S_w = 1.0 (blades angle 45°) and dimensionless transverse air jets velocity V_j= 1.0 (relative to u_0=20 m/s) are shown at figure 2.

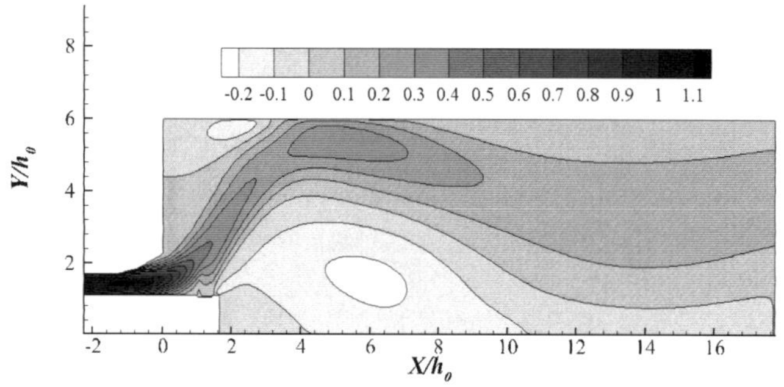

Figure 2. The averaged axial velocity field in the longitudinal section of the numerical domain for S_w = 1.0, V_j = 1.0.

The obtained velocity field used for modeling the propagation of a fuel jet into the swirling air flow and for modeling fuel jet disintegration into droplets. This simulation includes three stages: determining the shape of the liquid jet by solving the equations of mass and momentum conservation in a curvilinear coordinate system associated with the jet surface, using the primary disintegration model of the fuel jet, using the model of further fuel fragments disintegration and droplets formation.

IOP Conf. Series: Journal of Physics: Conf. Series **980** (2018) 012042 doi:10.1088/1742-6596/980/1/012042

The modeling of disintegration processes, details see [3], is carried out within the framework of the linear stability theory. It is assumed that the primary disintegration of the jet is caused by its instability in the field of aerodynamic forces of the air flow. In this case perturbations acting on the jet lead to the formation of waves on its surface, the amplitude of which increases and when a certain "critical" value is reached, the jet breaks into fragments. Similarly, the decomposition of fragments and droplets is occurred. We note that the technique for calculating the formation and breakup of liquid jets, described in [2], makes it possible to describe the shape of the jet in the flow of atomizing air, to determine the place of its breakup, and to obtain data on the droplet size in the spray cone. In the cases under consideration, the disintegration of the jet occurs at a distance of $3d$ to $4d$ (d - injection holes diameter) from the nozzle edge, and the average spray angle is 30° to the axis of the stabilizer. The average median sizes of the formed droplets are for the regime with $S_w = 0.6$, $V_j = 0.6$ $D_m = 40$ - 45 μm, and for regime with $S_w = 1.0$, $V_j = 1.0$, $D_m = 29$ μm. These data are the basis for numerical calculations of droplet motion and fuel distribution in the working volume of the combustion chamber. Note that the size of droplets formed in the flow for the second regime is in good agreement with the experimental data [1]: $D_{m,\ exp} = 30$ μm.

The method for calculating the fields of droplet fuel concentrations is based on models of the breakup of fuel jets onto droplets, droplets fragmentation in the air flow and analysis of the motion of individual droplets. It is assumed that the droplets sizes distribution obeys the Rozin-Rammler law with exponent $n = 3$: $F=1\text{-}\exp[-(D/D_m)^n]$, and the droplets distribution in the circumferential (along the angle φ) and longitudinal (along the angle ψ) directions - the normal law. Here F is the relative fraction of droplets whose diameter is less than D, D_m is the average median diameter of the droplets. The initial release velocities V_0 of the droplets are given by the velocity module V_0 and the values of the longitudinal and circumferential angles: φ_k and ψ_k. The initial coordinates of the entry points of droplets are determined from the solution of the problem of the breakup of the fuel jet injected into the air stream, the release angles - from experimental atomization data in the free space, the initial droplets temperature is assumed equal to 300 K. The mean median diameter of the droplets in the spray cone in the calculations varied depending on the regime from 30 to 50 μm, the fuel consumption is assumed equal to $G_f = 2.5$ g/s. Fuel - kerosene TS-1.

Figure 3 represents the results of the calculations of the average distributions of the mass concentrations of droplets obtained for regime $S_w = 1.0$, $V_j = 1.0$. The mass concentration of droplets C_k is referred to the air density, fuel injection was carried out in the radial direction.

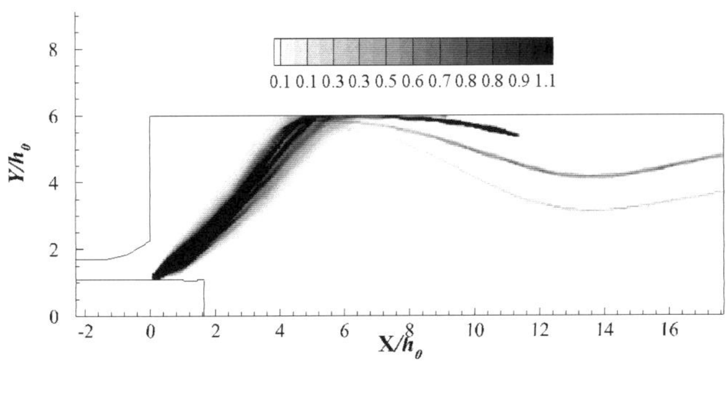

Figure 3. Average distribution of fuel concentration C_k in the longitudinal section of combustion chamber. $S_w = 1.0$, $V_j = 1.0$.

From the presented results it can be seen that significant part of the fuel falls on the outer wall of the chamber ($Y/h_0 = 0.6$). Analogous results were obtained for other operational regimes with sufficient high V_j ($V_j > 0.2$). We note that this phenomenon is determined exclusively by the unsteady nature of the flow. Similar results were obtained for the distribution of vapor fuel. The importance of taking into account transport processes nonstationarity is also evidenced by a comparison of the distributions of droplets fuel concentrations obtained in calculations with steady and unsteady

IOP Conf. Series: Journal of Physics: Conf. Series **980** (2018) 012042 doi:10.1088/1742-6596/980/1/012042

approach. Figure 4 and figure 5 show a comparisons of the droplets concentrations numerical profiles with experimental data for steady and unsteady calculations. It can be seen that the results of unsteady calculations are consistent with the experimental data in the entire flow region, while steady results are only in the internal region. Thus, the calculation from the steady model shows that the fuel drops in the cross section at *d* = 30 mm (here *d* is the distance counted from the end of the stabilizer, section correspond to dimensionless axial coordinate X/h_0=3.2 at figure 2, origin of Cartesian coordinate system placed at device symmetry axis and coincided in axial direction with frontal device outer vertical wall, see 2 at figure 1) are completely absent at radial distances greater than 20 mm, and in the section *d* = 50 mm at distances greater than 30 mm. At the same time, the calculation with the unsteady model demonstrates complete agreement with the experimental distributions. Thus, the presence of droplets in the vicinity of the walls of the flame tube of the combustion chamber observed in experiments can be explained only with unsteady flow model. It should be noted, that for the presented direct comparison of the results of experiments and calculations, a special method for processing the calculated concentration distributions has been developed: droplets mass concentrations were recalculated into volume one and local fuel distributions were averaged along the beam of the laser knife, in accordance with the experimental procedure. In addition, to investigate the process of breakup of fuel jets into droplets, a special technique was developed, based on visualization of the fuel cone in pulsed light. The experiments showed that the fuel jet in the considered regimes performs rotational-vibrational motions with an amplitude of spatial oscillations approximately equal to the output section of the swirler. The visually observed shape of the jet in non-pulsed lighting represents only the averaged distribution of droplets in a fuel-air cone.

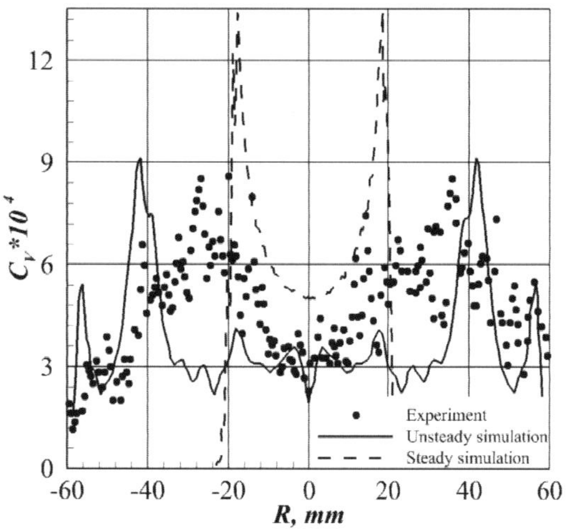

Figure 4. Average distribution of fuel concentration in the cross sections of the spray cone, *X*=30 mm, S_w = 0.6, V_j = 0.6. Steady, unsteady numerical models, experimental data.

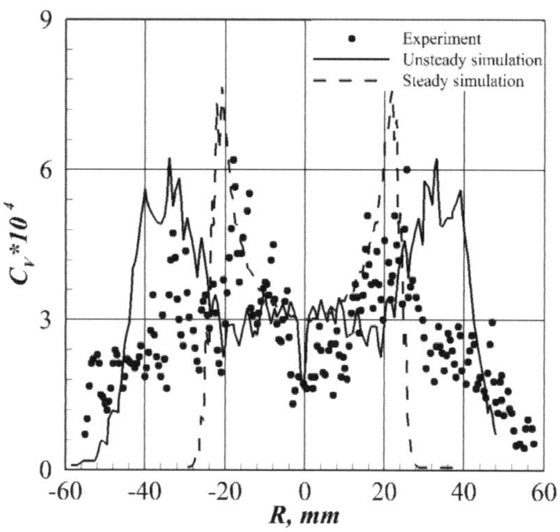

Figure 5. Average distribution of fuel concentration in the cross sections of the spray cone, *X*=30 mm, S_w = 1.0, V_j = 1.0. Steady, unsteady numerical models, experimental data.

In this paper, the study of the characteristics of the mixture formation is also carried out for a fixed air flow regime: S_w = 1.0, V_j = 1.0 with azimuthal fuel injection angle variation - φ_k. Calculations were carried out for three variants of fuel supply angles: φ_k = 0, ± 45°. It has been found that the degree of expansion of the spraying cone is substantially dependent on the direction of fuel jets and the swirl of the main air flow. The average size of the droplets formed in the flow also depends on these parameters. Thus, at φ_k = + 45° (in the direction of swirl of the main air flow), the spraying cone considerably narrows, and at φ_k = -45° (opposite to the direction of swirl of the main air flow) remains practically unchanged. In addition, in the case of φ_k = + 45°, there is no significant change in the droplets size, while at the injection angles of 45° there is a reduction in the droplet sizes from 50 to 30-40 μm.

3. Experimental measurements

For verification of jet breakup and droplets atomization mathematical model and testing of pressure atomizer performance experimental research under atmospheric conditions was performed. Experimental rig consist of air-fuel feed system, cylindrical operational region with mounted atomizer, diagnostic equipment and optically transparent window, exhaust section. High-speed video recording can be done for flow visualization, characteristics of disperse phase can be measured by small-angle laser scattering technique. Velocity field for single phase flow can be extracted by particle image velocimetry (PIV). Details of testing facility and measurement technique can be found in [4-6].

It should be emphasized that the presence of unsteady flow could principally change the characteristics of fuel jet breakup. Thus, measurements of the spraying cone characteristics showed that three maxima with droplet sizes in the range 20 ÷ 100 μm are distinctly observed in the concentration distribution near the axis (see figure 4, 5). Note that this result is obtained for free space jet. As for the combustion chamber, comparison between numerical and experimental data is complicated due to droplets deposition on the walls of the combustion chamber. It's difficult to organize local mass concentration measurements in the internal combustion chamber flow. Therefore, conclusions about the features of fuel distribution in the combustion chamber can be made easier by numerical calculations. It should be noted that the presence of flow boundaries significantly affects not only the hydrodynamics of the flow, but also all the characteristics of the atomization of the fuel. Experiments with fuel jet in air flow were performed to find main features of internal two-phase flow behind the frontal device of combustion chamber. Figure 6 shows photographs of the fuel jet injected into the swirling air flow in the combustion chamber.

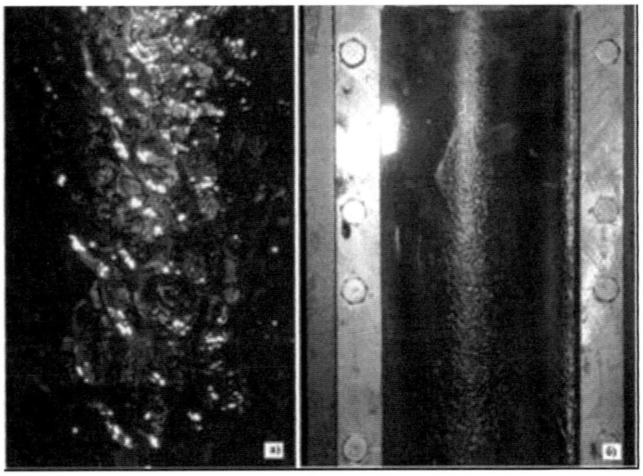

Figure 6. Photos of the fragment of the jet (a) and the whole jet (b) in the combustion chamber. ΔP_a = 3 kPa, G_f = 2.5 g/s. Flash lighting.

It can be seen from the figures that the presence of flow boundaries substantially changes the nature of the propagation of the jet and the mechanism of its breakup. Thus, in the breakup of a jet inside a chamber, waves formed on its surface are caused by the Rayleigh instability. The amplitude of these waves when interacting with a swirling air stream increases and the jet disintegrates into individual drops. At the same time, the jet does not make noticeable oscillations in the chamber, which is apparently due to the absence of significant large-scale precession-type motion in the combustion chamber.

Thus, the structure of the jet behind the swirler in the combustion chamber is radically different from its structure when it releases into free space. In the space confined to the walls of the chamber, the fuel jet does not make significant fluctuations, which are observed with similar fuel and air mass flow rates in the free space.

4. Conclusions

As can be seen from numerical and experimental research, distributions of drop-liquid phase in swirling flow in combustion chamber for typical regime conditions have essentially unsteady nature. This internal swirling flow is more receptive to perturbations than free space one. Some features of these flow, such as fuel droplets deposition on flame tube wall, can't be modeled in numerical simulation with simple steady approach. Additionally, special treatment need to be incorporated in numerical procedure to describe two-way coupling between continuum and discrete phases. Numerical studies of presented scheme of combustion chamber with gas-dynamic stabilizer show, that Sauter mean diameter of fuel droplets for wide range of dimensionless parameters S_w and V_j not exceed 50 µm for cold flow. Experimental research of liquid fuel atomization, which was held for verification of proposed mathematical models, show satisfactory agreement with numerical data.

Acknowledgments

Research has been supported by the RFBR grant No. 17-01-00213.

References

[1] Tretyakov V V, Sviridenkov A A and Toktaliev P D 2017 *Russian Aeronautics* **3** 106-12
[2] Armsfield S and Street R 1999 *Journal of Computational Physics* **153** 660-5
[3] Tretyakov V V and Sviridenkov A A 2016 *Herald of the Samara State University. Aerospace technique, technology and machinery* **15** (4) 162-73
[4] Tretyakov V V and Sviridenkov A A 2011 *Proc. Int. Conf. Heat and Mass Transfer and Hydrodynamics in Swirling Flows* vol 1 (October Moscow)
[5] Tretyakov V V 2006 *Herald of SSAU* **2** (10) 136-41
[6] Krasheninnikov S Yu, Mironov A K, Toktaliev P D, Zakharov D L and Maslov V P 2014 *Fluid Dyn.* **49** (1) 51–62

IOP Conf. Series: Journal of Physics: Conf. Series **980** (2018) 012043 doi:10.1088/1742-6596/980/1/012043

LDA Diagnostics of velocity fields inside the Ranque tube

I K Kabardin[1,2], V G Meledin[1,2], N I Yavorsky[1,2], M R Gordienko[1,2], M Kh Pravdina[1], D V Kulikov[1] and V I Polyakova[1], V A Pavlov[1]

[1] Kutateladze Institute of Thermophysics, SB RAS, Novosibirsk, 630090, Russia

[2] Novosibirsk State University, Novosibirsk, 630090, Russia

E-mail: kabardin@itp.nsc.ru

Abstract. Longitudinal and tangential components of the mean velocity field inside the Ranque tube of square cross section are presented for a central longitudinal section. Measurements are carried at central plane over the entire tube, with 1-2 millimeter step in transverse direction and with 10 millimeter step along the tube. The data is presented in the form of velocity maps.

1.Introduction

The experimental investigation of Ranque tube is carried out by means of LDA method. The nonintrusive diagnostics in the Ranque tube has currently demonstrated certain success. In [1] the large-scale vortex structure of double spiral form in the field of optical phase density is for the first time visualized by means of Hilbert optics. In [2] the same flow is diagnosed by scanning of a part of the central longitudinal plane section, using the LDA method with flow fumigation. In [3] a three-dimensional diagnostics is realized for water droplets in the nitrogen flow in the Ranque tube, and the data with good time resolution is obtained for velocity and size of particles. In [4] the mean velocity field is presented in the central plane of the tube, obtained by means of planar filtered Rayleigh scattering on gas molecules. In [3-4] the tube under investigation had the circle cross section, thus requiring the special correction of optical data out of the central plane. In [1-2] for the convenience of optical diagnostics all along the flow, the tube with square cross section, the so called square tube, was used. This particular tube is the subject of present research, also since it demonstrates large-scale hydrodynamic structure in the flow, observed by Hilbert optics in [1]. The influence of these structures on temperature separation has not been investigated yet. The careful comparison of circle and square tubes was made in [5] in the region of incoming overpressure from 1 to 6 Bar, showing a distinct Ranque effect of temperature separation in the square tube at nearly the same total flow rate, though being about 1.5 times less than that in a circular one with the tube diameter equal to the square side. In the present work the authors continue the investigation of the square tube [1-2, 5] in order to fit the LDA method for obtaining the two-dimensional mean velocity field all over the central plane as a first step. The more detailed data including on incoming pressure variation, measurements out of the center line and three-dimensional diagnostics are provided.

2. The experimental setup.

The actual Ranque tube (Figure 1) includes two-slit swirl apparatus, accelerating the vortex chamber with a planar top cover at one side and a profiled diaphragm at the other side, as well as the working square channel, adjusted to the diaphragm orifice. The cold exit from the tube is made as a cylindrical

Content from this work may be used under the terms of the Creative Commons Attribution 3.0 licence. Any further distribution of this work must maintain attribution to the author(s) and the title of the work, journal citation and DOI.
Published under licence by IOP Publishing Ltd

channel, passing through the swirler cover, and the hot exit is made as a radial diffuser. Air from the hot exit flows in the buffer volume that has an exit to the atmosphere, regulated by the valve.

The compressed air enters the vortex chamber through two tangential slits with total area of 40 mm^2 at the radius R_1=65 mm. The diaphragm is hyperbolically shaped that keeps the velocity circulation almost constant up to the entrance into the working channel of the tube. The radius of exit to the channel is R_2=17 mm. Thus, the tangential velocity at the entrance of the tube is higher than that at the slit exit. The channel length is 450 mm. Thermal insulation is absent. Sections of the square tube are made of duralumin; each section has two optical glass windows on opposite sides.

The cold fraction $\mu=G_c/G_{in}$ is regulated by the hot exit valve. The gap size between the disks of radial diffuser is 1.5 mm.

The cold exit channel is 10 mm in diameter. The cross section of the working channel is a square with the side of 34 mm.

The flow rates at the entrance and the exits were measured by calibrated [5] orifice flowmeters. The cold fraction μ, was kept equal to 0.24 at the incoming overpressure P being equal to 1 Bar and the incoming flow rate G_{in} equal to 0.02 kg/s. The temperature difference between hot and cold exits was 7K.

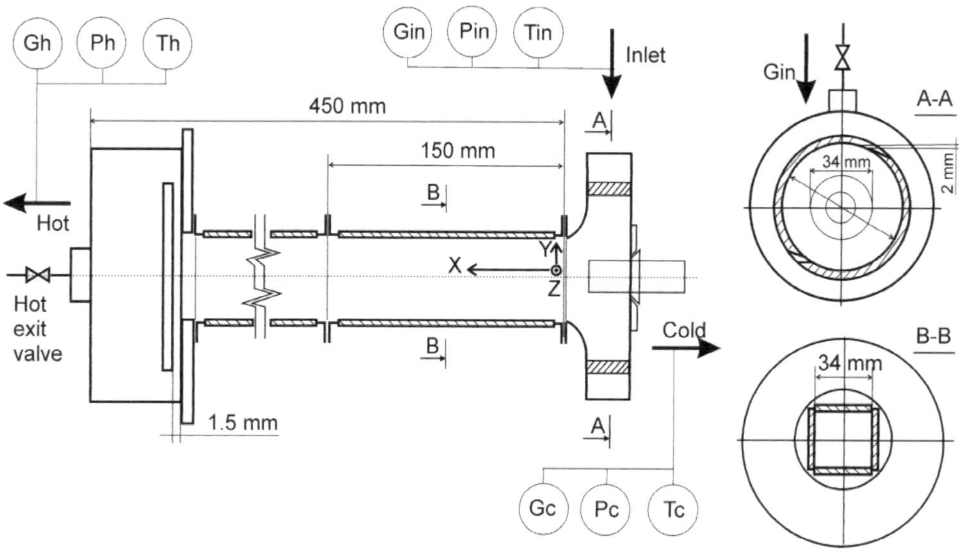

Figure 1. Vortex chamber geometry and coordinate system

The experimental flow kinematics inside the Ranque tube is studied using a laser Doppler anemometer (LDA) with an adaptive temporal selection of the velocity vector (LAD-05). The measuring device was designed and manufactured at the Kutateladze Institute of Thermophysics SB RAS, in Novosibirsk. The diagrams illustrating the LAD-05 complex operation principle and the LAD-05 photo are shown in Figure. 2. (a, b). The optical unit is mounted on a coordinate spacer (positioning device). The laser-optic velocity measuring circuit contains a laser, two orthogonally oriented acousto-optic modulators with traveling waves, a transmitting-receiving optical system, consisting of a series of mirrors and lenses arranged at the optical unit, and a photomultiplier tube (PMT). The photomultiplier tube transmits the signal to the quadroture mixer. The quadroture mixer passes the signal to the Doppler signal processor, which transmits the processed data to the computer. The test volume is scanned by moving the optical unit by a coordinate spacer (positioning device). LAD-05 operates as follows. After passing through the acousto-optical modulators the laser beam is diffracted into the zero and minus first orders. The split beams pass through a series of optical elements of the system and then are sent to the tested area of the flow by the lenses. Intersecting in that area, laser beams form an interference pattern with the known space-time periodic structure. The particle moving through the intersection area scatters the light. The photocurrent appears at the output of the detector, producing a frequency pulse in the radio range with the Doppler frequency shift, which is the known linear function of particle velocity. The signal duration is equal to the time of particle passage through the volume of intersection. The LAD-05 measures two projections of the velocity vector in the range of 0.001-300 m/s with a relative error of no more than 0.5%. The size of the intersection area is 0.1x0.1x0.5 (mm). The coordinate spacer moves the measurement unit in the area of 250 x 250 x 250 (mm) with an accuracy of 0.1 mm.

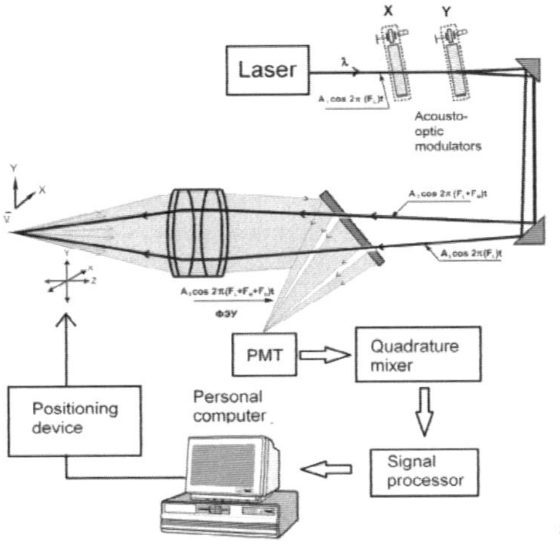

Figure 2.a. Diagram illustrating the LAD-05 complex operation principle.

Figure 2.b. Photo of the LAD-05 measuring complex.

3. The experimental results.

Velocity components V_x and V_y are measured in the central OZX plane, as shown in Figure 3. In these measurements, the module of V_y component coincides with tangential velocity.

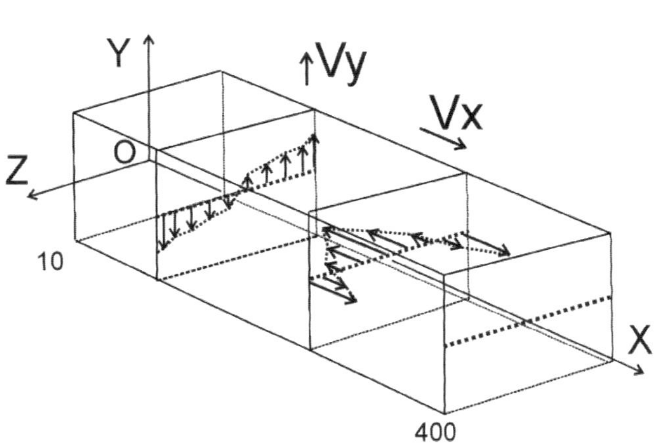

Figure 3. Illustrated measurement of two velocity components in channel cross section (several profiles).

The test points in each cross section were taken with the interval of 1-2 millimeters in Z direction. And the intervals between profiles in X direction were 10 millimeters. In each test point averaging was made on 300 particles for 3-15 seconds.

The results are presented in Figures 4, 5 in the form of velocity level maps.

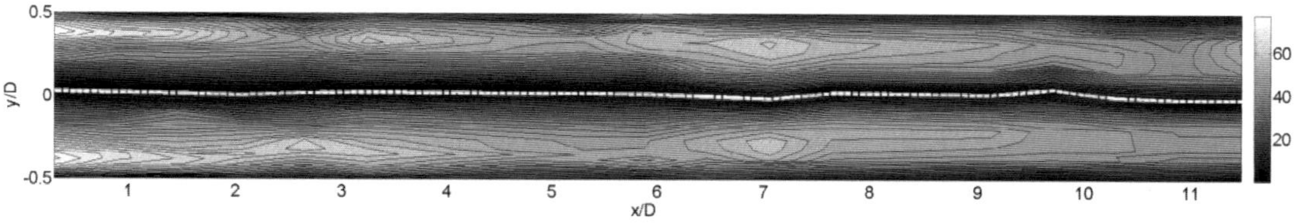

Figure 4. Map of module V_y (tangential velocity, m/s). The white dashed line denotes zero tangential velocity in the center.

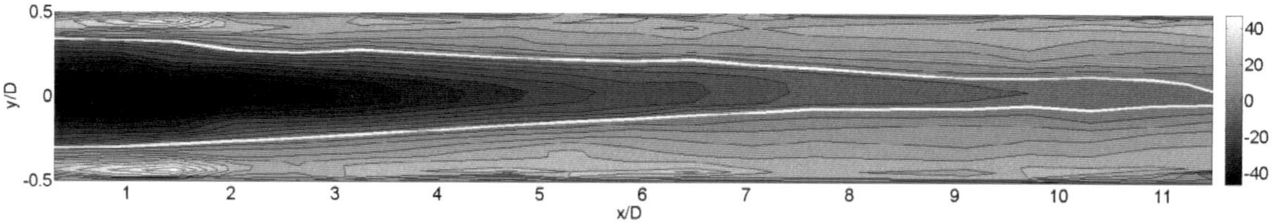

Figure 5. Map of V_x (m/s) with the white lines denoting zero longitudinal velocity and the boundary of the reverse flow area.

Conclusion

The longitudinal and tangential velocities were measured in the cross section of the square Ranque tube at one flow regime (overpressure of 1 Bar, incoming flow rate of 0.02 kg/s, cold flow fraction $\mu = 0.24$, the temperature difference in two exits of 7 K [5]). The maximum of tangential velocity was observed at a distance of one square side caliber from the inlet plane. The boundary of the reverse flow is distinctly shown in Figure 5 by white zero level lines for longitudinal velocity. The reverse flow area reaches the hot exit plane. No circulation areas were observed in the mean flow. The investigation of a variety of flow regimes should be continued. In particular, the three dimensional velocity fields should be obtained by using two LDA-meters simultaneously. A time development of velocity in certain points and PIV investigation of flow patterns will serve to verify and prove the presence of the large scale structures in the flow. The pressure pulsations will show if the rise and development of the large scale structures influence the mean characteristics and temperature separation.

Acknowledgements

Research was supported by the Russian Foundation for Basic Research RFBR-16-08-01120A.

References

[1] Arbuzov V A, Dubnishev Y N, Lebedev AV, Pravdina M H and Yavorsky N I 1997 *Technical physics letters* **23**(12) 938-940

[2] Dubnishchev Y N, Meledin V G, Pavlov V A and Yavorsky N I 2003 *Thermophysics and Aeromechanics* **10**(4) 587–598

[3] Doll U, Burow E , Beversdorff M, Stockhausen G, Willert C, Morsbach C, Schlüß D and Franke M 2017 *Proceeding of 9-th International Symposium on Turbulence and Shear Flow Phenomena* **1** 110-117
[4] Liew R, Zeegers J C H, Kuerten J G M, Michalek W R 2013 *Experiments in Fluids* **54** 1416-1432
[5] Kabardin I K, Meledin VG, Yavorsky N I, Pavlov V A, Pravdina M Kh , Kulikov D V and Polyakova V I 2017 *MATEC Web of Conferences* **115** 02022
[6] Raffel M, Willert C and Kompenhans J 2001 *Particle Imaging Velocimetry* (Berlin: Springer-Verlag)

IOP Conf. Series: Journal of Physics: Conf. Series **980** (2018) 012044 doi:10.1088/1742-6596/980/1/012044

Acoustic relaxation of the hydro-mechanical system under critical expiration of swirl flow

I G Pozdeeva, O V Mitrofanova

National Research Nuclear University MEPhI, Moscow 115409,
31 Kashirskoye sh.,Russia

e-mail: pozdeeva_irina@mail.ru

Abstract. The mechanism of generation of acoustic oscillations associated with the formation of stable vortex structures in the moving fluid was considered for the impact swirl flow. Experimental studies were carried out to determine the relationship between large-scale vortex motion and acoustic effects in hydro-mechanical systems. It was shown that a sharp change of the amplitude-frequency characteristic of the acoustic oscillations of hydro-mechanical system corresponds to the maximal flow rate of the swirl flow. The established connection between the generation of sound waves and geometrical and regime parameters of the hydro-mechanical system formed the basis for the developed method of diagnostics of the processes of vortex formation.

1. Introduction

Formation of vortex structures in the flow is connected with presence of viscosity in any real fluid. The presence of internal friction determines the fact that any change of channel geometry, e.g., contraction, expansion, bending, etc. leads to vortex formation. A detailed analysis of these effects is given in [1]. The main objective of the proposed methodology is the early identification of conditions of vortex generation to avoid dangerous operation modes of power units.

A self-regulatory effect of acoustic oscillations was revealed in a series of experiments, conducted to substantiate the developed diagnostic technique. The effect of self-regulation is expressed in a resonant amplification of the amplitude of natural frequencies of the hydro-mechanical system due to the absorption component of the spectrum of acoustic vibrations, generated by the vortex flow structure. In elements of thermal power equipment of nuclear power systems such effect can lead to a sharp increase in vibration and, consequently, to destruction of constructions. The comparison of amplitude-frequency characteristics of acoustic oscillations with visualization of the vortex structure of flows and numerical simulations allowed identifying the topological features of the impact swirl flow [2]. The connection of the internal vortex structure of the flow with the effect of the occurrence of acoustic resonances has been identified, using the acoustic method proposed in [3] and based on the measurement of frequency characteristics of sound waves in impact swirl jet. Theoretical analysis, using the approximation of acoustic flows [3] and the theory of screw flows [1], and comparison of experimental and calculated results are a justification of the proposed physical model of the flow, predicting the occurrence of acoustic resonances, caused by the topology of the vortex flow.

Content from this work may be used under the terms of the Creative Commons Attribution 3.0 licence. Any further distribution of this work must maintain attribution to the author(s) and the title of the work, journal citation and DOI.
Published under licence by IOP Publishing Ltd

IOP Conf. Series: Journal of Physics: Conf. Series **980** (2018) 012044 doi:10.1088/1742-6596/980/1/012044

2. Correlation of acoustic oscillations with vortex structure of the flow

The results of the preliminary studies, given in [3-5], allow correlating the generation process of vortex structures to the recorded acoustic oscillations with frequencies, concentrated mostly in the sound range. Detailed description of the experimental setup used in present studies was given in [2, 4].

The mechanism of the effect of self-regulation may be explained as follows: at the condition of the maximum flow rate of the working fluid, which corresponds to the condition of critical flow expiration, the flow obtains a clear vortex structure. The swirled jet, exiting the orifice of the vortex chamber in the region of twist, and the radial spread of the flow over the edge of the hole are divided into thin spiral vortices, which is confirmed by the picture of the wake vortex visualization (Figure 1g). In this case, the energy flux is sufficient to generate the vortex structure of the flow, characterized by two rotation frequencies (Figure 1): frequency f_1 = 63.5 Hz around the z axis and frequency f_2 = 273 Hz around the axis φ (Figure 1d).

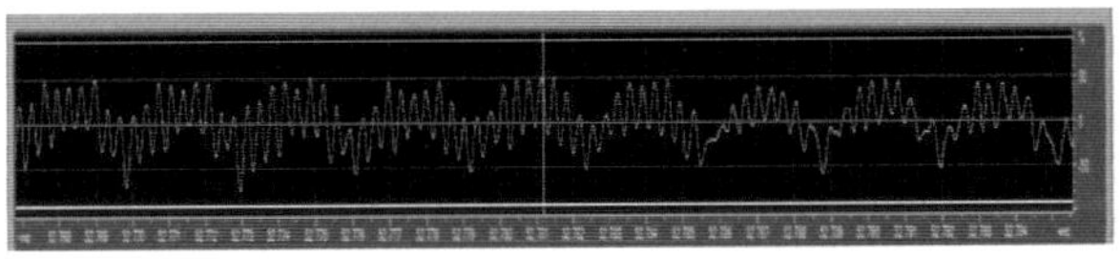

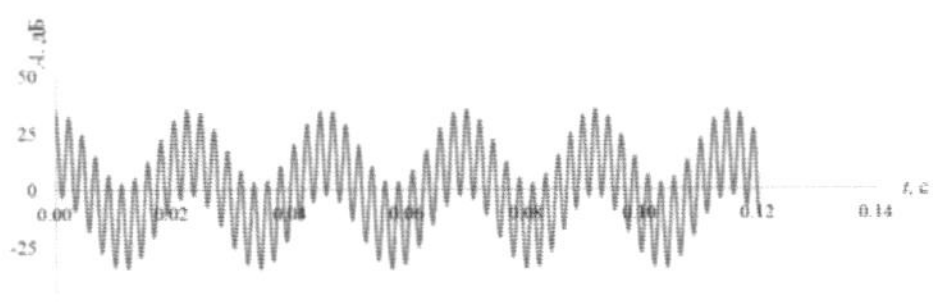

a) b)

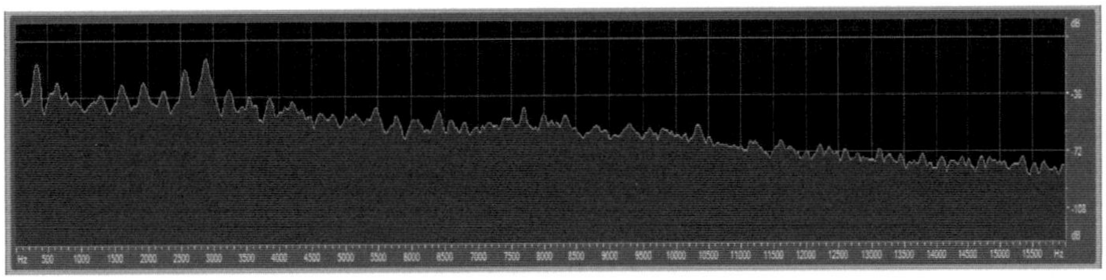

c)

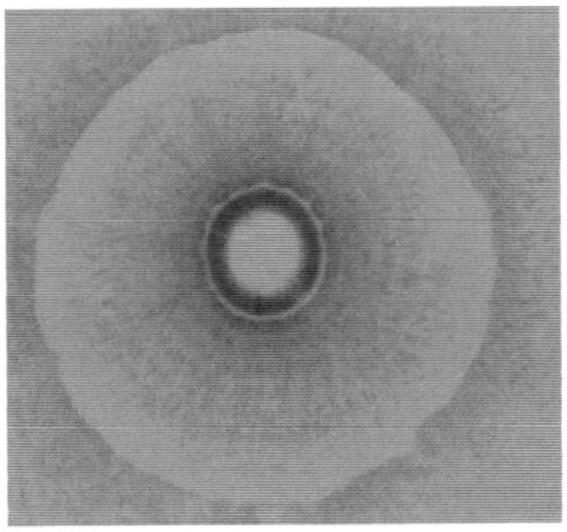

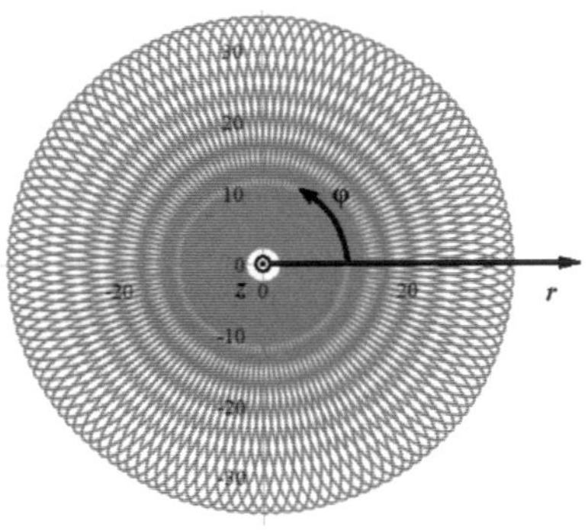

d) e)

Figure 1. Sub-critical expiration regime for the obstacle with diameter D = 70 mm at the output hole diameter d_0 = 5 mm and the flow rate $G = 1.15 \cdot 10^{-3}$ m^3/s: a) the time scan of the acoustic wave; b) the result of summing the harmonic oscillations with frequencies f_1 = 63.5 Hz and f_2 = 273 Hz; c) the amplitude-frequency characteristics; d) picture of the wake vortex on the lower surface of the obstacles; e) a Lissajous figure with frequency f_1 = 63.5 Hz and f_2 = 273 Hz in the cylindrical coordinate system (r, φ, z).

If the energy of individual vortex is not sufficient, then, reaching the edge of the obstacle and encountering the resistance of the atmospheric pressure of the external environment, a spiral vortex does not leave the system and returns to the center of the obstacle where it is again picked up by the effluent stream. Since the output flow is a swirled, jet, breaking into thin spiral vortices spreading in the radial direction, displaces on a certain angle with every turn of the stream.

Frequency f_1 is the harmonic oscillation frequency of rotation of the group of thin spiral vortices around the axis of the vortex chamber. The flow may go through several oscillatory cycles until its energy is sufficient for passing into the environment. The experimental diagram (Figure1a) shows the corresponding 1 smooth increase and decrease of sound pressure level. The resistance of atmospheric pressure depends on the area of the outlet section of the slit. So the number of cycles, for which energy necessary for the flow exit from under obstacle is accumulated, depends on the size of the obstacle D. In this case, the frequency f_2 must also depend on the diameter of the obstacles. The resulting vortex system tends to instability and rebuilding. Sources of additional energy are vibrations of structural elements of the vortex chamber.

The vortex chamber, from which the expiration air flow occurs, consists of a rigid metal construction with glued upper plexiglass surface with the central hole for the output of the swirl air flow. The bottom of the vortex chamber is a solid single surface, and the upper end surface of the chamber with a central hole of diameter d_0 is removable, that allows varying the diameter and the shape of the outlet hole edge.

In the considered case of excited self-oscillations at the swirl air outflow from the vortex chamber, its upper end surface is subjected to an asymmetric force impact that can cause bending vibration at a frequency f_{11}, which can be estimated by the formulas for a circular plate in accordance with the recommendations of [6]:

$$f_{11} = 2.09 \cdot f_{01}, \tag{1}$$

$$f_{01} = 0.9342 \frac{b}{2R^2} \sqrt{\frac{E}{3\rho(1-\nu^2)}}, \tag{2}$$

where R is the radius of the plate, b is the plate thickness, and E, ρ, ν are the Young's modulus, density, and Poisson's ratio of plate material, respectively.

In this case, (as it was illustrated in figure V. 2.1. in the work [6]), the fixed circumference is the edge of the round plate, and the nodal diameter is the line, connecting the region with occurring compensation of flow rate of inlet and outlet air flow under the obstacle. According to formulae (1, 2) frequency of bending vibration of plexiglass plate with radius R = 45 mm and thickness b = 5 mm is f_{11} = 2824, when values of $E = 41 \cdot 10^8$ Pa, ν = 0.394, and $\rho = 1.18 \cdot 10^3$ kg/m^3 Hz, which well agrees with the experimental values of the frequency f_3 and corresponds to the resonance condition. In a series of experiments with the upper end surface of the vortex chamber, which was made of duralumin alloy, a resonance effect was not observed, because the natural frequency of a duralumin surface was significantly higher and was outside the range of oscillations, generated by vortex structures.

Emerging from the vortex chamber, the flow of high kinetic energy excites vibrations of elastic surface, which are accompanied by fluctuations of the air volume under the obstacle. The coincidence of the frequency of one of the overtones of air vibrations with the natural frequency of the upper end surface of the camera f_{11} causes a resonance. This condition is necessary to form a thin structure of longitudinal helical vortices with frequency f_3 = 2796 Hz and rotation around the axis r (Figure 2d). The output flow leaves the space under the obstacle during the time of one cycle of rotation, not returning to the center of the obstacle. The visualization for this case is presented in Figure 2g. Figure 2a illustrates the corresponding experimental dependence of acoustic pressure on time, showing that under the resonance mode of the outflow the accumulation and discharge of energy takes place in one oscillation cycle.

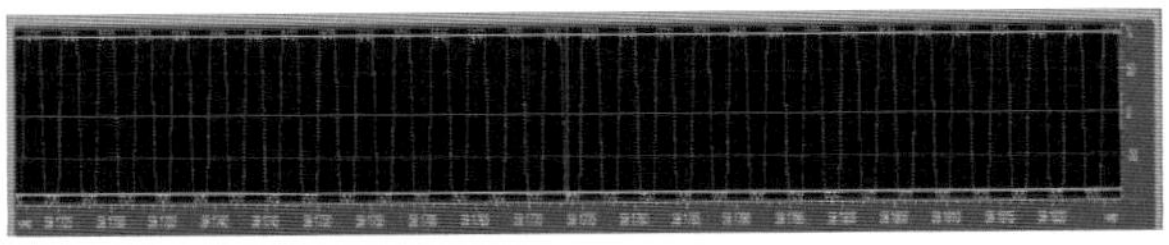

a)

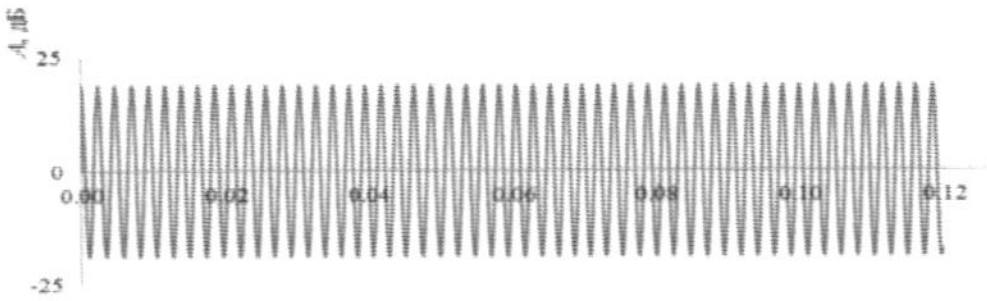

b)

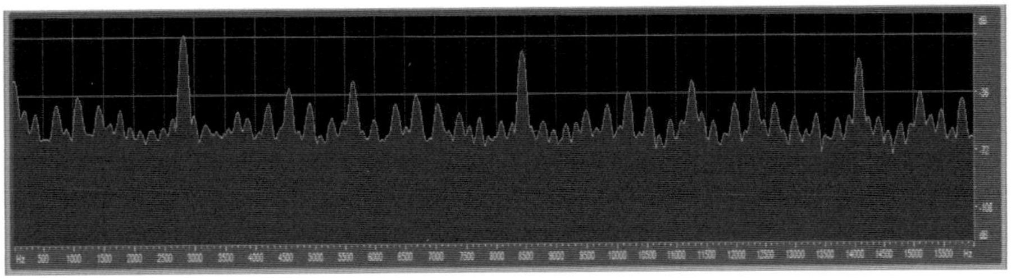

c)

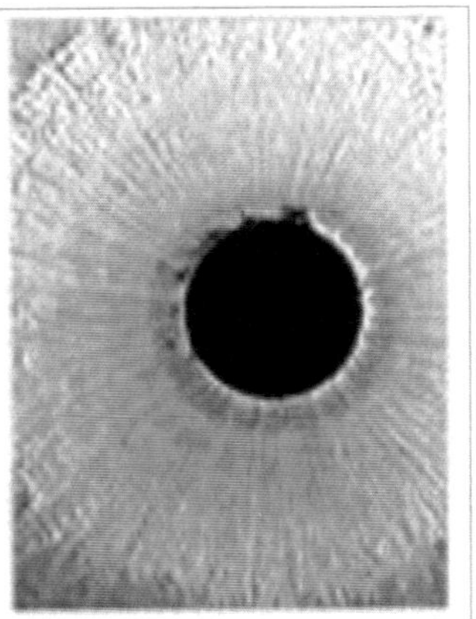

d)

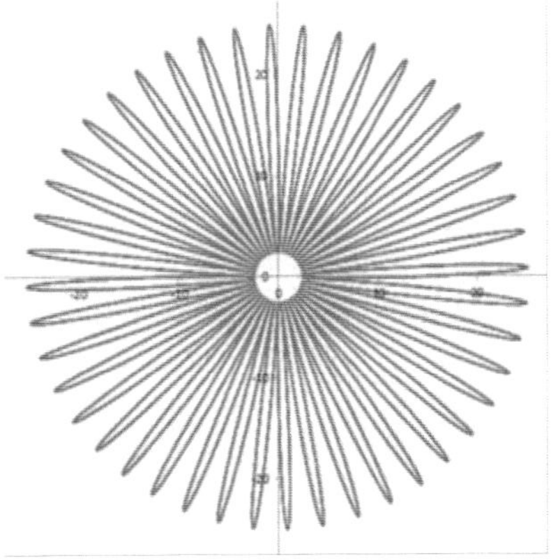

e)

Figure 2. Critical outflow regime for the obstacle with diameter $D = 70$ mm at output hole diameter $d_0 = 5$ mm and flow rate $G = 1.15 \cdot 10^{-3}$ m^3/s: a) the time scan of the acoustic wave; b) the result of summing the harmonic oscillations with frequencies $f_1 = 63.5$ Hz and $f_3 = 2796$ Hz; c) the amplitude-frequency characteristics; d) picture of the wake vortex on the lower surface of the obstacles; e) a Lissajous figure with frequency $f_1 = 63.5$ Hz and $f_3 = 2796$ Hz in the cylindrical coordinate system (r, φ, z).

3. The energy of acoustic oscillations

Three characteristic modes of impact swirl outflow were considered in the present work. These are subsonic, sonic sub-resonant and resonant modes. Figure 3 presents a graph, comparing the intensity of sound waves for these three expiration regimes of the flow.

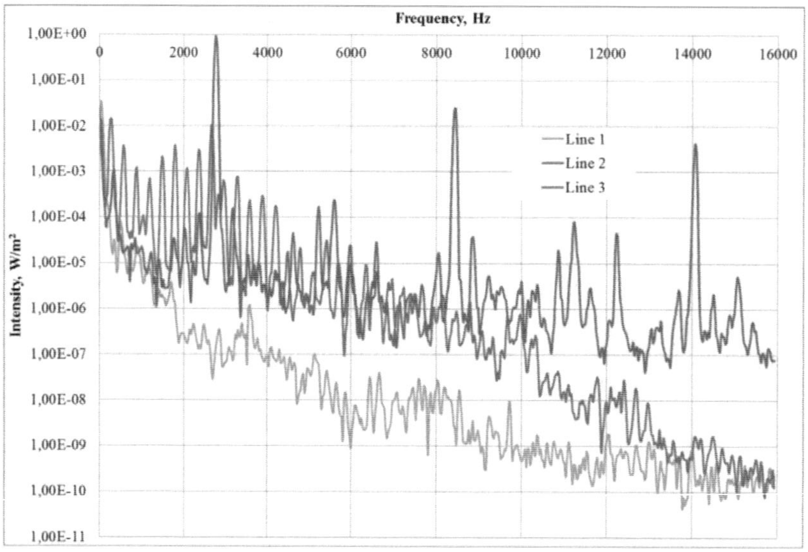

Figure 3. Frequency dependences of the sound wave intensity for the obstacle of diameter $D = 70$ mm (for $d_0 = 5$ mm and $G = 1.15 \cdot 10^{-3}$ m^3/s) for three flow regimes: 1 - subsonic mode, 2 - sonic sub-resonant mode and 3 - sonic resonant mode.

Subsonic regime occurs for small values of air flow rate in the vortex chamber. Under this regime corresponding to line 1 in Figure 3, a rotation of obstacle around its axis was observed while the sound effect was missing. In the subsonic mode, the energy of the air outflow from the vortex chamber goes for rotating the obstacle, and the air outflow is distributed evenly around the perimeter of the output gap.

The sonic sub-resonant mode occurs upon reaching the maximum value of the flow rate in the vortex chamber. The concept of limit of the fluid flow rate was introduced by academician I. I. Novikov in [7] to describe the hydrodynamical regime of a swirl flow, when the increase of pressure gradient in the work area does not increase the fluid flow rate. At this mode the cessation of the obstacle rotation around its axis was observed, and sound oscillations at low frequencies (line 2 in Figure 3) were identified.

The energy balance corresponding to the energy of the flow from under the obstacle for the sub-resonant regime (line 2 in Figure 3) consists of summing the kinetic energy of air flow arising from vortex chamber and energy of the sound waves, emitted by a hydro-mechanical system. This mode was recorded at the time of stopping the rotation of obstacles and the advent of sound. Such flow regime is unstable, and after some time it may spontaneously switch to mode 3.

The resonant mode is characterized by a sharp increase of sound oscillations at natural frequency of the vortex chamber (line 3 in Figure 3). In the experiment, the first of the achieved natural frequencies of the considered hydro-mechanical system was the frequency of oscillation of the lid, glued to the upper end surface of the vortex chamber. Figure 3 clearly shows that the energy of the acoustic oscillations increases considerably at the transition of the system to a resonant outflow mode (line 3 in Figure 3). A source of additional energy in the resonant system is the energy of vibrations in the elastic lid of the vortex chamber, which are excited due to the external force action of the vortex structure of the air flow.

4. Conclusion

As a result of this study it has been established that under the condition of critical outflow the swirl flow relieves excess energy through acoustic vibrations.

The identified vortex structure generates acoustic vibrations. Their frequency at the coincidence with the natural frequency of the structural elements of the work area, in turn, is the cause of excitation of the vibrating process and the redistribution of energy due to resonance.

On the basis of realized investigations the method of acoustic resonances is proposed. This acoustic method allows diagnosing the appearance of undesirable vortex structures and identifying threats of dangerous hydro-dynamical regimes, because the main natural frequencies of nuclear power installations relate to sonic and infrasonic ranges.

This work was supported by RFBR grant 16-08-00687 and the program of increasing the competitive ability of National Research Nuclear University MEPhI (agreement with the Ministry of Education and Science of the Russian Federation of August 27, 2013, project no. 02.a03.21.0005).

References

[1] Mitrofanova O V 2010 *Hydrodynamics and heat transfer of swirl flows in channels of nuclear power installations* (Moscow: Fizmatlit) p 288
[2] Mitrofanova O V and Pozdeeva I G 2015 Self-Adjustment Mechanism in Impinging Swirling Flows *Fluid Dynamics* **50**(5) 646–654
[3] Mitrofanova O V, Egortsov P P, Kokorev L S, Kruglov V B and Chernov A I An investigation of the mechanism of acoustic vibrations in swirl flows *High Temperature* **48**(2) 222–230
[4] Mitrofanova O V, Kruglov A B, Kruglov V B and Pozdeeva I G 2010 Investigation of the Topological Features of Impinging Swirling Jets *Teplovye Protsessy v Mekhanike* **10** 434-441
[5] Mitrofanova O V, Pozdeeva I G, Kruglov A B and Kruglov V B 2012 Comprehensive Investigation of the Effect of Generation of Large-Scale Vortex Formations in Nuclear Reactor Coolants. Part II. Experimental Investigations of Impinging Swirling Flows *Yadernaya Fizika Inzhiniring* **3**(2) 112-119
[6] Lependin L F 1978 *Acoustics* (Moscow: Vysshaya Shkola Publishers) p 448
[7] Novikov I I 1984 *Thermodynamics* (Moscow: Mashinostroenie) p 59